The Sound Studio

The Sound Studio

Sixth edition

Alec Nisbett

Focal Press
An imprint of Butterworth-Heinemann Ltd
Linacre House, Jordan Hill, Oxford OX2 8DP

℞ A member of the Reed Elsevier plc group

OXFORD LONDON BOSTON
MUNICH NEW DELHI SINGAPORE SYDNEY
TOKYO TORONTO WELLINGTON

First published 1962
Second edition 1970
Third edition 1972
Fourth edition 1979
Fifth edition 1993
Sixth edition 1995

British Library Cataloguing in Publication Data

Nisbett, Alec
  Sound Studio. – 6Rev.ed
  I. Title
  621.3893

    ISBN 0 240 51395 9

Library of Congress Cataloging in Publication Data

Nisbett, Alec.
  The sound studio/Alec Nisbett. – 6th ed.
  p.   cm.
  Previous ed. published under title: The technique of the sound studio.
  Includes bibliographical references and index.
  ISBN 0 240 51395 9
  1. Sound – Recording and reproducing. 2. Sound studios. 3. Radio
  broadcasting – Sound effects. 4. Television broadcasting – Sound
  effects. I. Nisbett, Alec. The technique of the sound studio.
  II. Title.
  TK7881.4..N5
  621.389′3 – dc20

Printed and bound in Great Britain

Contents

Introduction

For the fifth edition of this book much new material was added, the remainder was thoroughly revised to bring it in line with current practice, and the layout and order changed to increase clarity. Since then, further illustrations and updates have been introduced, particularly to the rapidly developing field of video editing (Chapter 18).

The principal remaining bastion of monophonic sound – television – has fallen (in many countries) to various coded forms of stereo. But the needs of those with obsolescent equipment must be respected. The shift towards digital recording and replay has continued, producing a marked reduction of background noise at previously weak points in the audio chain. As a result, the sound recordist's skills are all the more important, since errors and imperfections as well as more subtle effects can be heard even more clearly.

Even so, some compromises are still necessary, for the end product is no better than the weakest point in the chain. Radio transmission has improved little since the advent of FM, long ago, and will not do so until digital transmission, employing new wavebands, is well established. For both radio and television, the listening equipment and conditions may be poor, while the signal itself may have a useful dynamic range that is small in comparison with a record or recording that is being played. When they are broadcast, compact discs and digital cassettes may have to be compressed even more than is necessary for their predecessors, but mercifully this can be done without bringing up obtrusive background noise. It should never be forgotten that it is more important to satisfy the listener than to achieve the ultimate in technical perfection.

The increasing use of digital processing means more and more 'black boxes' – items of equipment to which the user has little access for repair in the event of failure. In many cases the user must become the dedicated user of a proprietary system, such as those used for digital video (or sound) editing. Here the main problem, as with so much computer-controlled equipment, is that new developments (which appear all the time) require new software – which can all too often lead to new bugs or new modes of breakdown. The best time for any user to move in is when a generation of equipment has become reasonably well established, rather than at the cutting edge of the next great advance – unless that advance is essential to the effects required, and the user is happy to help refine the system for achieving them. Intermittent users of this technology are liable to find it difficult to keep up with what is offered.

Where does the balance now lie between analogue and digital? Sound itself is 'analogue' as are (for all practical purposes) physical sound sources, studio acoustics and human hearing. Microphones and loudspeakers must also be analogue, or have analogue parts. Only the components between microphone and loudspeaker can be 'digital'. So analogue processes remain primary. Digital stages were introduced, initially, to overcome deficiencies in transmission, but they do also allow far greater flexibility in the manipulation of the signal. In popular music, in particular, it has made possible a shift in emphasis to more analytical techniques, toward perfection rather than performance.

Some reversals of opinions given in earlier editions have been made possible by such technical advances. 'Echo' devices have now reached the stage that microphones attained a generation before, where the choice of instrument was dictated by the nature of the original sound. Microphones, too, have improved a great deal since then. Some – at a price – now provide a substantially flat response throughout the audio frequency range. Here I hold to my opinion that there are advantages to matching the frequency characteristics of sound source and microphone.

The skills required are not primarily those of the engineer – which is just as well, since increasing quantities of equipment (now often digital) come as 'black boxes' which the user is not expected to interfere with. Even so, some degree of computer literacy is now desirable, if only enough to trouble-shoot and localize technical problems. It is also (at the very least) useful to be able to determine whether a fault is due to a poor electrical connection or something more serious. A recordist working alone on location will require more engineering skills than may be needed in a studio. These matters are outside the scope of this book.

Each new edition faces the same particular danger: the increasing complexity of technical equipment could all too easily be reflected in more complex technical descriptions. I have tried to avoid falling into this trap, for it would defeat the original purpose, which is to explain the subject simply, in terms acceptable both to the creative people in production – including writers and performers – and to those who, more technically orientated, deal with the balance and recording of sound and with its engineering aspects. By exploring the ground common to both groups, the work of each may be interpreted to the other.

Much of this account is derived from well over thirty years' experience of radio, television and film-making for Britain's public service broadcaster, the BBC, which has a range of output (central, regional and local) that has helped to create a body of techniques that is probably unique. Both the knowledge available to me and the opportunity to put it into practice have been as great as (indeed, possibly greater than) might have been provided in any other broadcasting organization or place of learning. I am deeply indebted not only to that organization, but also to the many specialists working in the medium who have generously shared their knowledge with me.

Where something is a matter of opinion, it is usually my own. In other cases the book reflects either a consensus or the views of an individual whose judgement I trust.

This book assumes a need and a desire for high standards, both of technique and programme content. These may be attained with apparent ease within the large network where there is time and money to spend, but some might question whether a knowledge and understanding of these high standards is also of value at the ground-roots level – for example, in a small-town, one-studio radio station, with a transmitter on a rack in the corner. In my own view it is, because high-quality work sets a standard by which all else may be measured. Sometimes, particularly for the freelance, it may make all the difference between success and failure: if you want to sell a product it helps if the quality is as high as the buyer can achieve by his own efforts. High standards are not only a matter of having good equipment: just as important is knowing how to use it.

The basic approach of this book has remained unaltered since the first edition: the emphasis has always been on general principles rather than rule-of-thumb, though the shorter routes to high-quality results are not neglected. It still insists on the paramount importance of subjective judgement – of learning to use the ears properly. Above all else, the sound man needs well-developed aural perception that is allied to a strong critical faculty. So studio and location techniques are described in terms of what each operation does to the sound and why, and *what to listen for.*

Of all the chapters in this book, that which has changed least is the last, on communication itself. Happily, in the years since it was first written, neither scientific study of this human capacity nor the evolution of the media themselves have invalidated the observations made then.

Wee have also Sound-Houses, *wher wee practise and demonstrate all* Sounds, *and their* Generation. *Wee have* Harmonies *which you have not, of* Quarter-Sounds, *and lesser* Slides *of* Sounds. *Diverse* Instruments *of* Musick *likewise to you unknowne, some* sweeter *then any you have; Together with* Bells *and* Rings *that are dainty and sweet. Wee represent* Small Sounds *as* Great *and* Deepe; *Likewise* Great Sounds, Extenuate *and* Sharpe; *Wee make diverse* Tremblings *and* Warblings *of* Sounds, *which in their* Originall *are* Entire. *Wee represent and imitate all* Articulate Sounds *and* Letters, *and the* Voices *and* Notes *of* Beasts *and* Birds. *Wee have certaine* Helps, *which sett to the* Eare *doe further the* Hearing *greatly. Wee have also diverse* Strange *and* Artificiall Eccho's, Reflecting *the* Voice *many times, and as it were* Tossing *it; And some that give back the* Voice Lowder *then it came, some* Shriller, *and some* Deeper; *Yea some rendring the* Voice, Differing *in the* Letters *or* Articulate Sound, *from that they receyve. Wee have also meanes to convey* Sounds *in* Trunks *and* Pipes, *in strange* Lines, *and* Distances.

From *The New Atlantis* by Francis Bacon

1624

Chapter 1

Audio techniques and equipment

Through years of intimate daily contact with the problems of the medium, through spending hours each day consciously and critically listening, the professional working in the radio or recording studio, or in television or film, builds a considerable body of aural experience. He becomes aware of subtleties and nuances of sound that make all the difference between an inadequate and a competent production, or between the competent and the exciting. He hears details that would pass unobserved by the untutored ear. At the same time he develops the ability to manipulate the equipment to mould sound into the shape that he requires.

The tools for recording sound continue to improve, and it becomes easier to acquire the basic skills; but because of this there is more that *can* be done. Microphones have become more specialized, creating new opportunities for their use. Studio control desks have become intimidatingly complex, but this only reflects the vast range of options that they offer. By both analogue and digital techniques, complex editing jobs can be tackled. Using lightweight portable recorders or tiny radio transmitters sound can be obtained from anywhere man can reach and many places he cannot.

The easy acquisition of tools, however, does not obviate the necessity for a well-trained ear. Each element in a production has an organic interdependence with the whole; every detail of technique contributes in one way or another to the final result; and in turn the desired end-product largely dictates the methods to be employed. This synthesis between means and end may be so complete that the untrained ear can rarely disentangle them – and even critics may disagree on how a particular effect was achieved. Faults of technique are nothing like so obvious as those of content. A good microphone balance in suitable acoustics, with the appropriate fades, pauses, mixing and editing can make all the difference to a radio programme – but at the end an appreciative listener's search for a favourable comment will probably land on some remark about the subject matter. Similarly, where the technique is faulty it is often, again, the content that is criticized. The sound man himself must develop a faculty for listening analytically to the operational aspects of a production, to judge technique independently of content while making it serve content.

In radio, television or film, the investment in equipment makes slow working unacceptable. Technical problems should be anticipated or solved quickly and unobtrusively, so that attention can be focused on artistic fine-tuning, sharpening performances and enhancing the material. Although each individual is concentrating on his own job, this is geared to that of the team as a whole. The major concern of everyone is to raise the programme to its best possible standard and to catch it at its peak.

This first chapter will briefly introduce the range of techniques and basic skills required, and set them in the context of the working environment, the studio chain or its location equivalent. In subsequent chapters the chain will be traced through in detail, beginning with the original sound source and the nature of sound itself, and including all of the vast array of operations that may be applied to the signal before it is converted back into sound and offered to the listener – who should be happily unaware of these intervening stages.

Studio operations

The operational side of sound studio work, or in the sound department of the visual media, includes:

- choice of acoustics: selecting, creating or modifying the studio environment
- microphone balance: selecting suitable types of microphone and placing them to pick up the most satisfactory sound from the various sources
- mixing: combining the output from microphones, replay machines, artificial reverberation devices, remote studios, etc.
- control: ensuring that the programme level (i.e. in relation to the noise and distortion levels of the equipment used) is not too high or too low, and uses the medium – recording or radio transmission – efficiently
- creating sound effects (variously called hand, spot or foley effects) in the studio
- playing records and recordings into programmes: this includes recorded effects, music discs or tapes, interviews and reports, pre-recorded sequences, etc.
- recording: ensuring that the resultant mixed sound reaches and is inscribed upon the recording medium without any significant loss of quality

In small American radio stations one additional responsibility may be added to this: supervising the operation of a transmitter which may also be housed in the studio area. Certain minimal technical qualifications are required.

The person responsible for balancing, mixing and control may be a balance engineer (in commercial recording studios) or sound or programme engineer; in BBC radio is called a studio manager; in television may be a sound supervisor with sound assistants; in film, a sound recordist or, at the re-recording stage, a dubbing mixer.

Whatever his designation, the sound supervisor has an overall responsibility for sound studio operations and advises the producer on technical problems. Where sound staff are working away from the studio, e.g. on outside broadcasts (remotes) or when filming on location, a higher standard of engineering knowledge is required. This is true also in small radio stations, where a maintenance engineer may not always be available. The sound balancer may also be responsible for recording items if no separate monitoring is required – though for complex recordings a separate recording engineer is employed. There may be one or several assistants: in radio these may deal with effects in the studio, and may play tapes and records. Assistants in television and film may also act as boom operators, positioning microphones just out of the picture.

Although in many parts of the world the person doing some of these jobs is an engineer, BBC experience indicates that in broadcasting – and for some roles in other media – it is often equally satisfactory to employ men and women whose training has been in other fields, because although the operator is handling technical equipment, the primary responsibilities are artistic.

The sound control room

In considering the layout of equipment that is needed for satisfactory sound control let us look first at the simplest case: the studio used for sound only, for radio or recording.

The nerve centre of any broadcast or recording is the control desk. It is here that all the different sound sources are modified and mixed. In a live broadcast it is here that the final sound is put together; and it is the responsibility of the sound supervisor to see that no further adjustments of any sort are necessary before the signal leaves the transmitter (except, perhaps, for any occasional automatic operation of a limiter to protect the transmitter from overloading).

The desk or console, in most studios the most obvious and impressive capital item, often presents a vast array of faders, switches and indicators. Essentially, it consists of a number of individual channels, each associated with particular microphones or other sources. These are all preset (or altered only rarely), so that their output may be combined into a small number of group controls which can be operated by one pair of hands. These in turn pass through a main control to become the studio output. That 'vast array' plainly offers

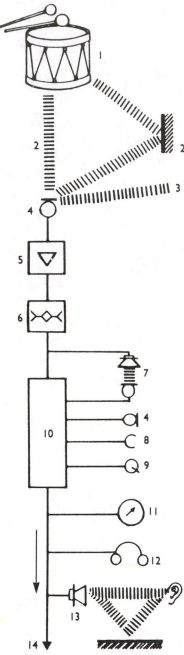

many opportunities to do different things with sound. Apart from the faders, of which the function is obvious, many are used for signal routing. Some allow modification of frequency content ('equalization') or dynamics (these include compressors and limiters). Others are concerned with monitoring. There are more still, for communications. Outside the desk, waiting to be 'inserted' at chosen points, are specialized devices offering artificial reverberation and further ways of modifying the signal. All these will be discussed in detail later.

In many desks, the audio signal passes first through the individual strips ('channels'), then through the groups, then to the main control and out to 'line' – but in some this process is partly or wholly mimicked. For example, a group fader may not actually carry the signal; instead, a visually identical voltage control amplifier (VCA) fader sends messages which modify the audio in the individual channels which have been designated as belonging to the group, before it is combined with all other channels in the main control. In some desks, the preset equalization, dynamics and some of the routing controls are located away from the channel strips, so that these can all be entered separately into the memory of a computer, thereby greatly reducing the (visual) complexity and total area of the layout.

In the extreme case, the signal does not reach the desk at all: the controls send their messages, via computer, to racks in which the real signal is modified, controlled and routed. What then remains of the original desk layout depends on the manufacturer's judgement of what the customer will regard as 'user friendly'. It should be possible for the majority of users to switch between analogue and fully digital systems (and back again) without having to go through a major course of re-education.

Apart from the microphones and control desk, the next most important equipment in a studio is a high-quality loudspeaker system. For a radio or recording studio is not merely a room in which sounds are made and picked up by microphone; it is also the place where shades of sound are judged and a picture is created by ear.

The main thing that distinguishes a broadcasting studio from any other place where microphone and recorder may be set up is that two acoustically separate rooms must be used: one where the sound may be created in suitable acoustics and picked up by microphone, and the other in which the mixed sound may be heard. This second room is often called the control room (although in BBC radio terminology it is a 'control cubicle': historically, the control room was a lines switching centre). Tape and disc reproducers, and the staff to operate them, are also in the control area, as are the producer or director (in BBC radio these two roles are combined) and the production secretary (who checks programme timing, etc.).

In the simplest television studios, sound and picture are controlled side by side, but more often there is an acoustically separate sound room with much the same physical layout as it would have in a radio studio (except that it will have picture monitors, including one for that

Components of the studio chain
Some of the symbols used in this book. 1, The sound source. 2, Direct and indirect sound paths. 3, Noise: unwanted sound. 4, Microphone. 5, Amplifier. 6, Equalizer. 7, Artificial reverberation, symbolically represented by acoustic path, but more commonly now by a digital device. 8, Tape head. 9, Record player. 10, Mixer. 11, Meter. 12, Headphones. 13, Loudspeaker: sound reaches the ear by direct and reflected paths. 14, Studio output.

which is currently selected and one for preview). The television sound supervisor does need to listen to programme sound in a separate cubicle because the director and his assistant must give a continuous stream of instructions on an open microphone, which means that their programme sound loudspeaker has to be set at a lower level than is necessary for the judgement of a good balance, and at times their voices obscure it. Their directions are relayed to the sound control room on an audio circuit, but at a reduced volume.

Studio layout for recording

It is no good finding a recording fault 10 minutes after the artist has left the studio. So an important rule in studio work is that every recording must be monitored *from the tape* as it is being made. On the simplest recordings – straight talk programmes in which there are no cues to be taken – the operational staff can do both jobs, operating the recorder and monitoring the output from a separate replay head.

When an analogue recording made on tape is monitored in this way, there is a delay of perhaps one-fifth of a second, 3 inches of tape between recording and reproducing heads at 15 inches per second, i/s (38 cm/s) or two-fifths of a second at $7\frac{1}{2}$ i/s (19 cm/s). This delay makes it unsatisfactory to combine the two jobs when the programme has any degree of complexity. One-fifth of a second can completely destroy timing. The perfect fade on a piece of music, for example, may have to be accurate to within one-twentieth of a second. These requirements mean that, ideally, there may be a third acoustically separate room in which the recording engineer can check the quality of the recorded sound. The most convenient plan is to group together the studio, control cubicle and recording room as a suite.

Many programmes are rehearsed and recorded section by section. Each few minutes of the programme is separately discussed, run through, and then recorded; and the results are edited together afterwards. However, this is expensive in both staff and space. In a

Radio studio layout
1, Studio. 2, Lobby. 3, Double-glazed window for acoustic separation. 4, Control console and monitoring loudspeakers. 5, Producer's position (behind balancer). 6, Recording and replay equipment. 7, Equipment racks.

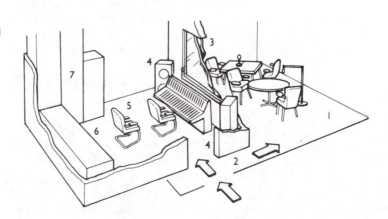

simpler arrangement, programme segments are recorded in the control cubicle by the play-in operator, then checked and rough-edited while the next piece is being set up.

An alternative is based on the idea that the dramatic contour of a programme is likely to have a better shape if it is fully rehearsed and then recorded as a whole, leaving any faults that occur to be corrected by retakes. In this case a more economic arrangement is to divide studio time into two parts, with the recording session limited to a relatively short period. This technique is based on the way studio time is organized for live transmissions. In such cases, the recording room may be remote from the studio.

There are various other possibilities, such as that commonly used for topical and miscellany programmes which go on the air in the form of a mixture of live and recorded segments. Here last-minute items are recorded in the control cubicle. The recording is monitored on headphones for an immediate check, and subsequently on the loudspeaker if time allows.

Studio communications

In both radio and television there are sound circuits that are used for local communication only.

The use of headphones carrying an independent talkback circuit between control cubicle and studio may help a recording session or radio transmission. Their use may range from the full-scale relaying of instructions (as occurs in television productions), through supplying a list of questions to an interviewer, to monitoring a feed of some separate event in order to give news of it. There may be both talkback (to the studio) and reverse talkback. The latter is an independent feed from the studio, sometimes associated with a microphone-cut key, so that a speaker can cut himself off the air and ask for information or instructions.

When a studio microphone is faded up it is usually arranged that the studio loudspeaker is cut off automatically (except for talkback purposes during rehearsal). Headphones or an earpiece may be used in the studio to hear talkback when the loudspeaker has been cut – which may be just before a cue.

In television the complexity of the communication system is much greater, because there are so many more people involved, and also because they need to occupy a range of acoustically separate areas. The director speaks to the floor manager over a local radio circuit; the floor manager hears him on a single earpiece; the cameramen and others in the studio hear him on headphones.

In his separate room the sound supervisor can hear high-level and high-quality programme sound. In addition, on a second much smaller loudspeaker (so that it is distinctly different in quality) he

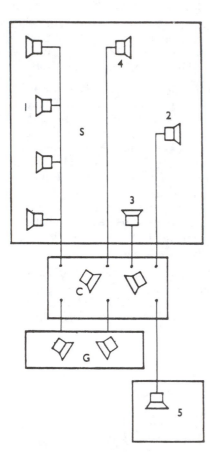

Loudspeakers in a television musical production
C, Sound control room. G, Gallery. S, Studio. 1, Audience (public address) system. 2, Foldback of recorded sound effects. 3, Special feed of vocalist's sound to conductor. 4, Special feed of rhythm group to vocalist. 5, Remote studio.

hears the director's instructions and reverse talkback from his own assistants on the studio floor.

Lighting and camera control also generally have a (third) acoustically separate area. They have their own talkback system to appropriate points on the studio floor; and cameramen in their turn have reverse talkback to them. In the control gallery itself the technical manager has in front of him what is in effect a small telephone exchange for communication with all points concerned technically with the programme. This includes control lines to remote sources, and also to the videotape recorder and telecine operators (who, in their own areas, hear the director's circuit as well as programme sound or their own output).

Television outside broadcast or *remote* trucks sometimes have sound control, director and camera control all in the same tiny area of the vehicle: this is best for sports and other unrehearsed events where side-by-side working helps the operators to synchronize their actions. But for rehearsed events, and programmes built on location sequence by sequence, good sound conditions are more important, and these demand an acoustically separate area. This may be either in the same mobile control vehicle or (for music, where more space is desirable for monitoring) in a separate truck or, alternatively, 'derigged' and set up in some convenient room.

In both radio and television outside broadcasts the sound staff provide communication circuits between the various operational areas (e.g. commentary points) and the mobile control room. This often takes the form of local telephone circuits terminating at special multi-purpose boxes which also provide feeds of programme sound and production talkback. (On television outside broadcasts the latter is particularly for the stage manager.) Additionally, field telephones may be required at remote points.

In broadcasting, the studio suite (studio and control cubicle) may be only the first link in a complex chain. In a station with mixed output, the next link is a continuity suite where the entire programme service is assembled. The output of individual studios is fed in live, and linked together by station identification and continuity announcements. The announcer in the continuity studio is also usually given executive responsibility for the service as a whole, and must intervene if any contribution under-runs, over-runs, breaks down, or fails to materialize. He cues recorded programmes, presents trailers (promotional material) and also commercials where appropriate.

In small specialized stations such as those of local radio, virtually the whole of the output emerges from a single room – though a separate studio may be available for news and discussion programmes or for recording commercials. In his 'electronic office' the presenter controls his own microphone and plays in the records, taped commercials, etc. For this, he should have a range of CD and record players, turntables, reel-to-reel tape decks and cartridge and cassette players – sufficient to allow commercials, promotional trailers, production effects and public service announcements to be played one after the other ('back-to-back'). Cartridges can be reproduced

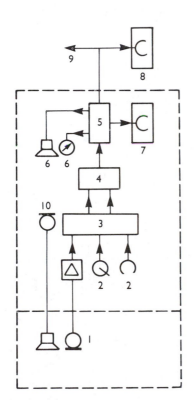

Elements of studio chain
1, Microphones: low level sources, fed through amplifier. 2, High level sources (replay machines, etc). 3, Microphone and line-input channels in control console. 4, Group channels in control console. 5, Main (or master) control. 6, Monitoring. 7, Studio recorder. 8, External recorder. 9, Studio output. 10, Talkback circuit.

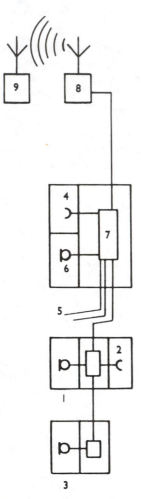

from carousels or stacks – though where large numbers of such sources are required it is now better to use hard disc storage. In addition to microphone and reproduction equipment the presenter may also have two-way communication with news or traffic cars or helicopters, which can also be arranged to record automatically, on tape, while on the air with other material.

In large networks the basic service is put together in the same way, but with greater emphasis on precision timing, and the output of the network continuity appears as a source in the regional or local continuity which feeds the local radio transmitter. The links between the various centres may be landlines or radio links.

In television the broadcasting chain is substantially the same as for sound radio, except that in the presentation area there may be both a small television studio and a separate small sound booth for out-of-vision announcements. Use of the sound booth means that rehearsals of more complex material in the main studio do not have to be stopped completely for routine, or unscheduled, programme breaks.

The broadcasting chain
The studio output may go to a recorder or it may provide an insert into a programme compiled in another studio – or it may be fed direct to a programme service continuity suite. Between studio, continuity and transmitter there may be other stages: switching centres, boosting amplifiers and frequency correction networks on landlines, etc. 1, Studio suite. 2, Recording area or machine room. 3, Remote studio. 4, Machine room. 5, Other programme feeds. 6, Continuity announcer. 7, Continuity mixer. 8, Transmitter. 9, Receiver.

Chapter 2

The sound medium

Some readers of this book will already have a clear understanding of what sound is. Others will know very little about the physics of sound, and will not wish to know, their interest being solely in how to use the medium. 'Do you have to know how paint is made to be an artist?' they might ask.

These two groups of readers can skip this section and go straight on to read about the operational techniques, the creative applications, and the practical problems and their solutions which are the main theme of this book.

The non-technical reader should have no difficulty in picking up as much as he needs to know as he goes along (perhaps with an occasional glance at the Glossary). Most of the terms used should be understandable from their context, or are defined as they first appear. But for the sake of completeness, and for those who prefer their raw materials – or ideas – to be laid out in an orderly fashion before they begin, here is a brief introduction to the physics of sound.

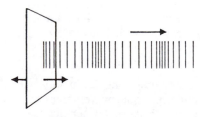

Sound wave
A vibrating panel generates waves of pressure and rarefaction: sound. A tuning fork is often used for this demonstration, but in fact tuning forks do not make much noise – they slice through the air without moving it. A vibrating panel is more efficient.

The nature of sound

Sound is caused by vibrating materials. If a panel of wood vibrates, the air next to it is pushed to and fro. If the rate of vibration is somewhere between tens and tens of thousands of excursions per second, the air has a natural elasticity which we do not find at slower speeds. Wave your hand backward and forward once a second and the air does little except get out of its way; it does not bounce back. But if you could wave your hand back and forth a hundred times every second, the air would behave differently: it would compress as the surface of the hand moves forward and rarefy as it moves back. In

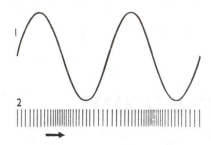

Waves
1, Sound waves are usually represented diagrammatically as lateral waves, as this is convenient for visualizing them. It also shows the characteristic regularity of the simple harmonic motion of a pure tone. Distance from the sound source is shown on the horizontal axis, and displacement of individual air particles from their median position is shown on the vertical axis. 2, As the particles are actually moving backwards and forwards along the line that the sound travels, the actual positions of layers of particles are arranged like this. It also shows the travel of pressure waves.

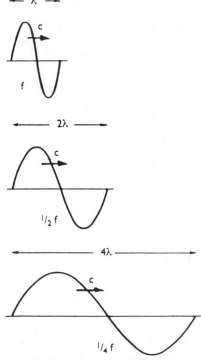

Frequency and wavelength
All sound waves travel at the same speed, c, through a given medium, so frequency, f, is inversely proportional to wavelength, λ.

such circumstances the natural elasticity of the air takes over. As the surface moves forward, each particle of air pushes against the next, so creating a pressure wave. As the surface moves back, the pressure wave is replaced by a rarefaction, and that is followed by another pressure wave, and so on.

It is a property of elastic media that a pressure wave passes through them at a speed that is a characteristic of the particular medium. The speed of sound in air depends on its temperature as well as the nature of air. In normal conditions sound travels about 1120 feet in a second (about 340 metres per second).

This speed is independent of the rate at which the surface generating the sound moves backwards and forward. The example was 100 excursions per second, but it might equally have been 20 or 20 000. This rate at which the pressure waves are produced is called the *frequency,* and this is measured in cycles per second, or *hertz:* $1 \text{ Hz} = 1 \text{ c/s}$.

Let us return to that impossibly energetic hand shaking back and forth at 100 Hz. It is clear that it is not a perfect sound source: some of the air slips round the sides as the hand moves in each direction. To stop this happening with a fluid material like air, the sound source would have to be much larger. Something the size of a piano sounding board would be more efficient, losing less round the edges. But if a hand-sized vibrator moves a great deal faster, the air simply does not have time to get out of the way. For very high frequencies, even tiny surfaces are efficient radiators of sound.

In real life, sounds are produced by sources of all shapes and sizes that may vibrate in complicated ways. In addition, the pressure waves bounce off hard surfaces. Many sources and all their reflections may combine to create a complex field of intersecting paths. How are we to describe, let alone reproduce, such a tangled skein of sound? Fortunately, we do not have to.

Consider a single air particle somewhere in the middle of this field. All the separate pressure waves passing by cause it to move in various directions, to execute a dance that faithfully describes every characteristic of all of the sounds passing through. All we need to know to be able to describe the sound at that point is what such a single particle is doing. In practice, the size of this particle is not critical so long as it is small compared with the separation of the successive pressure waves that cause the movement. It does not have to be *very* small; rather, we want to measure an average movement that is sufficiently gross to even out all the random tiny vibrations of the molecules of air – just as the diaphragm of the ear does.

Consideration of the ear leads us to another simplification. A single ear does not bother with all the different directional movements of the air particle; it simply sums and measures variations in air pressure. This leads us to the idea that a continuously operating pressure-measuring device with a diaphragm approaching the size of the eardrum would be a good instrument for describing sound in practical audio-engineering terms. For what we have described is the first requirement of an ideal microphone.

Wavelength

Pressure waves of a sound travel at a fixed speed for a given medium and conditions, and if we know the frequency of a sound (the number of waves per second) we can calculate the distance between corresponding points on successive waves – the *wavelength.*

Taking the speed of sound to be 1120 ft per second (340 m/s), a sound frequency of 1120 Hz (or cycles per second) has a wavelength of 1 foot (about 30 cm). Sometimes it is more convenient to think in terms of frequency and sometimes wavelength.

Because the speed of sound varies to a small degree with air temperature, any relationship between wavelength and frequency that does not take this into account is only approximate. But for most practical purposes, such as calculating the thickness of padding needed to absorb a range of frequencies, or estimating whether a microphone diaphragm is small enough to 'see' very high frequencies, an approximate relationship is adequate. Particular frequencies also correspond to the notes of the musical scale.

Note on piano   **Fundamental frequency. Hz**

The correspondence of frequencies to musical notes

Note on piano	Frequency (Hz)	Wavelength	
–	14 080	1 in	(2.5 cm)
–	7 040	2 in	(5 cm)
A^{iii}	3 520	$3\frac{3}{4}$ in	(9.5 cm)
A^{ii}	1 760	$7\frac{1}{2}$ in	(19 cm)
A^{i}	880	1 ft 3 in	(38 cm)
A	440	2 ft 6 in	(75 cm)
A_i	220	5 ft	(1.5 m)
A_{ii}	110	10 ft	(3 m)
A_{iii}	55	20 ft	(6 m)
A_{iv}	27.5	40 ft	(12 m)

Here, the speed of sound is taken to be 1100 ft/s (335 m/s), as it may be in cool air. Note that the metric equivalents (and shorter wavelengths) are approximate only.

The first thing to notice from this is the very great range in physical size these different frequencies represent. Two octaves above the top A of the piano, the wavelength is 1 in (25 mm). As it happens, this is close to the upper limit of human hearing; and the corresponding size is reflected in the dimensions of high-quality microphones: they must be small enough to *sample* the peaks and troughs of this. At the lower end of the scale we have wavelengths of 40 ft (12 m). The soundboard of a piano can generate such a sound, though not so efficiently as the physically much larger pipes in the lower register of an organ.

The second point to notice (and the reason for the first) is that for each equal interval, a rise of one octave, there is a doubling of frequency (and halving of wavelength).

Piano keyboard
Note and frequency of fundamental, octave by octave.

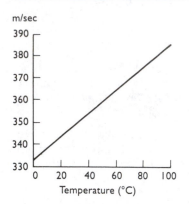

m/sec

Velocity of sound in air
The velocity of sound increases with temperature.

Much less obvious is the fact that as the pitch of a sound depends on the size of the object making it, and that size remains nearly constant as temperature increases, then because the velocity of sound increases with temperature, the pitch goes up. Stringed instruments go a little flat, while wind instruments go sharp. For strings, the remedy is simple: they can easily be tuned, but the vibrating column of air in most wind instruments cannot be changed so much. A flute, for example, sharpens by a semitone as the air temperature goes up 15°C, but flattens a little as the metal of the instrument itself warms up and lengthens. To compensate for weather, hall and instrument temperature changes, the head joint of the flute can be eased in or out a little – in an orchestra, matching its pitch to that of the traditionally less adaptable oboe.

Waves and phase

Still considering a pure, single-frequency tone, we can define a few other variables and see how they are related to the pressure wave.

First we have *particle velocity*, the rate of movement of individual air particles. As this is proportional to air pressure, the waveforms for pressure and velocity are similar in shape. But not only that: where one has a peak the other has a peak; where there is no excess of pressure, there is no velocity, and so on. The two waveforms are said to be *in phase* with each other.

The next variable is *pressure gradient*, the rate at which pressure changes with distance along the wave. Plainly, this must have a waveform derived from that of pressure, but where the pressure is at a peak (maximum or minimum) its rate of change is zero. Again it turns out that (for a pure tone or sine wave) the waveform is the same, but this time it has 'moved' a quarter of a wavelength to one side. The two waveforms are said to be a quarter of a wavelength *out of phase*.

Another variable, again with a similar waveform, is that for *particle displacement*. This is a graph showing how far a particle of air has moved to one side or the other of its equilibrium position. (This is, in fact, the equivalent in sound of the wave we see on a water surface.) Displacement is proportional to pressure gradient, and is therefore in phase with it. It is directly related to *amplitude*.

These terms are all encountered in the theory of microphones. For example, most microphones are *constant-velocity* or *constant-amplitude* in operation. These names are confusing: they actually mean that the electrical output of the microphone is equal to a constant *multiplied by* the diaphragm velocity, or to a constant *multiplied by* the displacement amplitude. Constant-velocity microphones include moving-coil and ribbon types: constant-amplitude microphones include electrostatic and crystal varieties. From the point of view of the operator, these descriptions are not

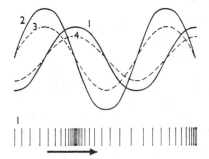

Waveform relationships
1, Pressure wave. 2, Displacement wave. 3, Pressure gradient. 4, Particle velocity.

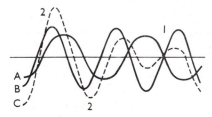

Adding sound pressures
At any point the sound pressure is the sum of the pressures due to all waves passing through that point. If the simple waves A and B are summed this results in the pressure given by the curve C. 1, Here there is partial cancellation. 2, Here the peaks reinforce. The resultant curve is more complex than the originals.

particularly important: there are high- and low-quality microphones in both categories: the terms are mentioned here only because they may be met in manufacturers' literature. They are, of course, important to the microphone design or service engineer. A great deal more will, however, be heard of pressure and pressure-gradient modes of operation, as these characteristics of microphones lead to important differences in the way they can be used.

Because of these differences, the *phase* of the signal as it finally appears in the microphone output is not always the same as that of the sound pressure measured by the ear. Fortunately the ear is not usually interested in phase, so it does not generally matter whether a microphone measures the air pressure or pressure gradient. The only problem that might occur is when two signals very similar in form but different in phase are combined. In an extreme case, if two equal pure tones exactly half a wavelength out of phase are added together, the output is zero. Normally, however, sound patterns are so complex that microphones of completely different types that are sufficiently far apart may be used in the same studio, and their outputs mixed, without much danger of differences of phase having any noticeable effect.

So far we have considered simple tones. If several are present together, a single particle of air can still only be in one place at one time: the displacements are added together. If many waves are superimposed, the particle performs the complex dance that is the result of adding all of the wave patterns. Not all of this adding together results in increased pressure. At a point where a pressure wave is superimposed on a rarefaction, the two – opposite in phase at that point – tend to cancel each other out.

Energy, intensity and resonance

There are several more concepts that we need before going on to consider the mechanics of music.

First, the *energy* of a sound source. This depends on the amplitude of vibration: the broader the swing, the more *power* (energy output per second) it can produce. The sound *intensity* at any point is then

Sound intensity
This is the energy passing through unit area per second. For a spherical wave (i.e. a wave from a point source) the intensity dies away very rapidly at first. The power of the sound source S is the total energy radiated in all directions.

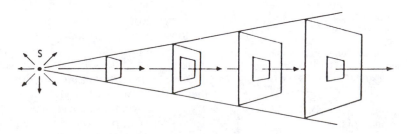

measured as the acoustic energy passing through unit area per second. But to convert the energy of a sound source into acoustic energy in the air, we have to ensure that the sound source is properly coupled to the air, that its own vibrations are causing the air to vibrate with it. Objects that are small (or slender) compared with the wavelength in air associated with their frequency of vibration (e.g. tuning forks and violin strings) are able to slice through the air without giving up much of their energy to it: the air simply slips around the sides of the prong or the string.

If a tuning fork is struck and then suspended loosely it goes on vibrating quietly for a long time. If, however, its foot is placed on a panel of wood, the panel is forced into vibration in sympathy and can transmit to the air. The amplitude of vibration of the tuning fork then goes down as its energy is lost (indirectly) to the air.

If the panel's natural frequency of vibration is similar to that of the tuning fork, the energy may be transferred and radiated much faster. In fact, wooden panels fixed at the edges may not appear to have clear specific frequencies at which they resonate: if you tap them they do not usually make clear musical sounds, but instead rather dull ones, though with perhaps some definable note seeming to predominate. In the violin, the wooden panels are very irregular, to ensure that over a whole range of frequencies no single one is emphasized at the expense of others. The string is bowed but transmits little of its energy directly to the air. Instead, it is transferred through the bridge that supports the string to the wooden radiating belly of the instrument.

At this stage an interesting point arises: the radiating panels are not only much smaller in size than the strings themselves, but also considerably smaller than some of the wavelengths (in air) that the strings are capable of producing. The panels might respond well enough to the lowest tones produced, but they are not big enough to radiate low frequencies efficiently. This leads us to consider several further characteristics of music and musical instruments.

Complex waveforms
Pure tones (1, 2 and 3) may be added together to produce a composite form (4). In this case the first two overtones (the second and third harmonics) are added to a fundamental. If an infinite series of progressively higher harmonics (with the appropriate amplitudes and phases) were added, the resultant waveform would be a 'saw-tooth'. When the partials are related harmonically, the ear hears a composite sound having the pitch of the fundamental and a particular sound quality due to the harmonics. But when the pitches are unrelated or the sources are spatially separated, the ear can generally distinguish them readily: the ear continuously analyses sound into its components.

Overtones, harmonics and formants

Overtones are the additional higher frequencies that are produced along with the *fundamental* when something like a violin string or the air in an organ pipe is made to vibrate.

On a string the overtones are all exact multiples of the fundamental, the lowest tone produced. It is this lowest tone which, as with most of the musical instruments, defines *pitch*. If the string were bowed in the middle, the fundamental and odd harmonics would be emphasized, as these all have maximum amplitude in the middle of the string; and the even harmonics (which have a node in the middle of the string) would be lacking. But the string is, in fact, bowed near one end so that a good range of both odd and even harmonics are excited.

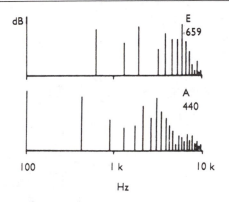
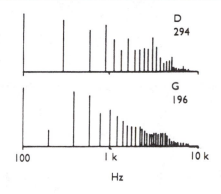

The open strings of the violin: intensities of harmonics

The structure of the resonator ensures a very wide spread, but is unable to reinforce the fundamental of the low G, for which it is physically too small. Note that there is a difference in quality between the lowest string (very rich in harmonics) and the highest (relatively thin in tone colour).

There is one special point here: if the string is bowed at approximately one-seventh of its length it will not produce the seventh harmonic. So, for the most harmonious quality this is a good point at which to bow, because as it happens the seventh harmonic is the first that is not musically related to the rest – though the sixth and eighth are both members of the same musical family, and there are also higher harmonics produced in rich profusion to give a dense tonal texture high above the fundamental. These are all characteristics of what we call string quality, irrespective of the instrument.

The quality of the instrument itself – violin, viola, cello or bass – is defined by the qualities of resonator; and most particularly by its size. The shape and size of the resonator superimposes on the string quality its own special *formant* characteristics. Some frequencies, or range of frequencies, are always emphasized; others are always discriminated against. Formants are very important: obviously, in music they are a virtue, but in audio equipment the same phenomenon might be called 'an uneven frequency response' and would be regarded as a vice – unless it were known, calculated and introduced deliberately with creative intent.

The first eight string harmonics

The fundamental is the first harmonic: the first overtone is the second harmonic, etc. They are all notes in the key of the fundamental, except for the seventh harmonic which is not a recognized note on the musical scale (it lies between G and G#). If the violinist bows at the node of this harmonic (at one-seventh of the length of the vibrating string) the dissonance is not excited. The eleventh harmonic (another dissonant tone) is excited, but is in a region where the tones are beginning to cluster together to give a brilliant effect.

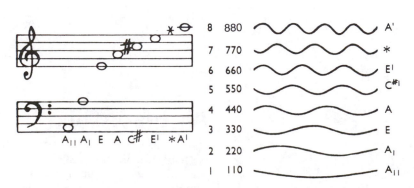

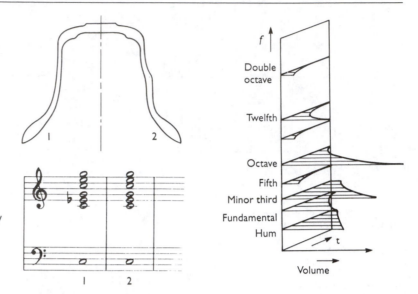

Bell overtones
Left: Profiles and tones produced by 1, a minor-third bell and 2, a major-third bell. *Right:* The relative volumes and rates of decay of the tones present in the sound of a minor-third bell.

On some instruments, such as bells, it may be a problem in manufacture to get the principal overtones to have any musical relationship to each other (the lowest note of a bell is not even distinguished by the name of fundamental). In drums, the fundamental is powerful and the overtones add richness without harmonic quality, while from triangles or cymbals there is such a profusion of high-pitched tones that the sound blends reasonably well with almost anything.

There are many other essential qualities of musical instruments. These may be associated with the method of exciting the resonance (bowing, blowing, plucking or banging); or with qualities of the note itself, such as the way it starts (the *attack*) or changes in volume as the note progresses (its *envelope*).

Vibration of a drumskin
(stretched circular membrane clamped at the edges)
The suffixes refer to the number of radial and circular nodes (there is always a circular node at the edge of the skin). The overtones are not harmonically related. If $f_{01} = 100$ Hz, the other modes of vibration shown here are: $f_{02} = 230$, $f_{03} = 360$, $f_{11} = 159$, $f_{12} = 292$, $f_{21} = 214$ Hz.

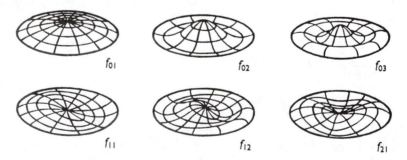

Air resonance

Air may have dimensional resonances very much like those of a string of a violin – except that whereas the violin string has transverse waves,

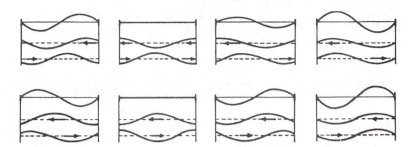

Stationary wave
Stationary waves are formed from progressive waves moving in opposite directions.

those in air, being composed of compressions and rarefactions, are longitudinal. And whereas radiating sound moves through the air forming *progressive waves*, dimensional resonances stand still: they form *stationary* or *standing waves*.

These stationary waves can again be represented diagrammatically as transverse waves. The waveform chosen is usually that for displacement amplitude. This makes intuitive sense: at *nodes* (e.g. at solid parallel walls if the resonance is formed in a room) there is no air-particle movement; while at *antinodes* (e.g. half-way between the walls) there is maximum movement of the air swinging regularly back and forth along the particular dimension concerned. Harmonics may

Vibration of air columns
Left: For an open pipe, f, the fundamental frequency is twice the length of the pipe. There is a node N at the centre of the pipe and antinodes at the open ends. The second and third harmonics are simple multiples of the fundamental: $f_2 = 2f$, $f_3 = 3f_1$, etc.
Left below: The third harmonic, half a cycle later. Air particles at the antinodes are now moving in the opposite direction, but the air at the nodes remains stationary.
Right: A pipe closed at one end (i.e. stopped). The wavelength of the fundamental is four times the length of the pipe. The first overtone is the third harmonic; only the odd harmonics are present in the sound from a stopped pipe: f, 3f, 5f, etc. In both types of pipe most of the sound is reflected back from an open end for all frequencies where the aperture is small compared with the wavelength.

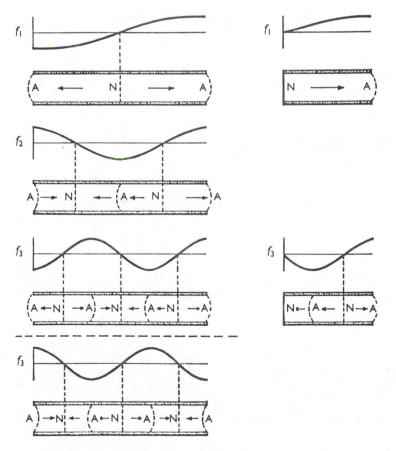

also be present. *Standing waves* are caused when any wave strikes a reflecting surface at right angles and travels back along the same path. A double reflection in a right-angled corner produces the same effect (though to a lesser degree, due to the greater loss in two reflections).

The air in a narrower pipe (such as an organ pipe) can be made to resonate. If it were closed at both ends there would be reflections, as at any other solid surface. But if the pipe is open at the ends, resonance can still occur. In a tube that is narrow in comparison with the wavelength, the sound has difficulty in radiating to the outside air.

The energy has to go somewhere: in practice, a pressure wave reflects back into the tube as a rarefaction, and vice versa. In this case the fundamental (twice the length of the pipe) and all of its harmonics may be formed.

If one end is open and the other closed, only the fundamental (the wavelength of which is now four times the length of the tube) and its odd harmonics are formed. The tone quality is therefore very different from that of a fully enclosed space, or again, from that of the pipe open at both ends.

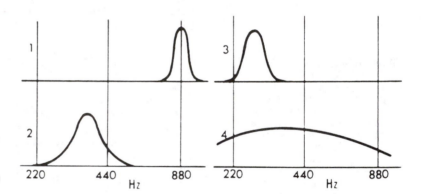

Wind instrumental formants

1, Oboe. 2, Horn. 3, Trombone. 4, Trumpet. The formants, imposed by the structural dimensions of parts of the instruments, may be broad or narrow. They provide an essential and recognizable component of each instrument's musical character.

Wind instruments in the orchestra form their sounds in the same way, the length of the column being varied continuously (as in a slide trombone), by discrete jumps (e.g. trumpet and horn), or by opening or closing holes along the length of the body (e.g. flute, clarinet or saxophone). Formants are varied by the shape of the body and the bell at the open end – though for many of the notes in instruments with finger holes the bell makes little difference, as most of the sound is radiated from the open holes.

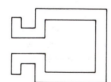

Helmholz or cavity resonator
This vibrates like a weight hanging on a spring. The volume of air inside the cavity is the 'spring'; the body of air inside the neck is the vibrating mass.

Another important way in which air resonance may govern pitch is where a volume of air is almost entirely enclosed and is connected to the outside through a neck. Such a device is called a *cavity* or *Helmholtz resonator.* It produces a sound with a single, distinct frequency, an example being the note obtained by blowing across the mouth of a bottle. In a violin the cavity resonance falls within the useful range of the instrument and produces a 'wolf-tone' that the violinist has to treat with care, bowing it a great deal more gently than

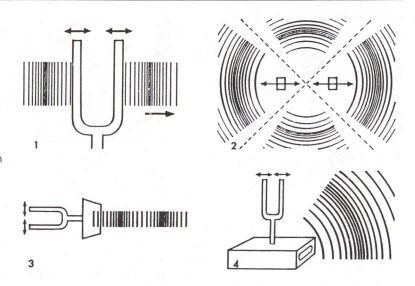

Tuning fork
1 and 2, Each fork vibrates at a specific natural frequency, but held in free air radiates little sound. 3, Placed against a wooden panel, the vibrations of the tuning fork are coupled to the air more efficiently and the fork is heard clearly. 4, Placed on a box having a cavity with a natural resonance of the same frequency as the tuning fork, the sound radiates powerfully.

other notes. Going back to the example of the tuning fork, boxes may be made up with volumes that are specific to particular frequencies. These make the best resonators of all – but each fork requires a different box.

The voice

The human voice is a great deal more versatile than any musical instrument. This versatility lies not so much in the use of the vocal cords to vary pitch as in the use of the cavities of the mouth, nose and throat to impose variable formant characteristics on the sounds already produced. It is as though a violin had five resonators, several of which were continuously changing in size, and one (the equivalent of the mouth) so drastically as to completely change the character of the sound from moment to moment.

These formant characteristics, based on cavity resonance, are responsible for vowel sounds and are the main vehicle for the intelligibility of speech. Indeed, sometimes they are used almost on their own.

For the effect of robot speech the formants are extracted from the human voice or simulated by a computer and used to modulate some simple continuous sound, the nature of which is chosen to suggest the 'personality' of the particular machine producing the sound.

In addition to the formants, a number of other devices are used in speech: these include sibilants and stops of various types which, together with the formant resonances, provide all that is needed for high intelligibility. A whisper, in which the vocal cords are not used,

Vocal cavities
1, Lungs. 2, Nose. 3, Mouth. This is the most readily flexible cavity, and is used to form vowel sounds. 4, Pharynx, above the vocal cords. 5, Sinuses. Together, these cavities produce the formants characteristic of the human voice, emphasizing certain frequency bands at the expense of others.

Human speech analysed to show formant ranges

1, Resonance bands. 2, Pause before plosive. 3, Unvoiced speech. 4, Voiced speech. These formants arise from resonance in nose, mouth and throat cavities. They are unrelated to the fundamental and harmonics, which are shown in the second analysis of the word 'see'. 5, The vocal harmonics are falling as the voice is dropped at the end of the sentence. But the resonance regions are rising as the vocal cavities are made smaller for the 'ee' sound.

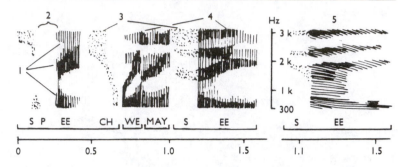

The ear

1, Sound enters via the outer ear and the auditory canal. This channel has a resonance peak in the region of 3–6kHz. At the end of the auditory canal is the ear drum, which vibrates in sympathy with the sound but is not well adapted to following large low-frequency excursions. 2, The sound is transported mechanically from the ear drum across the middle ear via three small bones which form an impedance-matching device converting the acoustic energy of the air to a form suitable for transmission through the fluid of the tiny delicate channels of the inner ear. The middle ear contains air: this permits the free vibration of the ear drum, and avoids excessive damping of the motion of the small bones. The air pressure is equalized through the eustachian tube, a channel to the nasal cavity. 3, The sound is pumped into the inner ear via a membrane called the oval window. 4, The inner ear is formed as a shell-like structure. There are two channels along the length of it, getting narrower until they join at the far end. Distributed along the upper channel are hair-like cells responding to particular frequencies: when the 'hair' is bent a nerve impulse is fired. The further it lies along the canal, the lower the frequency recorded. 5, A bundle of 4000 nerve fibres carries the information to the brain, where it is decoded.
6, Pressures in the cochlea are equalized at another membrane to the inner ear: the round window.

may be perfectly clear and understandable; in a stage whisper, intelligibility carries well despite lack of vocal power.

The vibrations produced by the vocal cords add volume, further character and the ability to produce song. For normal speech the fundamental may vary over a range of about 12 tones and is centred somewhere near 145 Hz for a man's voice and 230 Hz for a woman's. The result of this is that, as the formant regions differ little, the female voice has less harmonics in the regions of stronger resonance; so a woman may have a deep-toned voice, but its quality is likely to be thinner (or purer) than a man's.

For song the fundamental range of most voices is about two octaves – though, exceptionally, it can be much greater.

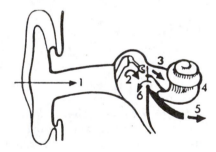

The human ear

The part of the ear that senses sound is a tiny spiral structure called the cochlea. It is a structure which gets narrower to one end of the coil, like the shell of a snail. But unlike the shell, it is divided lengthways into two galleries which join only at the narrow 'inside' end. The entire structure is filled with fluid to which vibrations may be transmitted through a thin membrane or diaphragm called the oval window. The acoustic pressures may then travel down one side of the dividing partition (the basilar membrane) and return up the other, to be lost at a further thin diaphragm, the round window.

All along one side of the basilar membrane are 'hairs' (actually elongated cells) that respond to movements in the surrounding fluid. Each of these standing in this fluid acts as a resonator system designed to respond to a single frequency (or rather, a very sharply tuned narrow band). The 'hairs' therefore sense sound not as an air particle sees it, as a single, continuous, very complex movement, but as a very large number of individual frequencies. The hairs are so arranged as to give roughly equal importance to equal musical intervals in the middle and upper middle ranges, but the separation is poor at very low frequencies: it is difficult to distinguish between very low pure tones. So (apart from indicating this lack of interest in low notes) a study of the mechanism of hearing confirms that we were right to be concerned with intervals that are calculated by their ratios rather than on a linear scale.

Growth of this sort is called *exponential*: it doubles at equal intervals, and a simple graph of its progress rapidly runs out of paper unless the initial stages are so compressed as to lose all detail. To tame such growth, mathematicians use a *logarithmic* scale, one in which each doubling is allotted equal space on the graph. The frequencies corresponding to the octaves on a piano progress in the ratios 1:2:4:8:16:32:64:128. The logarithms of these numbers are in a much simpler progression – 0,1,2,3,4,5,6,7 – and these equal intervals are exactly how we perceive the musical scale.

Frequency is not the only thing that the ear measures in this way: changes of sound volume follow the same peculiar rule.

Sound volume and the ear

The hairs on the basilar membrane vibrate in different degrees, depending on the loudness of the original sound. And they, too, measure this not by equal increases of sound intensity but by *ratios* of intensity: each doubling of intensity sounds roughly as loud again as the last. It is therefore convenient once again to use a logarithmic scale, and the measure that is used is the *decibel*, or dB. This is based on a unit called the bel, which corresponds to a tenfold increase or decrease. But the bel is inconveniently large, as the ear can detect smaller changes, and it is because of this that an interval a tenth of the size – the decibel – has been adopted.

The ratio of intensities in 1 decibel is about 1.26:1 (1.26 is approximately the tenth root of 10.) This is just about as small a difference in intensity as the human ear can detect in the best possible circumstances.

As it happens, the ratio of intensities in 3 dB is 2:1. This is convenient to remember, because if we double up a sound source we double the intensity (at a given distance). So if we have one soprano singing with gusto and another joins her singing equally loudly the sound level will go up 3 dB (not all that much more than the minimum detectable by the human ear). But to raise the level by another 3 dB two more sopranos are needed. Four more are needed for the next 3 dB – and so on. Before we get very far we are having to add sopranos at a rate of 64 or 128 a time to get any appreciable change of volume out of them. To increase intensity by increasing numbers soon becomes very expensive: if volume is the main thing you want it is better to start with something more powerful such as a pipe organ, a trombone, or a bass drum!

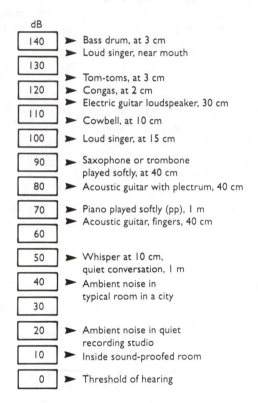

dB
140	➤ Bass drum, at 3 cm
	➤ Loud singer, near mouth
130	
	➤ Tom-toms, at 3 cm
120	➤ Congas, at 2 cm
	➤ Electric guitar loudspeaker, 30 cm
110	➤ Cowbell, at 10 cm
100	➤ Loud singer, at 15 cm
90	➤ Saxophone or trombone played softly, at 40 cm
80	➤ Acoustic guitar with plectrum, 40 cm
70	➤ Piano played softly (pp), 1 m
	➤ Acoustic guitar, fingers, 40 cm
60	
50	➤ Whisper at 10 cm, quiet conversation, 1 m
40	➤ Ambient noise in typical room in a city
30	
20	➤ Ambient noise in quiet recording studio
10	➤ Inside sound-proofed room
0	➤ Threshold of hearing

Sound pressure levels
Examples of sounds over the wide range that may reach microphone or ear.

Loudness, frequency and human hearing

The ear does not measure the volume of all sounds by the same standards. Although for any particular frequency the changes in volume are heard more or less logarithmically, the ear is more sensitive to changes of volume in the middle and upper frequencies than in the bass.

The range of hearing is about 20–20 000 Hz for a young person; but the upper limit falls with age to 15 000 or 10 000 Hz. Sensitivity is

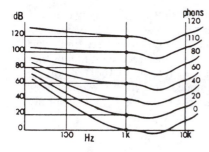

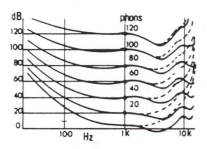

Equal loudness contours
Above: The classical curves of Fletcher and Munson, published in 1933. Intensity (dB) equals loudness in phons at 1 kHz.
Below: The curves measured by Robinson and Dadson take age into account and show marked peaks and dips in young hearing. The unbroken lines are for age 20; the broken lines show typical loss of high-frequency hearing at age 60. The lowest line in each set represents the threshold of hearing.

Eardrum and microphone diaphragm
The shortest wavelength that young ears can hear defines the size of diaphragm that is needed for a high-quality pressure microphone.

greatest at 1000 Hz and above: the auditory canal between the outer ear and the ear drum helps this by having a broad resonance in the region of 2000–6000 Hz.

Obviously *loudness* (a subjective quality) and the measurable volume of a sound are not the same thing – but for convenience they are regarded as being the same at 1000 Hz. Perceived loudness, in *phons*, can then be calculated from actual sound volume by using a standard set of curves representing average human hearing.

Figures for such things as levels of *noise* are also *weighted* to take hearing into account. This may refer to electrical as well as acoustic noise. For example, some types of microphone produce more noise in the low-frequency range: this is of less importance than if the noise were spread evenly throughout the entire audio range.

The lower limit is called the *threshold of hearing*. It is convenient to regard the lower limit of good human hearing at 1000 Hz as zero on the decibel scale. There is no natural zero: absolute silence would be minus infinity decibels on any practical scale. The zero chosen corresponds to an acoustic pressure of 2×10^{-5} pascals (Pa). One pascal is a force of one newton per square metre.

The upper limit is set by the level at which sound begins to be physically felt – a *threshold of feeling* at around 100–120 dB.

The ear and audio engineering

At all times when we are dealing with the techniques of sound we will bear in mind the capability of the human ear. The objective measurements of the engineer are meaningless unless we interpret them in this way. If we fall short of what the ear can accept we should be aware of this, and have good reasons. One is that the difference in cost may be too great for the difference in subjective appreciation. For example, an individual may be happy to spend money on equipment with a frequency response substantially level to 20 000 Hz, but for a broadcasting organization this may be too high. For most people hearing is at best marginal at 15 000 Hz; it is wasteful to go above this figure – particularly as the amount of information to be transmitted does not observe the same logarithmic law that the ear does. The octave between 10 000 and 20 000 Hz is by far the most 'expensive' in this respect.

A microphone is simply a device for converting sound into electricity in such a way as to retain the information content. The simplest way of doing this is quite adequate: *changes in pressure become changes in voltage; the electrical waveform is the analogue of the acoustical waveform.*

In the microphone, sound intensity is converted directly to electrical energy. But energy is proportional to the square of the voltage, and it is generally the voltage that we are interested in: fluctuations in voltage are amplified and carry the audio signal.

One result is that the full range of audible sound between the thresholds of hearing and feeling, a ratio in sound intensities of a million million to one, is reduced dramatically – to a mere million to one – in the corresponding voltages. In terms of decibels it means that the 120 dB of acoustic power are equivalent to only 60 dB of voltage change. Note that where the performance of microphones and other audio equipment is described in terms of 'decibels' that generally means the electrical level.

Electrical and mechanical analogues of sound
Early recording and transmission systems used analogues of the original sound wave. 1, Disc. 2, Tape. 3, AM radio. Digital techniques may replace all three, but the signal must eventually be converted back into the original form before it can be heard.

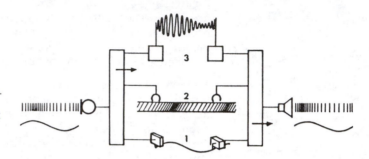

The electrical signal derived from the sound can in turn be used to magnetize particles on a moving strip of tape: it may be subject to volume control and other forms of manipulation, but remains an analogue system throughout. Alternatively, the signal may be sampled (at least twice as often as the highest required frequency) and the data generated is recorded digitally. This process may be partly or fully under computer control. However, a digital signal must, eventually, be returned to analogue form to drive the air which carries sound to our ears.

Note, however, that this book is not concerned with engineering aspects of the equipment used for recording or transmitting sound information except in so far as they affect operational techniques.

A special engineering problem that affects all aspects of operation is the efficient conversion of electrical signals back to sound energy – in other words, the design and construction of loudspeakers. Loudspeaker design (discussed briefly in Chapter 9) is probably the principal limiting factor in most sound systems, and should be studied by those with an interest in audio engineering, but is also aside from the main theme of this book.

Chapter 3

Stereo

The object of stereophony is to lay before the listener a lifelike array of sound – both direct sound and that which has been reflected from the walls of the original studio – which recreates some real or simulated layout in space.

In its simplest form, this *sound stage* is reproduced by a pair of loudspeakers, each of which has a separate signal fed to it. Sound reproduced by a single speaker appears to come directly from it. If, however, the sound is split and comes equally from the two speakers it seems as though it is located half-way between the two. Other directions on the sound stage are simulated by other mixtures being fed to the two speakers. It is usual to refer to the left-hand loudspeaker as having the 'A' signal and the right-hand one as having the 'B' signal. (In some manufacturers' literature they are called 'X' and 'Y'.)

A serious disadvantage of mono is that reverberation is superimposed on the same point-source of sound, reducing the clarity of music and intelligibility of speech. In stereo, the reverberation is spread over the full width of the sound stage – a vast improvement, although in two-channel stereo it still does not surround the listener.

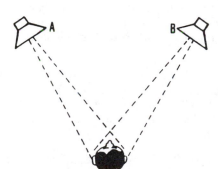

Hearing stereo
The position of apparent sources is perceived by time, amplitude and phase differences between signals received by the left and right ears, and (at high frequencies) by the effects of shadowing due to the head. For any particular frequency the phase at each ear is given by the sum of the sounds arriving from the two loudspeakers.

Two loudspeakers

A pair of stereo loudspeakers must work together in every sense. But the very minimum requirement is that if the same signal is fed to both, the cones of both loudspeakers move forward at the same time; in other words, the loudspeakers must be wired up to move in phase with each other. If they move in opposite senses an effect of sorts is obtained, but it is not that intended. For example, a sound that should be in the centre of the sound stage, the most important part, is not clearly located in the middle, but has an indeterminate position.

A second requirement is that the loudspeakers be matched in volume and frequency response. If the volume from one is greater than that from the other, the centre is displaced and the sound stage distorted. If the difference is as great as 20 dB, the centre is displaced to the extreme edge of the sound stage. The exact matching of frequency responses is less important to the average listener. Of a number of possible effects, probably the most important is that the differences could draw attention to the loudspeakers themselves, thereby diminishing the illusion of reality.

In the studio, operational staff need to work in fully standardized conditions, sitting on the centre line of a pair of matched high-quality loudspeakers placed about 8 ft (2.5 m) apart. It is recommended that the home listener should have his loudspeakers between 6 and 10 ft (about 2–3 m) apart, depending on size of room. With less separation the listener should still hear a stereo effect, but the width of the sound stage is reduced. On the other hand, the loudspeakers should not be too far apart. For normal home listening, if one person is on the centre line, others are likely to be 2 ft (say 60 cm) or more to the side, and the further the loudspeakers are apart, the greater the loss of stereophonic effect to the outer members of the group. If the listener is as little as one foot (30 cm) closer to one loudspeaker, phase effects that distort the sound stage are already marked (see later), and as the difference in distance grows, the nearer source progressively appears to capture the whole signal.

This second distortion of the sound stage – indeed, its near collapse – is induced by the *Haas effect*. This derives from the capacity of the ear to localize the source of a sound, distinguishing its direction from that of reflected sound, which is delayed by a few thousandths of a second (milliseconds, ms) even in a small room: in 1 ms sound travels just over 1 ft (34 cm).

The Haas effect is at its strongest for delays of 5–30 ms, which is produced by a difference in path lengths of some 6–18 ft (about 2–7 m). The capacity diminishes a little with increasing delay, but still remains effective, until eventually the reflection of a staccato sound can be distinguished as a separate echo.

But it is what happens at the lower end of this scale of delays and their corresponding distances that is more important for our perception of stereo. As the difference in distance from the two loudspeakers grows from zero to about 5 ft (1.5 m), the Haas effect grows rapidly and continuously to approach the maximum, where it would take an increase of 10 dB in volume from the farther loudspeaker source for it to begin to 'recapture' the source. Plainly, an optimum stereo effect is achieved only in a narrow wedge-shaped region along the centre line, and this is narrower still if the loudspeakers are widely separated.

Between the reduced spatial effect of narrow separation and the reduced optimum listening area created by a broad layout, it may

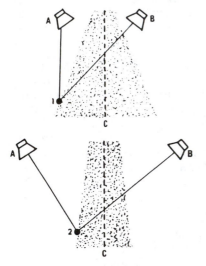

Stereo loudspeakers: spacing
With the closer loudspeakers, *above*, the spread of the stereo image is reduced, but there is a broad area of the listening room over which reasonable stereo is heard. With broader spacing, *below*, the stereo image has more spread, but for the benefit of a relatively narrow region in which it can be heard at its best. For listeners off the centre line C, the growing difference in distance rapidly 'kills' the stereo, due to the Haas effect. Position 1 (with a narrow spacing of speakers) has the same loss of stereo effect as position 2 (with the broader spacing).

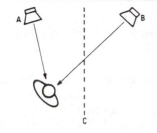

seem surprising that stereo works at all. The fact is, most casual listeners happily accept a sound picture that is spatially severely distorted. This is their prerogative: the sound man should be aware that his listeners often prefer their own seating plan to his clinical ideal.

Reduced loudspeaker separation

Haas effect
For a listener away from the centre line C, the difference in distances means that the sound from the farther loudspeaker arrives at the listener's ears later than that from the nearer one. The delay is 1ms for every 13.4 in (34 cm) difference in path length. For identical signals coming from different directions, a difference of only a few milliseconds makes the whole sound appear to come from the direction of the first to arrive. The effect can be reversed by increasing the volume of the second signal: the curve shows the increase in volume required to re-centre the image.

Many people accept a reduced loudspeaker separation in their own homes, either in portable or compact audio equipment, to which loudspeakers are integral or attached on either side. Television stereo is also usually heard through loudspeakers attached to the receiver itself. The sound stage in these cases is physically reduced in size to less than half or even a quarter of that available to the music balancer. The audio image will therefore be correspondingly reduced in size. For television, this will match the size of the visual image, but will limit any possibility of suggesting off-screen sound sources.

Fortunately there are compensations. One is that the Haas effect is reduced, permitting off-centre-line viewing – necessary for groups. There is still some separation of point sources and, equally important, reverberation is spread away from the original source, which increases intelligibility and is more pleasant to listen to.

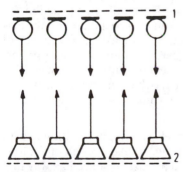

Stereo by reconstructed wavefront
1, Sound pressures at many points in a large surface sampled by many microphones. 2, The sound is reproduced at corresponding points in a similar surface. The biggest practical problem in this arrangement is the very large number of channels of communication required.

Alternatives to two loudspeakers

What alternative is there to the two-loudspeaker stereo layout?

One is to dispense with the loudspeakers entirely and to wear *stereo headphones* instead. These are convenient where people of different tastes are crowded together (particularly on an aircraft) and one advantage is that the spread of the image is the same for all listeners with normal hearing. But several characteristics of the two-speaker system are lost: one is that sound from *both* loudspeakers should reach *both* ears, and do so from somewhere to 'the front'; and another is that the sound stage is static – it does not swing around as the listener turns his head. As a result, listening with stereo headphones provides a different aural experience. But for most people this is less important than whether they are happy to spend substantial periods of time so cocooned and constrained, and whether they are prepared to accept, or perhaps prefer, the reduction in social contact that results.

The sound balancer should not use headphones to monitor stereo that will be heard by loudspeaker. For the location recordist this may be unavoidable, but the sound will usually be checked on loudspeakers at some later stage.

Stereo by wave reconstruction is a theoretical alternative to the two-speaker system in which the wavefront in the studio is sensed at many points so that it can be reconstructed at a similar number of points in the listening room. This requires a large number of microphones, transmission lines and loudspeakers, so for domestic use the idea is clearly impractical, but a modified multi-loudspeaker system is often used for stereo in cinemas – though for people sitting near the front or the side walls, individual or pairs of loudspeakers are bound to dominate the sound that they hear. The Haas effect is severe: most people are more than 6 ft (2 m) closer to some loudspeakers than to others.

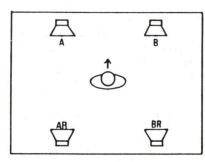

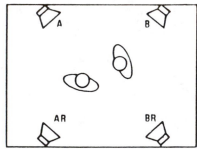

Four loudspeakers
Above: A and B are equivalent to two-channel stereo from the front, and AR and BR are placed at the back of the listening room (R indicates 'rear'). If most of the direct sound comes from the front and the spacing of the A and B loudspeakers is greater than for two-channel stereo, the position of the listener (close to the centre line) is correspondingly critical. If there is direct sound from all speakers the effect is multiplied: away from the centre of the room individual loudspeakers tend to dominate, and this is increased still further in the symmetrical layout, *below.*

Quadraphony

Quadraphony (or 'quad') is a system that has been tried as a potential compromise, but its technical merit did not attract a large enough public to make it a commercial success. Nevertheless, the application of well-developed quadraphonic techniques has left us a substantial library of recordings. There are still some specialist applications, especially as the starting point for the even more extensive spread used for certain big-screen films, so the system will be described briefly.

In quad, two further signals are fed to loudspeakers on the opposite side of the room from the normal A and B speakers. There are two ways in which this can be used. One is to surround the listener with direct sources, so that he is placed in the middle of the action. To bring individual sources forward they can also be 'placed' on the axis between pairs of diagonally opposed loudspeakers. The effect of being in the middle of the action can be dramatic and exciting. However, little existing classical music requires this, although notable exceptions include the Berlioz *Requiem* (which employs a quartet of brass bands) and early antiphonal church music by composers such as Gabrieli and Monteverdi.

The alternative use for the four channels is more conventional, with only the indirect sound, i.e. reverberation, spread over all of the four speakers, so that the listener is offered a sense of space in a way that cannot be achieved by the acoustics of a relatively small listening room. In fact, this in itself was sufficient to justify the system, without the need for more than an occasional *jeu d'esprit.* Indeed, many listeners prefer that their room has some orientation, with a 'front' and a 'back'.

A further complication arises (again) from the Haas effect. For the proper placing of sounds between loudspeakers, a speaker in each corner of the room renders the position of the listener very critical, except for sources that sit unequivocally in a particular loudspeaker. If two-channel stereo (or quadraphonics in which the rear speakers carry reverberation only) is reproduced on a four-loudspeaker system, it is better if the A and B channels are spaced as for conventional stereo. In this case the power handling capacity of the rear loudspeakers need not be as great as those in front and their spacing is less critical. Discrete quadraphony (in which the four channels are separately recorded) gives the best positional information at the sides and back.

Hearing and stereo

How, in real life, do our ears and brain search out the position of a sound and tell us where to look for its source? First, the Haas effect allows us to distinguish between direct and reflected sound (except in the rare case where the reflections are focused and their increased loudness more than compensates for the delay in their arrival). A second clue is offered by differences in the signal as it arrives at our two ears: a sound that is not directly in line with the head arrives at one ear before the other. At some frequencies the sounds are therefore partly out of phase, and this may give some indication of angle. Such an effect is strongest at about 500 Hz.

The head itself begins to produce screening effects for frequencies at 700 Hz and above, resulting in differences in the amplitude of signals reaching the ears from one side or the other. Sound-effective ridges in the outer ear help to distinguish front and rear. Further aids to direction finding are provided by movements of the head, together with interpretation by the brain in the light of experience. Even with all these clues, with our eyes closed we might still be uncertain: we can distinguish, separate out and concentrate our attention on the contents of particular elements of a complex sound field far better than we can pinpoint their components precisely in space. The clues tell us where to look: we must then use our eyes to complete and confirm the localization . . . if, that is, there is any picture to see. If there is none, the fact that in real life we rely rather heavily on the visual cues is perversely rather helpful to us: it permits us to continue to enjoy the spatially distorted image even if we are a little off the centre line. Two-speaker stereo measures up to this specification quite well.

Assume first that the source is in the centre of the sound stage and that the listener is on the centre line. The A signal reaches the left ear slightly before the B signal, and for middle and low frequencies the two combine to produce a composite signal that is intermediate in phase between them. For the right ear the B signal arrives first, but the combination is the same. The brain, comparing the two combined signals, finds them to be the same and places the source in the centre. If the amplitude of the sound from one speaker is now increased and the other reduced, the signals combine at the ears as before, but the resultant signals differ from each other in *phase* – in effect, it appears

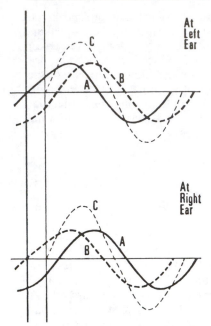

Two-loudspeaker stereo: central
The signals from the A and B loudspeakers are equal: the B signal takes longer to reach the left ear, and the A signal takes longer to reach the right. But the combined signal C is exactly in phase: the image is in the centre.

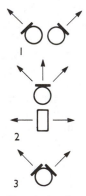

Stereophonic microphones
The method of picking up stereophonic sound that will be recommended in this book is the co-incident pair: two directional microphones very close together (1) or in a common housing, usually one above the other. The microphone elements will not necessarily be at 90° to each other. 2, Main or 'middle' and side microphones used as a co-incident pair. 3, Symbol used in this book for a co-incident pair, which may be either of the above.

as though the same signal is arriving at the two ears at slightly different times. This was one of the requirements for directional information.

If the listener moves off the centre line, or if the volume controls of the two loudspeakers are not set exactly the same, the sound stage is distorted, but there is still directional information that matches the original to some reasonably acceptable degree.

Microphones for stereo

In order to produce a stereo signal that can (after various processes) be fed to loudspeakers or headphones, the sound field has to be sampled by microphones in such a way as to produce two sets of directional information. This can be achieved in a variety of different ways. Given that loudspeakers are separated in space, it might seem that the appropriate technique would be to sample the field at two points spaced at a similar distance. However, the separation of the loudspeakers already produces phase cancellation effects, and separation of the microphones can only add to this.

Far greater control is achieved by placing them as close together as is physically possible, but pointing them in different directions. They are then virtually equidistant from all sound sources, and although (by their different directional characteristics) they may pick up different volumes, the two signals will be in phase throughout most of the audio-frequency range. Only at extreme high frequencies will the slight separation produce any phase differences, and because this region is characterized by erratic, densely packed overtones, even drastic phase changes will be relatively unimportant.

This first arrangement is called the *co-incident pair*, or *co-incident microphone* technique. There is a range of variations on it, but typically the two microphones feeding the A and B information are mounted one above the other or closely side by side and are directional, so that the A microphone picks up progressively more of the sound on the left and the B microphone progressively more on the right. Both pick up equal amounts of sound from the centre, the line half-way between their directional axes. For this to work well, the frequency characteristics of the two microphones must be virtually identical.

An alternative kind of co-incident pair uses radically different microphones, one behind the other. The front microphone covers most of the sound field, including all required direct sound. It produces the *middle, main* or '*M*' *signal* – which, on its own, is also mono. The second microphone favours sound from the *sides*, producing the '*S*' *signal*; it contains information required to generate stereo, but is not, by itself, a stereo signal. Sounds from the two sides are opposite in phase, so that if the M and S signals are added together they produce a signal that is biased to one side of the sound field – the left – and if S is subtracted from M it is biased the other way. Quite simply, M + S = A and M − S = B.

Whether by analogue or digital techniques, the conversion can be made very easily; indeed, its very simplicity can easily lead to confusion. For example, if the two tracks of a stereo recording are added to make a single-audio-track copy, this will work well if the originals were A and B – but if they were M and S, half the sound will disappear. (This could happen when a viewing copy of a videorecording is made if no proper sound instructions are given.)

Even so, there can be advantages to working with M and S signals, rather than A and B. In particular, analogue A and B signals taking similar but not identical paths may be changed differentially in volume or phase; if this happens to the M and S signals, the perceived effect is much less. Note also that because A + B will in general be greater than A − B, M is correspondingly greater than S, offering a useful visual check (by meter) that the original signal is not grossly out of phase.

If the M and S signals are produced in the microphone itself, its elements must operate by different principles and may have different frequency responses (see later), but it is unlikely that this will be perceived as a deficiency. The S signal will usually contain a higher proportion of reflected sound, and the ear expects this to be erratic.

Although co-incident pairs are the primary system recommended in this book, they are often augmented or may be replaced by spaced mono microphones. *Spaced microphone* techniques have been widely used, and in skilled hands have produced excellent results. Over the years, balancers who have been prepared to experiment with competitive techniques have, at best, improved the range of balances heard, but have also tended to produce a cycle of fashions in the kinds of balance that are currently most in favour.

The only test that can be applied is practical, if subjective: does the balance sound right? Experiments with a well-spread subject such as an orchestra soon demonstrate that one pair of widely spaced microphones do *not* sound right: there is plainly a 'hole in the middle'.

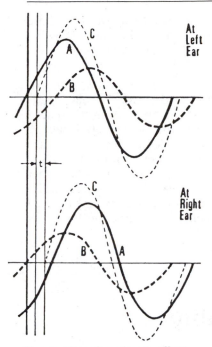

Two-loudspeaker stereo: offset
Signal A is increased. The effect of this is to make the peak in the combined signal C at the left ear a little earlier, and that at the right ear, later. There is therefore a slight delay (*t*) between the two signals C: this phase difference is perceived as a displacement of the image to the left of centre. In these diagrams it is assumed that the listener is on the centre line between A and B, and facing forwards. If, however, he turns his head this will cause the phase difference at his ears to change – but in such a way as to keep the image roughly at the same point in space. If the listener moves away from the centre line the image will be distorted, but the relative positions of image elements will remain undisturbed.

Spaced pair
Left: Two microphones. For subjects along the centre line the apparent distance from the front of the audio stage is greater than for subjects in line with the microphones; and the farther forward the subject is, the more pronounced the effect. It is called 'hole-in-the-middle'.
Right: The introduction of a third microphone in the centre does a great deal to cure this.

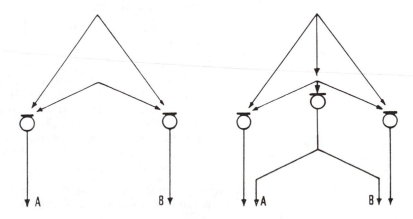

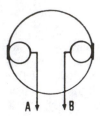

Dummy head
Microphones mounted in a solid baffle of head size. When used in combination with a two-loudspeaker system this cannot be justified on theoretical grounds. For headphone listening the effect is satisfactory.

Sources right at the front in the centre sound farther away than those at the sides which are closer to the microphones. In order to combat this, the usual technique is to add a third forward microphone in the centre, the output of which is split between the two sides.

Another technique that has been used is a *dummy head* with two pressure microphones in the position of the ears. The analogy is obvious, but is not related to the stereo we hear from spaced loudspeakers. There is a spatial effect, because the A and B signals are slightly different, especially in high frequencies at which the obstacle effect of a real head helps to discriminate position. But the test is whether it works in practice and it does not, in comparison with other stereo techniques (except when heard on headphones, as in airline in-flight entertainment). In all cases where loudspeakers may be used by at least some part of the audience, the earlier technique is recommended.

Stereo compatibility

If a stereo signal produces acceptable mono sound when reproduced on mono equipment, it is said to be *compatible*. As we have seen, the M signal $(A + B)$ may be compatible, but this is not automatic: it would be easy to place an M and S microphone in such a way that it is not. In order to achieve compatibility, the M component should include all the required direct sound sources in acceptable proportions, together with enough indirect sound (reverberation) to produce a pleasing effect, but not so much as to obscure it.

Domestic listening conditions vary greatly: in particular, the stereo component may be debased or absent. To cover this range, compatibility is desirable. Only where the listening conditions can be controlled is compatibility dispensable, but this means also that the material cannot be transferred for other uses without further, special processing. There may be good reason for incompatibility (such as cost, secrecy or copyright) but, if not, it should be built in to any system as a matter of course.

Historically, many analogue stereo systems have been so arranged, for example, on stereo records.

If a disc is recorded so that $A + B$ corresponds to lateral displacement, the mono signal can be picked up on a mono record player without difficulty. The $A - B$ signal is recorded in the vertical plane, by hill-and-dale signals: a mono stylus simply rides up and down but has no means of converting this information into an electrical signal. (In practice, however, the stylus tip radius, vertical stiffness and playing weight should all be less than for a traditional mono record, or the groove may be damaged.)

Similar advantages are obtained from using $A + B$ and $A - B$ signals in radio. When stereo was introduced, the $A + B$ signal occupied the place in a waveband previously used for the mono signal; the $A - B$ signal was modulated on to a sub-carrier above the main $A + B$ band.

The analogue method of recording stereo on magnetic tape is to put the A and B signals side by side on separate tracks. A full-track head reproduces both together, or alternatively the combined signal can be obtained electrically by adding the two outputs.

Digital systems have replaced or are replacing many of these intervening stages. Provided that the A and B or M and S signals are interlocked so they cannot be displaced, they can be converted back into an analogue system that preserves compatibility.

Chapter 4

Studios

A studio is a dedicated acoustic environment. This can be defined in terms of its purpose and layout, its acoustics and treatments that can be used to modify them, and its furnishings. These are described for radio and recording studios and also for television. Film studios, with less electronic equipment, and dedicated to successive individual shots, are similar to but perhaps simpler than television studios.

Studios also vary according to the economic needs and the size of the country, region or community they serve, and to their system of financing. Some must necessarily be for general (and perhaps limited) purposes, whereas others may specialize.

The range and function of studios

The cost of television is such that the number of studios that can be supported by any given population is limited. A regional population of 5–8 million is generally sufficient to sustain several centres, each with about three studios for major productions (including drama), for general purposes (including entertainment, with audience), and for speech only (presentation and news), plus an outside broadcast unit.

With a smaller population it may be difficult to sustain even this number: there may be only two studios or even one. But there would also be less use for an outside broadcast unit, which has in any case been partly displaced by the increased use of single videocameras (camcorders).

At the lower end of the scale, towns serving areas of a million or so may have single-studio stations (or more, in prosperous regions). But these studios will be small, big enough for only two or three setting areas of moderate size.

A national population the size of the UK or the USA is sufficient to sustain four or more national broadcast networks, augmented by many local and cable or satellite stations.

In radio (living, in most countries, in the shadow of television), a more confused situation exists. In the USA there is a radio station to every 25–30 000 population. Of the vast total number, half subsist on an almost unrelieved output of 'middle-of-the-road', 'conservative' or familiar 'wall-to-wall' music in some narrowly defined style. Then there are the top-forty, the country-and-western and the classical specialists, plus stations that offer full-time news, full-time conversation (e.g. telephone chat) or full-time religion. Educational stations have a broader range, broadcasting both speech and music. In addition, a few 'variety' stations still exist, segmented in the old style and containing an integrated range of drama, documentary, quiz, comedy, news and so on; and there are also ethnic stations which may be impelled to serve the broader interests of a real, defined community. Most of these survive with low operational costs by using records or syndicated tapes for much of their output, and their studio needs are necessarily simple: perhaps no more than a room converted by shuttering the windows against the low-frequency components of aircraft noise that heavy drapes will not impede. Acoustic treatment can be added on the inside of the shutters. For some stations, a second, general-purpose studio may be partly justified by its additional use for board or staff meetings.

But looking again at our region with 5–8 million people, we now have a population which can sustain a full range of studios for all of the specialized functions that the current purposes of radio permit. These are likely to be fewer in the USA, and other countries that have followed its example in dedicating commercial stations to a narrow range of output, than in public service broadcasting, where a regional centre may have the following studios:

- speech only (several)
- pop music
- light and orchestral music
- light entertainment (with audience)
- general purpose, including dramas and dramatized educational programmes

Smaller regions (100 000–1 million population) are capable of sustaining vigorous, but more limited, radio stations. National populations are capable of sustaining networks in about the same number as for television (though these are much more fragmented in the USA than in the UK, where they all reach almost the entire population). The BBC has about 60 radio studios in London alone, though many are for overseas broadcasting. Some are highly specialized – for example, those which are acoustically furnished for dramatic productions in stereo.

Noise and vibration

When designing or choosing buildings for the installation of radio or television studios, here are some points to consider:

● Do not build near an airport, railway or over underground lines.
● Prefer massive 'old-fashioned' styles of construction to steel-framed or modern 'component' architecture.
● In noisy town centres, offices that are built on the outside of the main studio structure can be used as acoustic screening. These outer structures can be built in any convenient form, but the studio must be adequately insulated against noise from *them*.
● The best place to put a studio is on solid ground.
● If a radio studio cannot be put on solid ground the whole massive structure, concrete floor, walls and roof can be floated on rubber or other suitable materials. The resonance of the whole floating system must be very low: about 10 Hz is reasonable. One inch of glass fibre or expanded polystyrene gives resonances at about 100 Hz, so care is needed. An alternative technique is to suspend the entire structure.
● Airborne noise from ventilation is a major problem. Low-pressure ducts are broad enough to cause resonance problems if not designed with care: they also require acoustic attenuation within them to reduce the passage of sound from one place to another. Grilles can also cause air turbulence and therefore noise.
● High-speed ducts have been suggested as an answer to some of the problems of airborne noise, but would also require careful design, particularly at the points where the air enters and leaves the studio.
● Double doors with pressure seals are needed.
● Holes for wiring should be designed into the structure and be as small as possible. They should not be drilled arbitrarily at the whim of wiring engineers. Wiring between studio and listening room should pass round the end of the wall between them, perhaps via a corridor or lobby.
● Windows between studio and control area should be double glazed and of different glass thicknesses, say $\frac{1}{4}$ in and $\frac{3}{8}$ in (6 and 9 mm), and the air space between them should be at least 4 in (100 mm). Contrary to what might be expected, it does not seem to matter if they are parallel. Triple glazing is not necessary, as the higher levels of sound will be the same in both places.
● Windows to areas outside the studio area *or to the control area if the loudspeaker is to be switched to replay during recordings* may be triple glazed if replay or noise levels are likely to be high. Measures to avoid condensation between the glass panes will be necessary only in the case of windows to the outer atmosphere, where there will be temperature differences.
● Any noisy machinery should be in a structurally separate area, and should be on anti-vibration mountings.

Although massive structures may have been recommended, *extremely* massive structures may be more expensive than their improvement warrants. *Doubling* the thickness of a brick wall gives an improvement of only 5 dB.

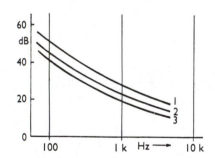

Permissible background noise in studios
1, Television studios (except drama and presentation). 2, Television drama and presentation studios and all radio studios except those for drama. 3, Radio drama studios.

For television studio roofs the limiting thickness for single-skin construction is about 10 in (25 cm): this gives an average attenuation of 60 dB over the range of 100–3200 Hz. BBC research indicates that for noisy jet aircraft flying at 1000 ft, a minimum of 65 dB attenuation is desirable. Even low-flying helicopters are less of a problem than this (provided they do not land on the studio roof), but supersonic aircraft producing a sonic boom with overpressures of 2 lb/ft² (9.6 kg/m²) would require 70 dB attenuation. Since a double skin is in any case desirable for more than 60 dB attenuation, it may be wisest to adopt the higher standard – 70 rather than 65 dB attenuation. This results, in fact, in a structure which is little heavier than a 60 dB single skin.

In BBC experiments there have been subjective tests using both types of aircraft noise attenuated as if by various roofs and added to television programme sound at quiet, tense points in the action. In the limiting case the action may be part of a period drama, in which aircraft noise would be anachronistic.

Obviously, permissible noise levels depend on the type of production. For example, in television drama background noise from all sources (including movement in the studio) should ideally not exceed 30 dB (relative to 2×10^{-5} Pa), at 500 Hz. For light entertainment the corresponding figure is 35 dB. However, in practice, figures up to 10 dB higher than these are often tolerated.

Reverberation

The studio structure is a bare empty box of concrete (or some similar hard and massive material). It will reflect most of the sound that strikes it; and this will be almost independent of frequency – though very broad expanses of flat thin concrete may resonate at low frequencies and so absorb some of the low bass.

When normal furnishings are placed in a room they include carpets, curtains and soft chairs. These (together with people) act as *sound absorbers*. Other furnishings include tables, hard chairs, and other wooden and metal objects. These reflect much of the sound striking them, but break up the wavefronts. They act as *sound diffusers*. Some things do both: a bookcase diffuses sound and the books absorb it.

The absorption or diffusion qualities of objects vary with frequency. In particular, the dimensions of an object condition how it behaves. A small ornament diffuses only the highest frequencies; sound waves that are long in comparison to its size simply pass round it. The thickness of a soft absorber affects its ability to absorb long wavelengths, and so does its position. At the hard reflecting surface of a rigid wall there is no significant air movement, so 1 in (2.5 cm) of sound-absorbing material reaches out into regions of substantial air movement (and damps them down) only for the highest frequencies; to have any effect on lower frequencies, the absorber must be well away from the wall.

Sounds in an enclosed space are reflected, many times, with some (great or small) part of the sound being absorbed at each reflection. The rate of decay of reverberation defines a characteristic for each studio: its *reverberation time*. This is the time it takes for a sound to die away to a millionth part of its original intensity, i.e. through 60 dB. Reverberation varies with frequency, and a studio's performance may be shown on a graph for all audio frequencies. Alternatively it may be given for, say, the biggest peak between 500 and 2000 Hz; or at a particular frequency within that range.

Reverberation time depends in part on the distance that sound must travel between reflections, so large rooms generally have longer reverberation times than small ones. This is not only expected but also, fortunately, preferred by listeners.

Coloration

In a large room an *echo* may be detected. If there is little reverberation in the time between an original sound and a repetition of it, and if this time gap exceeds about an eighteenth of a second – which is equivalent to a sound path of some 60 ft (18 m) – it will be heard as an echo.

In a small room, *coloration* may be heard. This is the selective emphasis of certain frequencies or bands of frequencies in the reverberation. Short-path reflections (including those from the ceiling) with less than 20 ms delay interfere with direct sound to cause cancellations at regular intervals throughout the audio-frequency range, resulting in a harsh *comb filter* effect.

In addition, there may be *ringing* between hard parallel wall surfaces which allow many reflections back and forth along the same path: at each reflection absorption always occurs at the same frequencies, leaving others still clearly audible long after the rest have decayed.

Eigentones, the natural frequencies of air resonance corresponding to the studio dimensions, are always present, and in small rooms are spaced widely enough for individual eigentones to be audible as ringing. If the main dimensions are direct multiples of, or in simple ratios to, each other, these may be further reinforced.

The rate of absorption in different parts of a hall may not all be the same, giving rise to anomalous decay characteristics. For example, sound may die away quickly in the partially enclosed area under a balcony, but reverberation may continue to be fed to it from the main hall. More unpleasant (though less likely) might be the reverse of this, where reverberation was 'trapped' in a smaller volume and fed back to a less reverberant larger area. Resonances may interchange their sound energies: this happens in the sounding board of a piano and is one of the characteristics which we hear as 'piano quality'. In a room it may mean that a frequency at which the sound has apparently disappeared may recur before decaying finally.

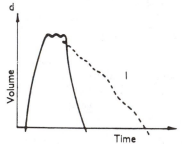

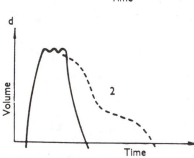

How sounds die away
1, In a good music studio the sound dies away fairly evenly. 2, In a poor music studio (or with bad microphone placing) the reverberation may decay quickly at first, and then more slowly.

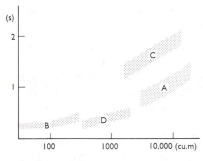

Radio studios
Typical reverberation times for:
A, Programmes with an audience.
B, Talks and discussions. C, Classical
music. D, Drama, popular music.

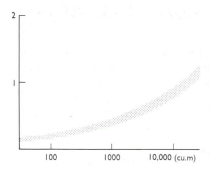

Television studios
Typical reverberation times for
general purpose studios.

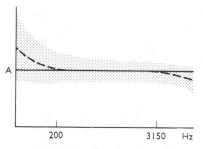

Range of reverberation times
A, Median for any particular type.
Below 200 Hz the reverberation in
talks studios and their control rooms
may rise; above 3150 Hz, for most
studios it should fall by 10–15%. For
a large music studio the tolerance is
about 10% on either side of the
average; for other studios and
control rooms it may be a little wider.

Not all of these qualities are bad if present only in moderation. They may give a room or hall a characteristic but not particularly unpleasant quality: something, at any rate, that can be lived with. In excess they are vices that may make a room virtually unusable, or usable only with careful microphone placing.

The answer to many of these problems – or at least some improvement – lies, in theory, in better diffusion. The more the wavefronts are broken up, the more the decay of sound becomes smooth, both in time and in frequency spectrum.

In practice, however, if the studio geometry has no obvious faults, acoustic treatment has been carried out and panels of the various types that may be necessary for adequate sound absorption have been distributed about the various walls and on the ceiling, little more generally needs to be done about diffusion.

Studios for speech

Studios that are used only for speech are often about the same size as a living-room in an ordinary but fairly spacious house. But where demands on space result in smaller rooms being used, the results may be unsatisfactory: awkward, unnatural-sounding resonances occur which even heavy acoustic treatment will not kill. And, of course, acoustic treatment that is effective at low frequencies reduces the working space still further.

Coloration, which presents the most awkward problem in studio design, is at its worst in small studios, because the main resonances of length and width are clearly audible on a voice containing any of the frequencies that will trigger them off. An irregular shape and random placing of the acoustic treatment on the walls help to cut down coloration. But in small rectangular studios treatment may be ineffective, and this means working closer to the microphone. With directional microphones, bass correction must be increased and the working distance restricted more than usual.

But coloration, as already indicated, is not necessarily bad; a natural sounding 'indoor' voice always has some. But it has to be held to acceptable limits, which are lower for monophonic recordings. Even on a single voice there is a world of difference between mono and stereo: with mono all of the reverberation and studio coloration is collected together and reproduced from the same apparent source as the speech. Note that even when the final mix is to be in stereo, pre-recorded voices are usually taken in mono, so the acoustics of (for example) a commentary booth should allow for this.

At first sight the answer might appear to be to create entirely 'dead' studios, leaving only the acoustics of the listener's room. But listening tests indicate that most people prefer a moderately 'live' acoustic – whether real or simulated. The experience of the BBC has suggested that, for normal living-room dimensions, 0.35–0.4 s is about right for speech when using directional microphones. A smaller room would need a lower time – a quarter of a second, perhaps.

Remembering that reverberation time is the time it takes for a sound to die away through 60 dB, you can use a hand-clap to give a rough guide to both duration and quality. In a room that is to be used as a speech studio the sound should die away quickly, but not so quickly that the clap sounds muffled or dead. And there must certainly be no 'ring' fluttering along behind it.

Listening rooms and sound-control cubicles should have similar acoustics, with a maximum of 0.4 s at frequencies up to 250 Hz, falling to 0.3 s at 8000 Hz.

General-purpose sound studios

In Britain there is still a need for studios in which a variety of acoustics is used for dramatic purposes. In some other countries such facilities can hardly be justified by their use except, perhaps, to add interest to commercials. They may also be used for special projects to be issued as records. This studio type is therefore still important, though much less so than when radio was the principal means of broadcasting.

In a dedicated drama studio there are two main working areas, usually separated by no more than a set of curtains. One end, the 'dead' end, has a speech acoustic that is rather more dead (by about 0.1 s) than would be normal for its size. The other end is 'live', so that when used for speech the acoustic is very obviously different from that of the dead end. If the curtains are open and microphones in the dead end can pick up sound from that direction the resonance of the live end can sometimes be heard on speech originating in the dead end. This parasitic reverberation generally sounds unpleasant unless there is a good reason for it, so the normal position for the curtains is such as to cut off the live end (i.e. at least partly closed). However, useful effects can sometimes be obtained near the middle of the studio.

The live end can also be used for small musical groups working with speech in the dead part of the studio.

There may also be a very dead area; though if this is small it may sound like the inside of a padded coffin rather than the open air it is supposed to represent. For good results the type of treatment used in anechoic chambers for sound testing would be best for the walls and ceiling (these have wedges of foam extending out about 3 ft (1 m) from the walls). On the floor a normal carpet is sufficient. However, some performers find it unpleasant to work in such surroundings.

Studios for stereo speech are more dead than other general-purpose studios. The reason for this will be seen later: where co-incident microphones are used, voices must be placed physically further back than for mono; otherwise, movement is exaggerated.

The music studio

The foremost characteristic of a studio with live acoustics, such as is used for classical music, is its reverberation time. Preference tests indicate that, for any given size of studio, music requires longer reverberation than speech does. These preferences are not absolutes, but are valid for the conventional music and instruments played in them during the tests. However, both music and instruments as we know them in the West today have developed in a very specialized way: in particular, the orchestral and chamber music is developed from that played in rooms of large houses in the Europe of the seventeenth and eighteenth centuries. The original acoustics has defined the music – and that music now, in turn, defines the acoustics needed for it.

In contrast, at one extreme, much Eastern music is designed for performance in the open air, while church music is at its best in highly reverberant surroundings. Moreover, some modern music is written for drier acoustics than those which sound best for classical works. Nevertheless, the acoustics that we have been accustomed to have led to the design of instruments for which music of great beauty and power has been composed; so the studios we look for have certain specific qualities which can be judged by these subjective tests.

Listening tests are not always the best indication of true value: they measure what we like best at the time of the test, and not what we could learn to like better if we took the trouble to change our habits. Nevertheless, several clear results appear when such tests are applied to music studios. It turns out that the ideal (i.e. preferred) reverberation time varies with the size of the studio: small studio – short reverberation; large studio – long reverberation. For the very smallest studios (e.g. a room in a private house set aside as a music room) the ideal is between three-quarters and one second. This is what might be expected if the room has no carpet and few heavy furnishings. Some authorities suggest that there should be more reverberation in the bass.

An interesting sidelight on reverberation is that for mono listening, reverberation times about a tenth of a second lower than those for live music are preferred. The emergence of the reverberation from the same apparent source as the direct sound means that less is required. And in any case the reverberation of the listening room is added.

Apart from this, the principal quality of a good music studio is good diffusion, so that the sound waves are well broken up and spread. There should be no echoes due to domes, barrel vaults or other concave architectural features.

Dead pop music studios

The studio acoustics play no part in much of today's popular music. The internal balance of the orchestra is deliberately surrendered in exchange for electronic versatility. A celeste, piano or flute may be given a melody and may be expected to sound as loud as brass, and various other effects may be sought which can be achieved only if the sound of the louder instrument does not spill to the microphones set at high amplification for quiet instruments. Microphones are placed as close as may be practicable, and their directional properties are exploited to avoid picking up direct sound from distant loud sources, and to discriminate against reverberation from the open studio. Reflected sound is just a nuisance.

The solution to these problems has been to make the studio dead – more dead than is required for a speech studio. This requires substantial attenuation even at a single reflection from any wall, so that much less of the sound of the brass (for example) will strike the walls and reach the flute or celeste microphone from directions that cannot be discriminated against.

An aspect of acoustics that is very important to pop music, not for its presence but again for its absence, is attenuation due to air transmission. At frequencies below about 2000 Hz, absorption by the air is negligible. But for 8000 Hz, air at 50% relative humidity has an absorption of 0.028 per foot (30 cm). This means there is something of the order of a 3.5 dB loss at a distance of 100 ft (30 m) for high frequencies. For damper air the loss is a little lower, while for very dry air it rises sharply and begins to affect lower harmonics (down to 400 Hz) appreciably. This may make a great deal of difference between two performances using a distant balance in the same concert hall, but it does not affect close-balanced pop music, except to make instrumental quality *different* from that which may be heard normally: substantially more of the higher harmonics of each instrument are heard.

Television and film studios

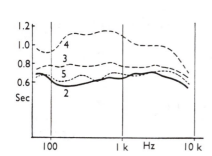

Four television studios: frequency response
Variation in reverberation time with frequency.
BBC Television Centre Studios 2 and 5 are both 116 000 ft³ (3320 m³); studios 3 and 4 are 357 000 ft³ (10 230 m³). Studio 4 was designed primarily for musical productions.

Studios for television and film (and particularly the larger ones) are usually made very dead for their size. The primary reason is that if they are not to appear in vision, microphones must be farther away from their subjects than in other types of balance. But it also helps to reduce inevitable noise from distant parts of the studio. Noise arises in a variety of ways. For example, studio lamps produce heat, and the necessary ventilation cannot be completely quiet. Scenery may have to be erected or struck during the action, or props set in. Performers and staff must walk in and out of the studio, and from set to set. Cameras move about close to the action, dragging their cables. The floor of a television studio must allow the smooth movement of heavy equipment, so it is acoustically highly reflective.

Sound that radiates upwards may be reflected several times before it can reach a microphone, but sound travelling horizontally may be reflected only once. For this reason the most effective place for the absorption of sound is the lower parts of the walls; more of the treatment is concentrated here than is placed higher up.

The set itself may reduce the overall reverberation a little: braced flats absorb low frequencies, while soft furnishings and heavy drapes reduce the highs. In addition the set may have unpredicted effects on the local acoustics: hard surfaces may increase high frequencies. A cloth cyclorama makes little difference, but if constructed of wood or plastered brick it reflects sound strongly (and curved sections focus it).

Television studios have acoustics approaching those used in 'dead' music studios; they are therefore suitable for music techniques of this type. For classical music, however, they are basically unsuitable, and artificial reverberation must be applied liberally to compensate for this. But this does not solve the problem of the musician who finds the surroundings acoustically unpleasant. In particular, string players need reflected sound to satisfy themselves that they are producing a good tone quality at adequate power. In its absence they are apt to bow harder, changing the internal balance of the orchestra and making its tone harsh and strident. The timing of ensemble playing also suffers. An orchestral shell (acoustically reflective surface behind and above) helps all of the players, but reinforces the brass players (who do not need it) more than anyone else: a full shell turns a balance inside out and should not be used. In fact, the best solution is to design the setting to have reflecting surfaces for the benefit of the strings. This does nothing for the overall reverberation time, but helps the musicians to produce the right tone.

Film studios have in the past tended to be even deader than television studios. In addition to the other problems, film cameras are, by the nature of their intermittent action, noisy and difficult to silence really effectively.

Television sound control rooms and film dubbing theatres have the same characteristics as their counterparts in radio: 0.4 s at 250 Hz falling to 0.3 s at higher frequencies. The acoustics of the main production and lighting control areas should be relatively dead, in order that the director and his assistant and the technical director shall be clearly audible on open microphones in an area that also has programme sound. In practice, however, absorption is limited by the necessarily large areas of hardware and glass.

Other technical areas, such as those for telecine and videotape, need acoustic treatment: machines other than the smaller video formats can be noisy in operation. If machines are grouped in partly open bays, acoustic partitioning is desirable: aim for 20 dB separation between one bay and the next.

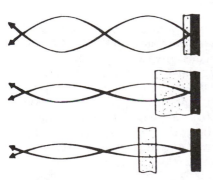

Sound absorbers
Padding which is close to a reflecting surface is efficient only for sufficiently short wavelengths. Screens with a thin layer of padding are poor absorbers at any but the highest frequencies. For greater efficiency in absorbing a particular wavelength, the absorber should be placed at a quarter wavelength from reflecting surfaces.

Acoustic treatment

Three basic types of acoustic treatment are available:

Soft absorbers. These are porous materials applied to walls: their action depends on loss of sound energy, as air vibrates in the interstices of the foam, rockwool or whatever is used. The method

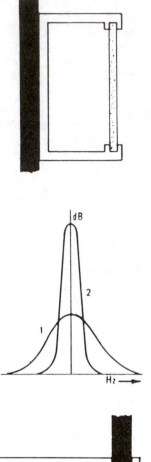

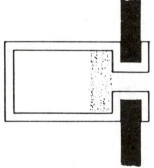

Absorbers
The membrane absorber, *above*, is a box covered by a panel which is fixed at the edges and weighted with some material such as roofing felt. It absorbs a broad range of frequencies (1). The Helmholz or damped cavity absorber, *below*, is in effect a bottle of air resonating at a particular frequency, with damping material within it. It absorbs at the frequency of resonance (2).

works very well at high and middle frequencies, but for efficiency at low frequencies absorbers need to be about 4 ft (1.2 m) deep. So it is reasonable to lay these at thicknesses that are efficient down to about 500 Hz and then become less so. Many different types are commercially available. Control of high frequencies (for which absorption may be excessive) is provided by using a perforated hardboard surface: with 0.5% perforation much of the 'top' is reflected; with 25% perforation much of it is absorbed. Different surfaces may be alternated.

Helmholtz resonators. Cavities open to the air at a narrow neck resonate at particular frequencies. If the interior is damped they absorb at that frequency. Such resonators are useful for attacking dimensional resonances in sound radio studios. This is not a problem that should be overstated, however. At one stage it was calculated that only about seven out of 100 BBC studios seemed to be affected. In any case, it may not be the ideal solution, as it requires treatment to be fitted either at particular places that may be already occupied or some of the only places that are not.

Membrane absorbers. These are often used to cope with the low frequencies. But if the low frequencies are a problem, so too can be the absorbers themselves. The idea is simple enough: a layer of some material or a panel of hardboard is held in place over an air space. The material moves with the sound wave like a piston, and this movement is then damped, so that low-frequency sound energy is lost.

A typical *combination absorber* might consist of units of a standard size – say, 2 ft × 2 ft × 7 in deep (60 × 60 ×18 cm) – which are easy to construct and to install; building contractors can be left to do this without constant supervision by acoustic engineers. A level response for a studio can be obtained by making (literally) superficial changes on this single design.

The box for this is made of plywood, with a plywood back. The interior is partitioned into four empty compartments by hardboard dividers 6 in (15 cm) deep. Over this airspace, and immediately behind the face of the box, is 1 in (2.5 cm) of heavy density rockwool. The face itself may be of perforated hardboard or similar. If the open-area perforation is very small there is a resonance peak of absorption at about 90 Hz. If the perforation area is large (20% or more), wideband absorption is achieved, but with a fall-off at 100 Hz and below. By using similar boxes with two kinds of facing, it is possible to achieve almost level reverberation curves.

Using sound absorbers

If the acoustics of a sound studio are fully designed, a range of different absorbers is likely to be used. To help evaluate them, their *absorption coefficient* is measured. That most commonly used is

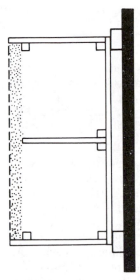

All-purpose absorber
A plywood box mounted on battens fitted to the wall and covered with 3 cm-thick rockwool and perforated hardboard. The response is tuned by varying a single factor, the amount of perforation in the hardboard surface layer. Combinations of boxes with 0.5% and 20% perforation can be used as required.

averaged over all angles of incidence, and over the range from 100 Hz to 3150 Hz.

For total reflection the absorption coefficient is zero; for total absorption it is unity. In practice the figure should therefore lie between zero and one. For example, free-hanging heavy drapes may be measured at 0.8 for 500 Hz and above. But note that for relatively small areas absorbers may have a coefficient that is greater than one – an apparent impossibility that can be produced by diffraction effects.

Larger objects have sometimes been described as the number of 'open window units' (their equivalent in the number of square feet of total absorption) that they represent. Individual, separated people rate about four and a half on this scale (at lower-middle to high frequencies), and a padded sofa may have 10 times the effect of the person sitting on it.

For seated audiences, this would give too high a number, as sound does not approach each individual body from all angles, so it is necessary, and in any case more convenient, to calculate their absorption as though they are part of a flat surface: this gives a coefficient of 0.85.

Tables of these values are available: in addition, they give values for the studio structure itself and for the volume of air within it. The calculations are complex, but here are some rule-of-thumb ideas that can be applied when attempting to correct the performance of a given room:

● apply some of each type of absorber to surfaces affecting each of the three dimensions
● try to ensure that no untreated wall surfaces remain to face each other (note that a surface that is treated at one frequency may be reflective and therefore effectively untreated at another)
● put the high-frequency absorbers at head level in speech studios.

The cheapest and best treatment for the floors of speech studios is a good-quality carpet and underlay (though the roof will need compensation in the bass). Wooden blocks are good on the floor of a (live) music studio.

As for the general furnishings of sound studios, it is plainly unwise to get the acoustics right for the studio when it is empty and then bring in a lot of items (particularly large soft absorbers) that subsequently change it. Furniture, decoration and people must be included in the calculations from the start. But apart from this, it is not worth worrying too much about problems until you actually hear them. For example, if there is enough treatment in a room to get the reverberation low and level, and if the main rules for the layout of absorbers have been observed, the diffusion of sound is probably going to be all right anyway. So you can arrange the furniture with aesthetic and other considerations in mind: for how it looks best, makes the occupants feel happy, and gives the most convenient layout of working areas.

The use of screens

Very often the acoustics that are built into a studio are not exactly those that are wanted for a particular programme, so ways of varying studio quality have been devised. Sometimes the acoustic treatment of a studio is set in hinged or sliding panels, so an absorber can be replaced by a reflecting surface. But a more flexible approach relies on layouts of studio screens (in America called flats or gobos).

If padded screens are to be easily movable (on castors), they must be light in weight (which in practice means they cannot be very thickly padded) and not too broad, often about 3 ft (0.9 m) wide. For stability they are mounted on cross-members. Individual free-standing screens are effective only for sound of short wavelengths: low-frequency sound flows round and past them as though they were not there. But grouping screens together improves things somewhat. At best, thin screens may reduce the sound flow between neighbouring areas by some 10–15 dB, while heavier screens with several inches of absorber (say 4 in, 10 cm thick) will do substantially better.

A common type of screen consists of a panel of wood with a layer of padding on one side: used dead side to the microphone, it will damp down the highs: with the bright (reflective) side forward, the highs are emphasized, and the ambient sound in the studio is somewhat reduced by the absorbent backs. In music studios with dead acoustics, large free-standing sheets of perspex or wood are used in addition to the thickly padded screens that separate instruments and their associated microphones from each other.

Whatever the type, the effect of all screens is the same in several important respects:

- There are shorter path lengths between reflections of the sound in the studio: there are more reflections and more losses in a given time. This means, in general, a lower reverberation time for the studio as a whole.
- Sound from distant sources is reduced; but sound that does get round the screens is usually lacking in top. This includes any appreciable reverberation from the open studio.
- Coloration can be introduced by careless layout. If there are two parallel screens on opposite sides of the microphone a standing-wave pattern will be set up. On the other hand, such distortion effects may be used constructively – though this is more often done by placing a microphone fairly close to a single screen.
- For the purposes of multimicrophone music balances, the separation of a large number of individual sources that are physically close together can be made more effective. Important enough in mono, this is vital in stereo.

In any studio it helps to have a few screens around. They may be wanted to produce subtle differences of voice quality from a single acoustic, or to screen a light-voiced singer from heavy accompaniment, or in classical music for backing and reinforcing horns.

Altering the acoustics of concert halls

Traditionally, the design of concert halls has been a fallible art rather than an exact science. In many existing halls of a suitable size it is difficult to pick up good sound, and even where after careful experiments a reasonable compromise is found for microphone placing, many members of the audience (whose audible enjoyment may contribute to the sense of occasion of a broadcast) may suffer poor sound in the hall itself. Fortunately it is now possible to do something about many such halls (or studios). Among those that have had their acoustics physically altered are several of the main concert halls of London.

Ever since it was built in Victorian times, the Royal Albert Hall inflicted severe echoes on large areas of seating and its unfortunate occupants. This was partly due to the shape of the hall and partly its size. Being very large, for many seats the difference in path length between direct and reflected indirect sound was of the order of 100 ft (30 m), resulting in a delay of nearly 0.1 s. Some delay between direct sound and the earliest reflections is a characteristic of all halls. Whether this is acceptable or not depends on the nature of those reflections.

The Albert Hall was designed as a near-cylinder surmounted by an elegant dome. Such concave surfaces produce focused reflections, and some of them coincided with the seating. Apart from the gallery, where the sound was clean, though distant, there were (regular concert-goers claimed) only about two places in the hall that were satisfactory. Staccato playing, including percussion and piano, were special problems. Also, the natural reverberation time of the hall is long. As a result, the range of music that could be played to good effect was limited virtually to one type: romantic music with long flowing chords, and even this lacked impact – it is simply not loud enough because the hall is too big.

Attempts to improve the acoustics without changing the appearance of the hall failed. The problem was eventually tackled by suspending flights of 'flying saucers' 6–12 ft (2–3 m) in diameter: 109 of these were hung just above gallery level. Taken all together, they defined a new ceiling, slightly convex towards the audience and filling about 50% of the roof area.

These polyester-resin-impregnated glass-fibre diffusers (some of them with sound-absorbent materials on the upper or lower surfaces) did largely solve the problem of echo from above (if not from the oval-sectioned walls) and at the same time reduced the reverberation at 500 Hz to more reasonable proportions for a wider range of music. Even so, the design has been superseded by another, in which diffusers are integrated into the ribs of the dome.

In London's post-war Royal Festival Hall the error was in the opposite direction. Musicians had been asked in advance what acoustics they would like to have, 'tone' or clarity. The answer came, 'tone' – but unfortunately what they actually got instead was a quite remarkable degree of clarity. The sound was much more dead – particularly in the bass – than had been expected, largely due to the construction of too thin a roof shell. Minor measures, such as taking up carpets from walkways and making surfaces more reflective, had insufficient effect. While a cough sounded like a close rifle shot, the bass reverberation remained obstinately at about 1.4 s. How, short of taking all the seats out and sending the audience home, do you lengthen a reverberation time?

The answer here was *assisted resonance.* Cavity resonators were used, but not in their normal form: these contained microphones, which were thereby tuned to respond to narrow bands of frequencies present and to re-radiate sound through individual loudspeakers. A hundred channels could each have a separate frequency; allocating each a band of 3 Hz, the whole range of up to about 300 Hz could be boosted. In this way, resonance was restored at surfaces at which, previously, sound energy had begun to disappear. A reverberation time of 2 s was now possible, but it could be varied and actually tuned at different frequencies.

Assisted resonance has since become available as a commercial package, prolonging a selected range of about 90 frequencies between 50 and 1250 Hz by half as long again. At the Festival Hall this dealt very effectively with the balanced overall duration of reverberation, but offered less improvement in other ways: in particular, loud brass or tympani may still overwhelm the other instruments. Modern works that require a dry acoustic remain the most successful.

In both of these places, however, broadcasters had been able to get good or fairly good sound even before the alterations by suitable choice of microphone and careful placing; and the balance problems (other than for reverberation *time*) have hardly been affected.

Forty years later, another British concert hall (in Birmingham) was designed to have much greater physical control of its acoustics. It has a movable canopy over the orchestra and concrete-walled chambers with adjustable openings around the periphery of the hall – the value of both of which are debated – and 'banners', large movable areas of drapes, which offer a more generally accepted control measure.

Recording companies working with London's many first-rate orchestras have found their own solutions. Audiophiles all over the world may not know it, but the acoustics *they* often hear on record are not those of the well-known concert halls, but obscure suburban town halls at places like Walthamstow, Wembley, Watford and Hammersmith – in wood-panelled, rectangular boxes which, by chance rather than musically dedicated design, have much in common with Amsterdam's Concertgebouw, a favourite of musicians. The moral of this story is that architecture has not always done so well as happy chance, the fortuitous combination of circumstances that gives perfect blending. Search until you find it.

Acoustic modelling of studios

A more scientific approach to the design of new studios – or redesign of defective ones – can be achieved by the use of scale models. If the model is made to a scale of one-eighth linear, the sound path lengths are reduced in the same proportion. With the absorbers scaled accordingly, the reverberation is one-eighth of that for the full-size studio. Surfaces are one sixty-fourth life size and the total volume about one five-hundredth.

Variations in design can be explored at modest cost. A special acoustically dead selection of musical recordings is replayed into the model at eight times normal speed, and re-recorded with the acoustics of the model added. This can be replayed at normal speed, and the results of various configurations and forms of treatment compared.

The design of a loudspeaker to fill the model with sound at the required 400 Hz to 100 kHz is necessarily unorthodox. One used for this scale had three elements: a 11 cm thermoplastic cone (400–3000 Hz), a commercial 20 mm domed plastic diaphragm (3–21 kHz), and a cluster of 45 small electrostatic units mounted on a convex metal hemisphere giving wide-angle high-frequency radiation (21–100 kHz). Microphones to reproduce the same raised frequency band required a very small diaphragm. One of 6 mm was adequate, if rather large for the highest frequencies. The microphones were omnidirectional, so for a stereo pick-up a spaced pair had to be used. The tape recorder was required to cover 50 Hz to 12.5 kHz at normal speeds and the 400 Hz to 100 kHz band when speeded up: for this a commercially available machine was modified.

Structural components and miniature furnishings to be used in the model were first acoustically tested in a specially constructed reverberation chamber. Materials with textures similar to the full-scale object were used where possible; for example, velvet for carpets. Perforated absorbers were represented by scaled-down absorbers of similar construction. It proved essential to include miniature musicians too; an orchestra is widely spread and mops up a great deal of the sound it creates. Indeed, the early experiments were unsuccessful until the players were adequately modelled in cutouts of expanded polystyrene with 3 mm of felt stuck to their backs.

The biggest problem was the air itself – for this must absorb eight times more sound in transit. This was achieved by drying it to 4% relative humidity (compared with a typical, though very variable, 45% in normal air). Silica gel was not good enough for this, but an artificial zeolite dried the air in the model in only half an hour.

The first application of the technique was in the redesign of the BBC's main orchestral studio, a former ice-rink that had served the BBC Symphony Orchestra reasonably well for many years, but which had certain nagging deficiencies. A vast number of experimental modifications were tried out on a model of this hall. Absorbers were changed on the end wall, and added on the roof, then covered with a low-frequency reflective material; the choir rostra were changed, then switched to a different position, and subsequently replaced by a new design; while various shapes of wooden canopy were suspended over the orchestra. The results were compared with each other and (as was possible in this application) with the sound produced by playing the same test material through the real studio. The experiment saved more than its own cost by one result alone: it indicated that the big reflective canopy would have been a waste of money. The chosen design was built, without the canopy, and came within 10% of the performance predicted – an unusually good result in this problem-ridden field.

The second application was to the construction of a completely new music studio (in Manchester), where the question asked was: would a simple design work, using the 'combination' absorbing boxes described earlier to tailor its response? The answer was yes, and, surprisingly, that far fewer such boxes were required than simple extrapolation from results in smaller studios would have led the designers to expect. They had modified the absorbers a little, making them a little deeper in order to handle the low frequencies better. In the model it was found that scattered boxes absorbed more than 100% of the sound falling on their equivalent area of wall: evidently their sides made an important contribution, as well as the front. Again the initial experiments proved extremely valuable.

Scale models that are smaller still (1:50) are now sometimes used, though plainly with greater limits on the testing that is practicable. This is too small for listening tests, but a spark can simulate a gunshot and reverberation is measured using microphones with a $\frac{1}{8}$ in (3 mm) diaphragm. There are substantial savings of time and money in preliminary design, and a better chance of getting good results later. It should be obvious that the best approach to any design is to model the acoustics first and build the structure around it, but unfortunately this is not yet standard architectural practice.

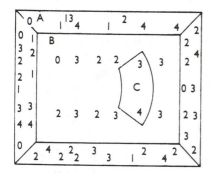

Ambiophony

A, Gallery level. B, Grid level.
C, Orchestral area below.
0–4, Loudspeakers with different delay times: O, No delay. 1, 30 and 60 ms delay. 2, 90 and 150 ms. 3, 120 ms. 4, 180, 210 and 240 ms. This is an example in which the delays were produced by multiple replay heads on a tape loop. This can now be achieved electronically with greater flexibility.

Ambiophony

A general-purpose television studio has quite the wrong acoustics for orchestral music. The addition of artificial 'echo' is only a half-solution, because of the effect of the acoustics on the players themselves: the string and some other players may tend to force their tone to hear themselves as they think they should be, thereby marring both the internal balance of the orchestra and the overall orchestral quality. Another problem for the musicians is that more distant sections that can normally be heard clearly may not be, so that ensemble playing is more difficult.

One early technique that was tried in order to combat this was called *ambiophony*. This replaced the missing reflecting surfaces by a large number of loudspeakers – 50 or more in a studio of moderate size – fed by special microphones through a tape delay system. The additional delay is essential because the acoustic path lengths are shorter than the actual dimensions of the studio would suggest: the total distance from musician to ambiophony microphone, plus that from wall or roof loudspeaker to listener (or the studio microphone that represents him), is less than first reflection path via the same point on the wall. With tape delay the apparent reflective surface can be made to recede to a point even beyond the actual limits of the studio. In practice, the best result – simulating the well-diffused and open textured reverberation of a good, large hall – was achieved by introducing not just one delay system but four (plus some loudspeakers relaying undelayed sound). The four systems were derived from combinations of the outputs of eight replay heads situated at intervals on a tape loop. The sound passed repeatedly through the studio, diminishing as it did so.

Ambiophony microphones needed to be placed close to the sound source – preferably not more than 6–8 ft (2–3 m) away – so that they discriminated strongly against the relatively distant wall loudspeakers. Where close spotting microphones are used on individual instruments or sections, such as the string and woodwind sections, as part of the microphone balance for broadcast or recording, an ambiophony feed could be split off using the public address controls. Otherwise, for wider shots, the special microphones necessarily appear in vision: small electrostatic microphones hanging by their own cables are the least obtrusive.

The system was tuned by taking each loudspeaker in turn, raising its volume until a howlround occurred, then setting back a little from this. The overall master ambiophony feedback volume control must, of course, be left at the same setting throughout the process, and at the end was set back a further 4 dB. This then became the maximum level at which the system could be used.

Ambiophony created conditions in which musicians could work comfortably, and offered a fair simulation of music studio acoustics. But the studio layout, once optimized, was inflexible.

Acoustic holography

The ultimate all-purpose auditorium has an intrinsically short reverberation time, but with reinforcement that can be varied to simulate the acoustics of any hall required, provided only that the new reverberation time is longer. This can be done electronically – and is now relatively inexpensive, provided that there is no need to generate the enhanced acoustic within the original auditorium. Electronic artificial reverberation and its uses are described later. Far more expensive, but also more rewarding to perform and listen in, is a hall in which the ideal sound field for any purpose can be recreated electronically, by means of *acoustic holography*. One such arrangement is the acoustical control system (ACS). The audience is between lines of loudspeakers along the side and rear walls. The

Acoustic control system (ACS)
Signal from microphones above the performers is fed to a complex digital system which mixes, delays and feeds the recombined signals to loudspeakers (more than shown here) around the hall. Feedback between loudspeakers and microphones is kept as low as possible. The aim is to simulate the sound field of a good music studio, but with variable reverberation time.

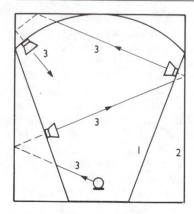

'Acoustic holography' principles
In this example, the original walls,
1, carry speech rapidly to the back of
the fan-shaped auditorium, so that
people at the rear can hear individual
speech syllables clearly. At the back
wall, sound must be absorbed so
that it does not reduce clarity at the
front. 2, Ideal music auditorium
simulated by the acoustical control
system. 3, Simulated sound path
using three of the loudspeakers
lining the walls at sides and rear. In
practice, there will be many
loudspeakers in a line along these
walls.

sound waves from such horizontal lines radiate up and down, so that
those at the side nearest the stage radiate only weakly in that
direction. In addition, the microphones in the stage area discriminate
against sound from the direction of the hall. The system is designed to
recreate an ideal wavefront in the audience area. This requires so
much computer processing that it only became practical in the mid-
1980s.

ACS employs a matrix of electronic delays and filters which
interconnect all microphones and loudspeakers. Acoustic feedback
can be reduced to a safe, low level and plays little part in the final
sound, so the coloration that accompanies it is avoided. In
ambiophony, coloration limited the increase in reverberation time to
no more than double; here it can be increased sixfold, to make a small,
relatively dead theatre sound like a cathedral. An auditorium
designed for speech can be used for many different kinds of music,
and can be instantly reset by the control computer between pieces, or
even between announcements and music.

In a hall that is treated in this way, sound is balanced between direct
and reinforced sound as though the acoustics were natural, taking
care only to keep microphones well away from individual
loudspeakers.

Chapter 5

Microphones

A microphone converts sound energy into its electrical analogue – and, in principle, should do this without changing the information content in any important way. A microphone should therefore do three things:

- for normal sound levels it must produce an electrical signal that is well above its own electrical noise level.
- for normal sound levels the signal it produces must be substantially undistorted.
- together with its associated equipment, it should ideally, for a particular sound source, respond almost equally to all significant audio frequencies present.

With today's high-quality microphones the first two objectives are easily met – though to achieve a signal that can be transmitted safely along the microphone lead, a transformer or amplifier is often required so close to the head as to be regarded as part of the microphone itself. The third requirement is more modest, and deliberately so. Indeed, until the advent of FM broadcasting there was no call in radio for microphones that could pick up the full audio range, and even today the practical limit of many distribution systems may be 15 kHz. To have demanded more would have been a waste of money. One is deeply sceptical of advertising copy or commercial enthusiasm which claims that sound above the range of human hearing somehow 'feels' better. This smacks of mysticism. Even apart from that, for a source of restricted range, a full audio-frequency range microphone response is hardly necessary: it may pick up spill from neighbouring sources with a more extended range or unwanted high-frequency noise.

Microphones differ from one another in the way that air movement is converted into electrical energy. The most important types in current professional use are electrostatic (also called condenser or capacitor microphones, and including electrets), moving coil ('dynamic')

Microphone sensitivities are now usually given in dB relative to 1 volt per pascal. This corresponds to a (linear) scale of millivolts per pascal, as follows:

dB rel. to 1 V/Pa	mV/Pa
−20	100
−25	56
−30	31.6
−35	17.8
−40	10.0
−45	5.6
−50	3.16
−55	1.78
−60	1.00
−65	0.56

Other units that may be encountered are dB relative to 1 volt per dyne/cm²: this gives figures which are 20 dB lower. One mV/Pa is the same as 1 mV/10μbar or 0.1 mV/μbar.

microphones and ribbons. A wide range of other principles has also been used. These include crystal (piezoelectric), carbon, inductor, moving-iron, magnetostriction and ionic microphones, none of which is described here; several have practical disadvantages that have limited their development.

Nearly all microphones employ a *diaphragm*, a surface that is mechanically shifted in response to sound waves. But another possible principle is that of the pressure-sensitive semiconductor, forming part of an integrated circuit.

Microphones also differ from each other in the way in which air pressures are converted into movement of the diaphragm. These variations show themselves as differences in directional characteristics and are of great importance to the user. Although the output of a microphone may be deliberately designed to change with the angle from a given axis, the frequency response, ideally, should not. Unfortunately, however, this is an almost impossible demand, so it is usual to define the 'useful angle' within which this condition is, more or less, met; or to devise ways in which the deficiency can be turned to advantage.

Microphone properties

The range of properties affecting choice of microphone includes:

Sensitivity. This is a measurement of the strength of the signal that is produced. In practice, most professional microphones of a particular type have output voltages within a range of perhaps 10 dB. Taking moving coils as 'average', ribbons tend to be lower, while electrostatic types can be higher because they need a head amplifier anyway.

Microphone performance figures sometimes quote maximum sound pressure levels (usually for 1% harmonic distortion). A conversion to corresponding linear pressure levels in pascals is:

SPL (dB)	Pressure (Pa)
140	200
134	100
128	50
94	1
0	0.000 02

The zero approximates to our threshold of hearing.

Characteristically, rather more than 70 dB (further) amplification is required to reach a level suitable for input to a high-level mixer or for transmission by line between studio and radio transmitter. The upper limit to sensitivity is dictated by overload level, that at which unacceptable harmonic distortion is generated. Moving-coil microphones are relatively free of this, unlike electrostatic types, which also produce more electronic noise (hiss). The limits for both of these may be described in terms of sound pressure levels. For example, a microphone with a stated upper limit of 130 is claiming low distortion up to sound levels that are close to (or over) the threshold of pain, though this may still be less than that encountered close to the mouth of a loud singer. A lower limit (also called 'equivalent noise level') of 20 dB means that the electrical noise from the microphone is no higher than the 20 dB (sound pressure level) of a quiet recording studio.

Robustness. Moving-coil microphones in particular stand up well to the relatively rough treatment a microphone may accidentally suffer when used away from the studio. Electrostatic requires greater care.

Sensitivity to handling, e.g. by pop singers. Moving-coil and some electrostatic types score highly. Some designs have mechanical insulation between the casing and the microphone capsule and associated electronics. A reduced low-frequency response helps. Ribbon microphones are unsuitable, as their ribbon diaphragms typically have a resonant frequency of about 40 Hz, which is easily excited by any form of movement, even on a boom. Cables and connectors should also be checked for handling noise.

Sensitivity to wind. Gusting produces erratic pressure gradients, so microphones relying on this principle are at a disadvantage. Ribbon microphones also have a relatively floppy diaphragm and so are doubly at risk. Shape and the use of windshields are important.

Shape, size and weight. Appearance matters in television, and also size and weight in microphones that are hand-held or attached to clothing.

Cost. Electrets (permanently polarized electrostatic microphones) can be cheap, and good moving-coil microphones moderately so; some high-quality electrostatic microphones are more expensive.

Suitable impedance. Nominally a microphone should be rated appropriately for the impedance of the system into which its signal is fed, otherwise the interface between components reflects some of the signal and may affect the frequency response. In practice, for complete safety the impedance of a microphone should be less than a third of that of the system it is plugged into. After a transformer or head amplifier (where necessary), many professional microphones have an impedance of about $200\,\Omega$; mixers typically have an input of $1200\,\Omega$. Sometimes a minimum input impedance is specified. Cable runs longer than 200 ft (60 m) require low impedance microphones.

Sensitivity to temperature and humidity variations. Location filming may sometimes be undertaken in extreme climatic conditions, and heat and humidity can also cause trouble high up among the lights in a television or film studio. Water vapour condensing on to a lightweight diaphragm increases its inertia and temporarily wrecks its response. An electrostatic microphone that has been stored or transported in the cold is especially vulnerable, and also to any moisture that impairs its insulation or head amplifier.

Transient response. Speech and music contain transients which are 10–100 msec in duration. High-quality electrostatic microphones, with a response of less than 1 msec are excellent, and ribbons are also good. Moving coils are limited by the mass of the voice-coil.

Flexibility of response. Some microphones have a variable polar response, and the switches may be located on the microphone head, or on a separate control unit, or remotely on the control desk. Another switch may control bass response.

In a given situation any one of these points may be dominant, severely restricting or even dictating the choice of microphone. Many commercially available microphones emphasize particular characteristics at the expense of others, and are manufactured with some particular function in mind. Popular music attracts a wide range

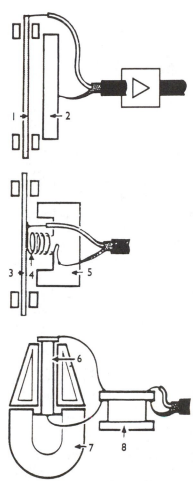

Types of microphone
Above: Electrostatic (condenser). 1, Foil diaphragm. 2, Backplate. An amplifier is needed so close to the head as to be considered part of the microphone. *Centre:* Moving coil. 3, Thin diaphragm, typically of Mylar. 4, Voice-coil fixed to diaphragm. 5, Permanent magnet. *Below:* Ribbon microphone. 6, Light, conductive foil ribbon, typically 0.08 in (2 mm) thick. 7, Permanent magnet, with pole pieces extending above. 8, Transformer. Ribbons, less popular for some years because of their limited high-frequency response and archaic appearance, have since returned to greater favour. But the magnet is heavy, and must also be kept away from recorded tapes.

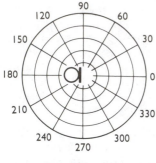

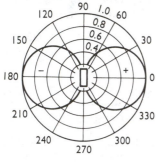

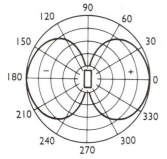

Polar diagrams
Field patterns of *omnidirectional* and *bidirectional* microphones.
The scale from centre outwards measures sensitivity as a proportion of the maximum response, which is taken as unity. *Above:* A perfect omnidirectional response would be the same in all planes through the microphone – the diagram is a section through a sphere. *Centre:* A perfect bidirectional response is the same in all planes through the 0–180° axis. At 60° from the axis the output is reduced to half, and at 90° to zero. The response at the back of the microphone is opposite in phase. This diagram has a linear scale, reducing to zero at the centre, and the figure 'eight' is perfect. *Below:* The figure-of-eight, shown on a more conventional decibel scale, is fatter, but the scale itself does not have a zero at the centre.

of different types, some to cope with extreme physical conditions, others on account of frequency characteristics which are appropriate to individual instruments or for what manufacturers call their 'perceived' tonal qualities.

Directional response

Microphones fall into several main groups, according to their directional characteristics. Their field patterns are best visualized on a *polar diagram*, a type of graph in which the output in different directions is represented by distance from the centre.

Omnidirectional microphones. These, ideally, respond equally to sounds coming from all directions. Basically, they are devices for measuring the *pressure* of the air, and converting it into an electrical signal. Many designs of moving-coil and electrostatic microphones work in this way. The diaphragm is open to the air on one side only.

Bidirectional microphones. These measure the difference in pressure (*pressure gradient*) at two successive points along the path of the sound wave. If the microphone is placed sideways to the path of the sound, the pressure is always the same at these two points and no electrical signal is generated. The microphone is therefore *dead* to sound approaching from the side and *live* to that approaching one face or the other. Moving round to the side of the microphone, the response becomes progressively smaller and a graph of the output looks like a *figure-of-eight* (a term often used to describe this type of microphone). The *live angle* is generally regarded as being about 100° on each face. For sound reaching the microphone from the rear, the electrical output is similar to that for sound at the front, but is exactly opposite in phase. Ribbon microphones, which were once widely used, employ this principle. They respond to the difference in pressure on the two faces of a strip of aluminium foil. Single-diaphragm electrostatic figure-of-eight microphones work in this way if the air pressure can reach both sides of the diaphragm equally.

Cardioids have a heart-shaped response. This is obtained if the output of a pressure-operated microphone is added to that of a pressure-gradient microphone with a response of similar strength along its forward axis. Various ways of combining the two principles are possible: early cardioid microphones actually contained a ribbon and a moving coil within a single case. The output of the two was added together for sounds at the front; but at the back the two were antiphase and cancel. At the side there was no output from the ribbon, but the omnidirectional element retained its normal output. So there was a substantial pick up of sound on the front and right round to the side, but beyond this relatively little.

Supercardioids and hypercardioids. If the pressure and pressure-gradient modes of operation are mixed in varying proportions, a range of polar diagrams is produced, passing from omnidirectional

Directivity

The range from pure pressure operation (omnidirectional) to pure pressure gradient (bidirectional) operation.

Curve 1 shows the ratio of sound accepted from the front and rear. The most unidirectional response is provided by the supercardioid, A, for which the microphone is, essentially, dead on one side and live on the other. Curve 2 shows discrimination against ambient sound. With the hypercardioid response, B, reverberation is at a minimum and the forward angle of acceptance is narrower. In the range between omnidirectional and cardioid, direct sound from the front and sides is favoured, so a microphone of this type could be placed close to a broad source. (Note that the close balance will already discriminate against reverberation, and that this microphone will reduce long path reverberation still further.)

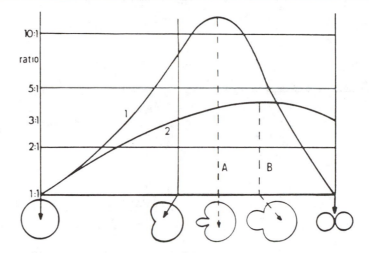

through cardioid, then supercardioid to hypercardioid, to bidirectional. The directional qualities within this range may be described in two ways. One is in terms of the degree of *unidirectional response*, indicating the ratio of sound accepted at front and back of the microphone (i.e. before and behind a plane at right angles to its directional axis). This is 1:1 for omnidirectional and bidirectional responses and maximum, reaching a peak ratio of 13:1, at an intermediate setting. Another attribute of a microphone's directivity is its *discrimination* against indirect sound: in practice, this too is expressed as a ratio, here describing the proportion of the total solid angle over which the microphone is effectively sensitive to sound. It is 1:1 for omnidirectional microphones, 3:1 for both cardioid and bidirectional pick-up, and is higher (a little over 4:1) for a response half-way between the latter two.

The combination of the pressure and pressure-gradient principles in different proportions is very useful for cutting down the amount of reverberation received and for rejecting unwanted noises. Some microphones can be switched to different polar diagrams; others are designed for one particular pattern of response. When choosing between different field patterns, note that the breadth of pick-up narrows as the pressure-gradient component gets stronger. A *hypercardioid* or *cottage loaf* pattern is most evident at the point where the discrimination against ambient sound is at its greatest; and the term 'undirectional' may be taken to refer to the response between cardioid and hypercardioid.

Cardioid response

Left: The sum of omnidirectional and figure-of-eight pick-up when the maximum sensitivity of the two is equal. The front lobe of the 'eight' is in phase with the omnidirectional response and so adds to it; the back lobe is out of phase and is subtracted. The supercardioid, *right*, is a more directional version of the cardioid microphone. It is one of a continuous range of patterns that can be obtained by combining omni- and bidirectional response in various proportions. As the bidirectional component increases further, the rear lobe grows further, to make a cottage-loaf shaped or hypercardioid response.

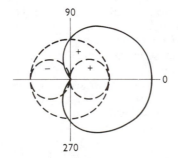

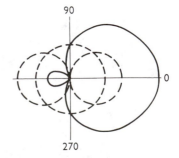

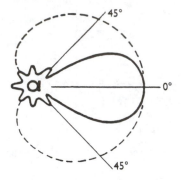

Highly directional microphones. These are substantially dead to sound from the side or rear. They are characterized by their size – for some purposes that of a large parabolic dish used to concentrate sound for a microphone placed near the focus or, more commonly, a long tube extending forward from the main unit. The frequency response is a narrow forward lobe at high frequencies, degenerating into a broader response at wavelengths greater than the major dimension of the microphone.

Highly directional microphone
The unbroken line shows a polar response for medium and high audio frequencies. At low frequencies (broken line) the response degenerates to that of the microphone capsule being used (in this case cardioid).

The frequency response of practical microphones

Practical, high-quality professional microphones do not have a response that is perfectly flat. Peaks and troughs of the order of 3–4 dB are common and are to be found even on relatively expensive microphones. This is not bad workmanship: at one time it was difficult to design a microphone for a level response at all audio frequencies, and even today some deviations may be helpful. In addition, an individual sample may vary by 2 dB or so from the average for its type. Fortunately, few people notice variations of this order: in a reverberant balance they will probably be overshadowed by variations due to the acoustics, anyway. Sometimes a high-frequency peak is permitted on-axis, to allow for a more level response at an angle to it.

High-frequency effects on a large diaphragm
The response R at 0° (i.e. on axis) is the sum of several components including: 1, Obstacle effect (which sets up a standing wave in front of the diaphragm). 2, Resonance in cavity in front of the diaphragm. 3, Inertia of the diaphragm as its mass becomes appreciable compared with high-frequency driving forces.

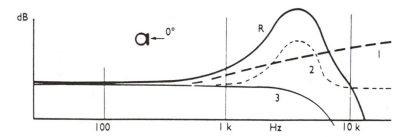

Some microphones have a response that falls away in the bass enough to make them unsuitable for particular orchestral instruments (and certainly for a full orchestra). But much more common is loss of top – from about 12 kHz on some professional microphones. Others have an effective high-frequency response that goes well beyond this, but with these the angle at which the microphone is used may be critical: for some the axis of the diaphragm should point directly at the sound source. A common defect of microphones – and a serious one for high-quality work – is an erratic response in the upper-middle range. Except in individual cases where a continuous trace of the frequency response is provided with each microphone,

Medium quality moving-coil omnidirectional
Trade-off between sensitivity (achieved here by size of diaphragm) and frequency response.
In this example, the difference between the axial response and that at 90° is vast. The most level response is at about 45°. The presence of a windshield (0° W) further modifies the response, but may not unduly degrade speech quality unless it is already sibilant.

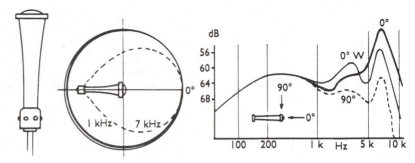

manufacturers tend to publish data and diagrams that somewhat idealize the performance of their products. So this defect may not be apparent from publicity material.

The size of a microphone affects its frequency response. In general, the larger the diaphragm and the larger the case, the more difficult it is to engineer a smooth, extended top response (but the easier it is to obtain a strong signal).

Proximity effect

The pressure-gradient mode of operation, by its very nature, produces a distortion of the frequency response called *proximity effect* or *bass tip-up*: when placed close to a source the microphone exaggerates its bass. The distance at which this begins to set in

Ribbon microphone: live arc
At one time this type of microphone was very widely used on account of the directional properties which are a direct result of the pressure gradient principle employed. An arc of approximately 100° on each side provides a 'live' working area. The frequency response at 60° to the axis is halved, but begins to depart from the common curve at about 2000 Hz, where high-frequency effects begin. Those for variation in the vertical plane (v) are more pronounced than those for the horizontal plane (h).

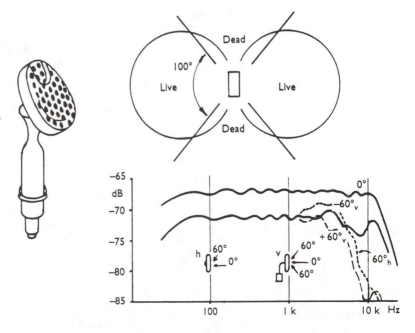

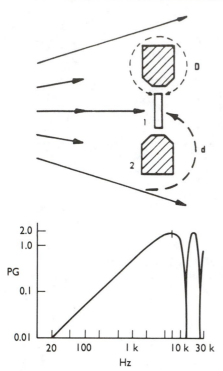

Ribbon microphone: dimensional effects

Above: The sound wave travels farther to reach the back of the ribbon (1) than it does to the front. The effective path difference d is roughly equivalent to the shortest distance D round the pole-piece (2).
Below: Variation of pressure gradient (PG) with frequency where effective path difference, D, from front to rear of ribbon is 2.5 cm. In practice, however, at this wavelength the ribbon and pole pieces create a shadow, and the principle switches over to pressure operation. Ribbons are designed to have a low resonant frequency.

Ribbon microphone: vertical response

At oblique angles there is partial cancellation of very short wavelengths (i.e. high frequency) sounds along the length of the ribbon. In the horizontal plane there is little variation with wavelength.

depends on the difference in path length as the signal reaches first the front then the rear of the diaphragm. Bass tip-up occurs when the sound source is close enough for the path difference to be a significant proportion of that distance, and is due to the fact that, as it radiates outward, sound dies away – in fact, as we have seen, for spherical waves it diminishes in proportion to the square of the distance.

Pressure gradient is the sum of two components. One is produced by the phase difference at two points on a wave separated by the distance from front to back of the diaphragm. The other is the change in intensity with distance. For low frequencies to be picked up at all by a pressure-gradient microphone, it is arranged that it is sensitive to very small phase differences. As a result, when a source is close, the drop in intensity may be large in comparison. In fact, the distance at which this begins to dominate depends on wavelength. For middle and high frequencies the distances at which it becomes important are too small to be of practical importance; it is only for low frequencies that the effect reaches out to a normal working distance from the microphone. The greater the path difference, the greater the frequency range over which a significant amount of bass tip-up occurs.

Bidirectional ribbon microphones offer an extreme example of this, as they require a substantial path difference to achieve adequate sensitivity. In some cases the effect can be heard as far out as 20 in (50 cm), becoming increasingly marked as the distance is reduced. But all directional microphones that employ pressure gradient are affected in some degree.

Some directional microphones have arrangements to switch in bass roll-off to compensate for the increase that occurs at a particular working distance. Movement back and forth from this chosen position produces changes in the ratio between bass and middle frequencies that cannot be compensated by simple control of volume; to make matters worse, such movements also introduce changes in the ratio of direct to indirect sound.

When two bidirectional microphones are placed close together a check should be made that the two are in phase. If they are not, cancellation of direct sound occurs for sources equidistant from the two (although normal reverberation is heard) and severe distortion may be apparent for other positions of the source. The cure is simply

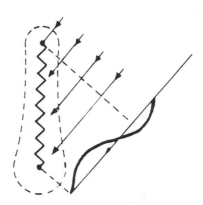

Proximity effect: variation with distance
This increases as a directional microphone approaches a sound source. For a bidirectional ribbon the curves 1–4 may correspond to distances of 24 to 6 in (60 to 15 cm).

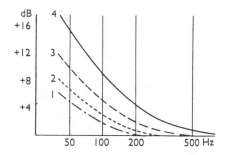

to turn one of them round (or to reverse its leads). When two microphones are needed to pick up separate sources, bidirectional types may be used, with each placed dead-side-on to the other source. This gives better control in mixing.

Compensation for proximity effect
1, Some microphones have a reduced bass response which automatically compensates for the tip-up (2) due to close working.
3, With the microphone at a certain distance from the source, its response to direct sound is level.

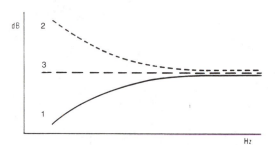

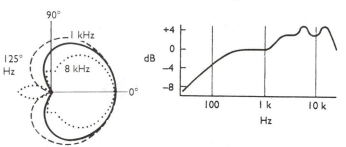

Singer's microphone
The supercardioid response with unswitchable bass roll-off makes this suitable for close working only. The moving coil is rugged and will withstand high sound pressure levels. The double peak is designed to give strong presence on some voices.

Cardioid and hypercardioid microphones

With true cardioid microphones the useful angle is a broad cone including at least 120° on the live side – but which may be taken as extending to a total included angle of about 180°, though so far round to the side the overall response is noticeably lower and in some designs the high-frequency response may be lower still. Continuing round to the back of the microphone, the output should in theory simply diminish gradually to nothing, but in practice, though at low level, it may be uneven, with an erratic frequency response.

Coverage being much broader than with a figure-of-eight, a cardioid microphone can be used where space is limited. When working

Double-ribbon hypercardioid microphone

Polar and frequency response. Its low sensitivity, bass roll-off and fairly smooth mid- and upper-midrange response make it suitable for use with percussion and brass in popular music balance.

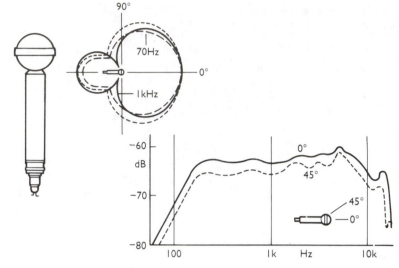

Double moving coil cardioid microphone

Polar and frequency response. Inside, a forward diaphragm responds to high frequencies, while that behind it is more sensitive to low frequencies. Entry ports lower down the stem allow sound to reach the rear of the second diaphragm: this partially compensates for any increase in bass owing to proximity effect.

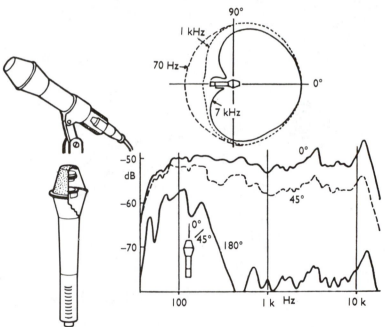

sufficiently close there is again some bass tip-up – in fact, half as much as with a similar figure-of-eight microphone. This reflects the combination of polar responses that makes a cardioid.

The most obvious way of producing a cardioid, by connecting up omnidirectional and bidirectional units in parallel, is no longer used, as the combination of different types with disparate effects produces an erratic response.

A second method is to have both front and back of the diaphragm open to the sound pressure, but to persuade the signal to reach the back with a change of phase. For this a complicated phase-shifting

Moving-coil cardioid microphone
A complex system of acoustic labyrinths is built into the microphone housing: the air reservoir R (shown in the simplified sectional diagram) provides the damping for one of three resonant systems which are used to engineer the response.

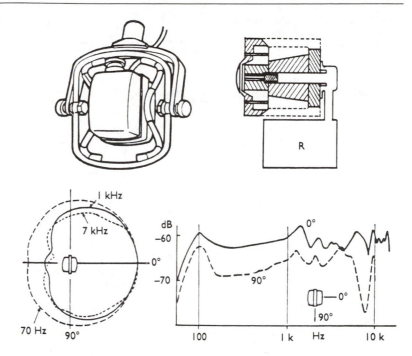

network of cavities and tunnels is engineered. Moving-coil, electrostatic and ribbon microphones have all been adapted to this principle. Such a system can be described mathematically in terms of acoustic resistances, capacitances and inductances. Using these, the pressures on both sides of the diaphragm can be calculated for a range of frequencies arriving from different directions, so that working back from the desired result – for example, a cardioid response reasonably independent of frequency – it has been possible to design an appropriate acoustic network. Even so, such a complex design cannot readily lead to an even response.

An electrostatic microphone with a pair of diaphragms on either side of a central base-plate will also work as a cardioid. In this case there is a polarizing charge between one of the diaphragms and the base-plate; but the other plate has the same charge as the base, so only one diaphragm is being used as a normal electrostatic microphone element. Sound waves cause both diaphragms to vibrate; and the resulting pressures in the cavity behind the second diaphragm are transmitted through holes perforating the base-plate: these holes are of such a size and number that they produce a phase change, and this in turn produces the cardioid response.

In recent years designers have produced a wide range of professional microphones based on these and similar ideas: the majority, today, are electrostatic, but some are moving-coil. Not all of these microphones are pure cardioid: they can, in principle, be anywhere in the spectrum between omnidirectional and bidirectional.

For extended high-frequency response, electrostatic microphones have particular advantages. The diaphragm has an extremely low mass, so it responds readily to minute changes of air pressure. In

Cardioid response of high-quality electrostatic microphone
The axial response is substantially flat to 15 kHz and the response at 90° almost so. The irregularities in the 180° curve are at low level and therefore unimportant.

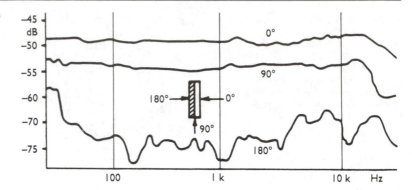

addition, as it has to have one stage of amplification near the microphone capsule (in order to convert the signal to a manageable form), the diaphragm itself can be made small. The precision engineering and head amplifier add to the cost, but on account of its quality (in particular, in frequency response) and sensitivity, it has increasingly dominated the market for studio microphones.

Cost and size can both be reduced in a variation on the electrostatic principle, called an *electret*, in which the diaphragm has an electrostatic charge sealed within it during manufacture. This eliminates the power supply that would otherwise be required to charge the condenser. A small d.c. battery or phantom power supply is still needed for an amplifier close to the diaphragm, either within the head itself or nearby. Early versions have suffered some loss of high-frequency response as they aged, so this may need to be checked. Later variants have been designed to hold their polarizing charge better, and the technology (of sealing the surface) continues to improve.

One professional use is in personal microphones, where the small size of the capsule allows it to be both inconspicuous and extremely light. Another is on musical instruments, to which it may be clipped, usually on a small arm or gooseneck arranged to angle in towards the source. So close to it, the reduced sensitivity and therefore higher noise level associated with a smaller diaphragm is acceptable.

The electret, by its simplicity, is ideal for engineering to a fixed polar response: and this represents one line of development of the condenser principle. Moving in another direction, the polarizing voltage has been retained and turned to new advantage – in order to make the microphone more versatile.

Electret circuitry
Compared with earlier electrostatic microphones this is very simple.
1, Integrated circuit amplifier.
2, Low-voltage d.c. power supply and resistor. Phantom power may also be used. 3, Output (cable).

Switchable microphones

In a switchable electrostatic microphone one diaphragm is constantly polarized and the other can have its polarization changed. If the two diaphragms are polarized in the same sense the microphone operates

Switchable electrostatic microphone

Left: The early version of a high-quality microphone with variable polar response, and, *above centre*, a later, smaller version that replaced it. Its side-fire operation is more obvious. The response is varied by changing the polarizing voltage to one diaphragm. 1, Front diaphragm. 2, Rigid centre plate (perforated). 3, Rear diaphragm. 4, Multi-position switch and potentiometer. 5, Polarizing voltage. 6, High resistance. 7, Head amplifier. 8, Microphone output terminals. When the polarization is switched to position O the voltage on the front and back diaphragm is the same, and above that of the centre plate: the capsule operates as an omnidirectional (pressure) microphone. At B, the capsule measures pressure gradient and is bidirectional. At C, the polar response is cardioid. *Above right:* Two such capsules in a single casing, making a co-incident pair.

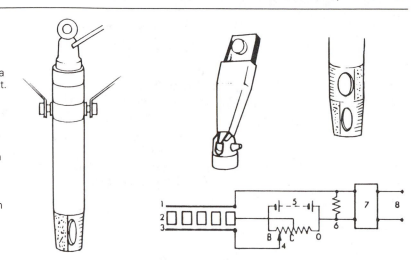

as a pressure capsule, as would two single-diaphragm cardioid electrostatic microphones placed back to back and wired to add outputs. But if the polarizing current on one diaphragm is decreased to zero (and the centre plate has suitable perforations) the polar response becomes cardioid.

Taking this process a stage further, if the polarizing current is now increased in the opposite sense, so that the voltage on the centre plate is intermediate between those of the two diaphragms, the mode of operation is fully pressure-gradient and the output bidirectional. This is now very similar to the symmetrical layout in which a central diaphragm is balanced between two oppositely polarized perforated plates.

Even with a high-quality double-diaphragm microphone of this type some compromises have to be made. In a particular case the cardioid response is excellent and the intermediate positions useful; the omnidirectional condition is good, but the bidirectional condition has a broad peak in its high-frequency response.

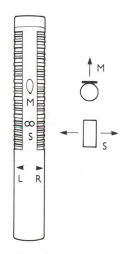

MS microphone

This microphone combines capsules employing two different principles in a single housing. The forward M or middle microphone is end-fire; behind this is a bi-directional S or side microphone. The M response has a 'presence' peak which is suitable for speech or actuality; S is flatter, but matches the effect of the peak. Stereo is obtained by combining the two signals.

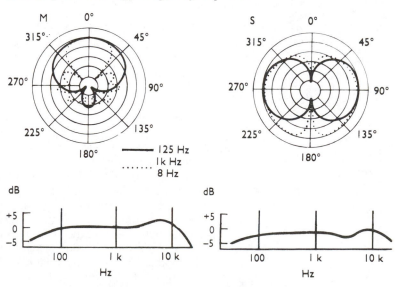

A combination of two types is acceptable for direct middle and side stereo, and may be favoured for television stereo sound, including that for television video and film. Unlike an A and B pair, the M and S capsules do not have to be precisely matched and could even have different transducer principles, although in practice electrostatic components have been used for both. The 'M' component will generally be cardioid or supercardioid.

The most exotic combination microphone has four identical electrostatic capsules, set close together in a tetrahedral arrangement. Named the Soundfield, its great flexibility in use allows a wide range of applications. For example, its response can be altered during a live event to favour a source from some unexpected direction, which may be outside the plane set for a standard stereo microphone pair.

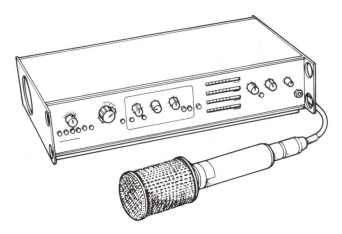

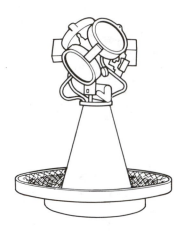

Soundfield microphone and the array of capsules it contains
The variable polar response is switched remotely, as are azimuth (horizontal response) and the angle up or down. A control for 'dominance' narrows the response, giving a zoom effect, within the usual limits (i.e. there is no highly-directional response available). The signals from the four capsules can be recorded separately, then fed back through the control box so that the directional response can be modified later.

Highly directional microphones

For a more highly directional response, a microphone capsule (which may have any of the characteristics described above) is placed within an enhanced or focused sound field. The most common way of achieving this is by the *interference* method employed in *gun* or *shotgun microphones*, also called *line microphones*.

Early designs (offering the simplest way of understanding the principle) had many narrow tubes of different lengths forming a bundle, with the diaphragm enclosed in a chamber at the base. When these tubes are pointed directly at a subject, the sound pressures passing along the different tubes all travel the same distance, so the tubes have no significant effect on the sound. But for sound approaching from an oblique angle many different path lengths are travelled, and as the pressures recombine in the cavity before the diaphragm there is cancellation over a wide range of frequencies. The further round to the side or rear, the more efficient this cancellation becomes; for sound from the rear the path lengths vary by twice the length of the 'barrel' of the 'gun'. But there is a limit to effective cancellation, and therefore to the directional response itself. The length of a gun microphone gives an immediate visual indication of the longest wavelength (and so the lowest frequency) at which its directional properties are effective.

In today's gun microphones the many tubes have been replaced by a single narrow barrel, to which the sound has access at many points along the length. The effect is the same.

One extreme example of 'rifle' microphone has a tube 6 ft 8 in (2 m) long, and is therefore highly directional. As an acoustic engineering aid it has been used to pinpoint the sources of echoes in concert halls. Similar microphones have been used for picking up voices at a distance; for example, reporters in a large audience at a press conference. But its operation can be aurally distracting, as it is very obvious when it is being faded up or down, or panned to find a person who is already speaking. The speech it picks up has a harsh, unpleasant quality – but is, at least, clearly audible.

Commonly-used gun microphones have interference tubes about 3 ft (0.9 m), 18 in (46 cm) and 10 in (25 cm) long, each facing a unidirectional electrostatic capsule. The off-axis cancellation is limited to the shorter wavelengths, with the switch-over in the three cases occurring at approximately octave intervals, depending on length of tube. At wavelengths longer than the tube, the directional response reverts to the first-order effect of the capsule itself, with no great loss of quality. Gun microphones have found favour with many film and television recordists for location work. They are sensitive to low-frequency noise from the sides, but if bass-cut can be applied (as it often can be for exterior work) traffic rumble is reduced considerably. Some have a pistol grip below the microphone, and are pointed by hand. A windshield is often used for outdoor work, but its bulk makes it very obtrusive when there is a danger that it may appear in vision. A strong, gusting wind can cause break-up of the signal even when the windshield itself is protected by the body of the recordist and other screens. In this respect, however, there is little to choose between this and 'personal' microphones (which are described later); but when used out of doors the gun microphone is preferable for its quality.

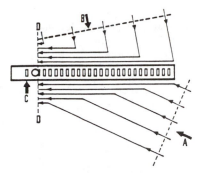

Gun microphone: interference effect
Wave A, approaching from a direction close to the axis, reaches the microphone front diaphragm D by a range of paths that differ only slightly in length. Cancellation occurs only at very high frequencies. Wave B reaches the diaphragm by a much wider range of path lengths. Cancellation is severe at middle and high frequencies. Undelayed sound reaches the back of the diaphragm via port C. The polar diagram of the capsule without the acoustic delay tube in front is cardioid; with the tube it becomes highly directional.

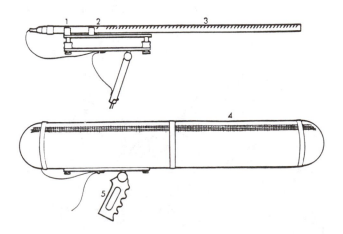

Gun microphone
1, Housing for head amplifier.
2, Electrostatic capsule (cardioid operation). 3, Acoustic interference tube (with entry ports along upper surface). 4, Basket windshield for open-air use (a soft cover may be slipped over this for additional protection in high wind). 5, Hand grip.

When used indoors or in acoustically live surroundings the gun microphone suffers partial loss of its directional qualities because the reverberation is random in phase and therefore not subject to systematic cancellation. Indoors, therefore, the response may be no better – nor significantly worse – than that of the capsule itself. Unless the microphone is brought close to the subject (which defeats the object of using the 'gun') off-axis sound is likely to be unpleasantly reverberant. It can, however, be used on a boom in television or film

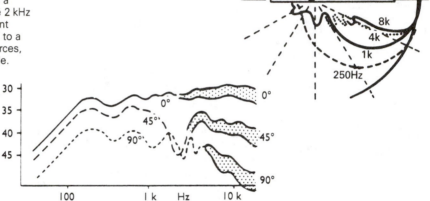

Response of gun microphone
This has 18 in (45 cm) interference tube. For high frequencies the cone of acceptance encloses 30°; movement outside this will pass through regions of marked and changing coloration. The axial response is reasonably level, with some roll-off. One version has a broad peak in the range above 2 kHz to allow for air losses on distant sources; this can be switched to a flatter response for closer sources, or where sibilance is noticeable.

studios, which have relatively dead acoustics, and this allows it to be somewhat farther back than a first-order directional microphone when a wide, clear picture is required. The increased discrimination in favour of high-frequency sound may improve intelligibility.

The short (10 in, 25 cm) gun can also be used indoors as a hand microphone. But here it may reintroduce an older problem of directional microphones, proximity effect. Fortunately, in this design

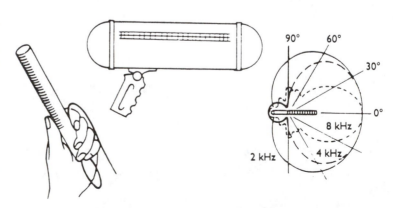

Short interference microphone
Overall length about 10 in (25 cm). At 2000 Hz and below, the response is cardioid. The minimum distance from capsule to mouth is controlled by the length of the tube, so that there is little proximity effect (bass tip-up) even when held close to the mouth. Cancellation due to the interference tube makes the microphone highly directional only at high frequencies. It has excellent separation from surrounding noise when hand-held by solo singers or reporters, or when used as a gun microphone for film sound recording.

it is controlled in part by the distance between mouth and microphone that is created by the length of the tube itself, and the remainder can be compensated by switching in the appropriate bass roll-off. Some examples are sold in a kit which permits the use of the capsule either by itself as an electrostatic cardioid or with an interference tube. One of these is based on an inexpensive electret with a tiny head amplifier powered by a long-life miniature battery.

Gun microphones have not entirely replaced an earlier method of achieving a highly directional response: *the parabolic reflector.* The principle whereby this achieves its special properties is obvious; equally obviously these, too, degenerate at low frequencies, for which the sound waves are too long to 'see' the dish effectively. For a reasonably manoeuvrable reflector of 3–4 ft (1 m) in diameter the directional response is less effective below 1000 Hz. Also, the microphone is subject to close unwanted noises at low frequencies, unless a bass cut is used.

For picking up faint sounds, the parabolic reflector has one great advantage: its directional properties rely on acoustic amplification of the sound before it reaches the microphone, achieving perhaps a 20 dB gain. This offers a far better signal-to-noise ratio than could be obtained from a microphone that has a similar capsule but which relies on phase cancellation. The reflector has therefore found favour for recording birdsong and for other specialized purposes where its bulk and poor low-frequency pick-up are acceptable. The sound can be concentrated almost to pin-point accuracy at high frequencies, or defocused slightly by moving the microphone along the axis to give a broader lobe of pick-up: the latter is generally preferred. An advantage over gun microphones is that reflectors work just as well indoors as out. Tested in highly reverberant conditions (an underground garage), a parabolic reflector picked up intelligible speech at twice the distance that could be achieved by the better of two gun microphones.

To allow for the contribution from the rim, the microphone capsule itself may have a cardioid response, pointing into the bowl. But listen for phase cancellation effects at frequencies related to the size of the dish, where the reflected signal is weak and out of phase with the direct one.

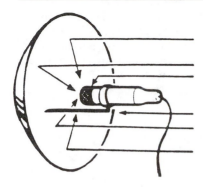

Microphone in parabolic reflector
The bowl may be of glass fibre or metal (coated on the outside with a material to stop it 'ringing'). Alternatively, if it is made of transparent plastics the operator can see through it to follow action, e.g. at sports meetings. In another design the reflector is segmented and can be folded like a fan for travel and storage.

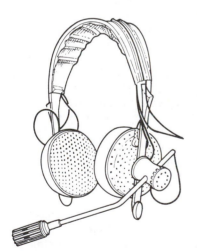

Headphone/microphone combination
The omnidirectional electret has a frequency response that is engineered to suit the direction of the mouth. Discrimination is otherwise by distance, but loud nearby voices will still be audible and may be distracting. Loud sound effects may be satisfactory if the speaker projects to balance them.

Noise-cancelling microphones

There are several ways of getting rid of noise: one is to put the microphone user – a sports commentator perhaps – into a soundproof box. But commentary boxes tend not to be soundproof against the low-frequency components of crowd noise, and in any case are bound to introduce unpleasant resonances if they are small. Another technique is to use an omnidirectional microphone close to the mouth; but in very noisy surroundings this may still not provide the ideal balance between the voice and the background sound, and if the background is still too loud, there will be no good means of correcting the mixture. For best results the voice and background need to be almost completely separated. The one broadcast-quality microphone that can do this in almost any circumstances, even for a commentator speaking in a normal voice with the roar of the crowd all around him, is a *lip ribbon.*

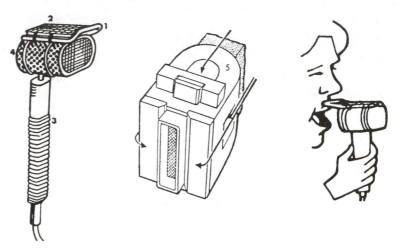

Lip-ribbon microphone
1, The mouth guard is placed against the upper lip. 2, A stainless steel mesh acts as a windshield below the nose. 3, The handgrip contains a transformer. 4, Case containing magnet assembly. 5, Yoke of magnet is towards the mouth, and the ribbon away from it. The sound paths are marked.

Here, the ribbon is very close to the mouth of the speaker. Working in its pressure-gradient mode, it is subject to very heavy bass tip-up for sound from such a close source, but there is no bass tip-up on ambient sound arriving from distant sources. If a specific distance is chosen, say a little over 2 in (5 cm), for the separation between lip and ribbon, and sufficient equalization is introduced to make the microphone response flat for a sound source at such a distance, the ambient bass noise level is also reduced by the same amount. Obviously, the closer to the mouth the ribbon is, the better – so long as the explosive sounds in speech, and streams of air from the mouth and nose, can be controlled. With a distance of $2\frac{1}{8}$ in (54 mm) – fixed by placing a guard against the upper lip – these factors can be controlled by windshields (which, as they are so close to the mouth or nose, need to be made of stainless steel or plastic), and by cupping the ribbon itself behind the magnet.

Some versions have a variety of equalization settings for different degrees of bass; but in general the reduction of noise is about 10 dB at 300 Hz, increasing to 20 dB at 100 Hz. A high-frequency response of up to about 7 kHz is all that is needed for speech in these conditions, so above that frequency the sensitivity of the microphone falls away. These figures are, of course, to be added to those for discrimination due to the extremely close working, and (for lateral noise) the figure-of-eight response.

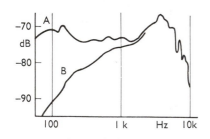

Response of lip-ribbon microphone
A, Response to spherical wave (user's voice) at $2\frac{1}{8}$ in (54 mm).
B, Response to plane wave (ambient sound), on axis. This effect is achieved by equalization, without which B would be flat and A would be high at low frequencies, due to extreme bass tip-up at such a close working distance.

It is possible that even more efficient noise-cancelling microphones could be devised. But this one – though not visually attractive because it obscures the mouth – has proved sufficient for almost any conditions of noise yet encountered in radio, as well as for out-of-vision announcements over loud applause in television.

Microphones for use in vision

Everything within a television or film picture, as much as everything in the sound, adds to the total effect; and a microphone which may be in the picture or a series of pictures for a long time is an important visual element. Accordingly, many modern microphones have been designed for good appearance as well as high-quality performance. They may be mid- or light-toned, often of silver-grey, or black, but always with a matt finish to diffuse the reflection of studio lights. If the capsule is not intrinsically small it should be slender, and performers will more easily understand how they should work to do it if any directional properties it may have are in line with the long axis (*end-fire*): in use it then points towards them. A microphone looks better entering frame from the bottom rather than from the top. A microphone suspended in vision is particularly untidy and calls attention to itself.

Table microphones. The electrostatic microphone lends itself to suitable design. A pencil shape perhaps half an inch or more (1.5–2 cm) in diameter has room for a directionally sensitive capsule pointing along the axis; the body of the pencil houses the head amplifier. The mounting, if it is to be relatively heavy and stable, and is to protect the microphone from vibration, cannot also be made small. But this, too, can be given clean lines and can often be concealed from the camera in a sunken well in the table.

Microphones on floor stands. For someone standing up, a microphone may be placed at the top of a full-length stand, which again should be reasonably strong and have a massive base. It will normally be telescopic. Whatever happens, in a long shot there is going to be a vertical line very obviously in the picture, but the cleaner this is, the better. Any clip holding the microphone itself should be neat but easy to release (e.g. by a performer in vision). Some microphones have only an electrostatic capsule at face level, and are plugged in to one of a selection of thin rigid extension tubes

Headset (without headphones) for singers
This is designed for light weight and minimal disturbance of hairstyle. The electret microphone is in a fixed position, and at this distance must be beyond the corner of the mouth in order to avoid popping.

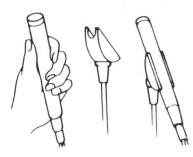

Easy-release clip on floor stand
This padded clip allows the performer to take the microphone from the stand with one hand while in vision.

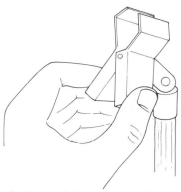

Spring-loaded clamp
This requires two hands to release but is more secure than that above.

Electrostatic microphone kit: I
This is supplied with alternative capsules for cardioid and omnidirectional operation.
1, Head amplifier and capsule.
2, 3, Extension pieces (these are fitted between head amplifier and microphone capsule).
4, 5, Windshields. 6, Power supply unit.

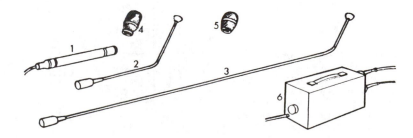

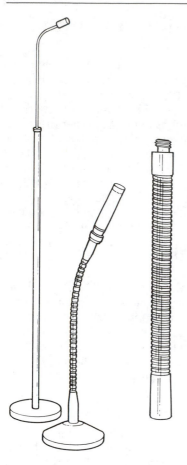

connecting them to the head amplifier: the capacitance of the wiring is taken into account in the design. Some microphone kits come with a choice of capsules for omnidirectional or cardioid response and different lengths of column, including a flexible 'goose-neck' section, so that the capsule can be angled precisely without the need for a separate arm.

Hand microphones. These generally have a 'stick' shape so that they can be held easily. If the diaphragm is at the tip, its distance from the user's mouth cannot be predicted, so in order to avoid variable proximity effects (bass tip-up) it requires an omnidirectional response at low frequencies. An additional advantage of this is that by its pressure operation at these frequencies the microphone is less sensitive to wind noise. While a hand microphone that becomes directional at middle or high frequencies should have an end-fire response (as it is natural to point the microphone toward the mouth) the high-frequency pick-up lobe must be broad enough to accommodate a good range of angles. If the axis is angled more towards the mouth, there may be an increase in the high-frequency response. This effect may be employed creatively by a skilled user.

Robustness and lack of handling noise are essential, but sensitivity is less so: the microphone is often held close to the lips, which is visually intrusive but improves separation from other sound.

Flexible gooseneck connectors
These are stiff enough to allow precise positioning of the microphone.

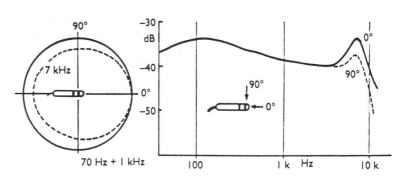

Electrostatic microphone kit: II
Above: Response curves for omnidirectional head fitted directly to head amplifier (i.e. without the extension pieces). *Below:* Response obtained when cardioid head is used in combination with extension piece.

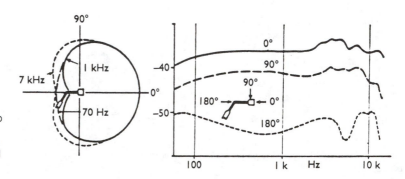

Boundary microphones

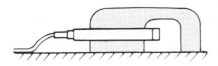

Microphone mouse in section
Polyurethane foam is moulded with a tubular slot for the insertion of microphone, and a cavity for the capsule. It is placed on the floor.

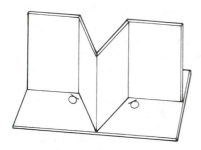

Floor-mounted boundary microphone stereo array
One of many options for the application of the boundary principle in stereo, this American home-made layout stands on a 2 ft × 2 ft (60 × 60 cm) base, with reflectors 1 ft (30 cm) high. The hinged 'V' forms two sides of an 8 in (20 cm) equilateral triangle.

Pressure Zone Microphone with flanges
These may be fixed to a stage surface by tape of a matching colour, or by double-sided adhesive tape or adhesive putty. In this case the response is near-cardioid, and the response at low frequencies depends on the distance of the source. Spaced at intervals of 3–4 m along the apron of a stage, these have been used successfully for opera.
The polar response (for a source at an angle of 30° above the stage) is unidirectional. There are controls to boost bass, B, and to cut it, C.

Generally, microphones are kept away from reflective surfaces owing to the interference fields these produce. However, as the diaphragm is moved closer to the surface this effect moves up through the sound frequency spectrum, until at a distance of half an inch (1.2 cm) it occurs only in the highest octave of human hearing, so that on speech and song the effect becomes unimportant. At closer spacings still the full range is restored. This leads to a design called the *boundary microphone*, or, in one commercial range, *Pressure Zone Microphone, PZM* (which sounds better with the American 'zee' pronunciation).

For a sound wave approaching a rigid, solid surface at right angles a microphone so placed registers only pressure variations, not the pressure gradient, but for a sound travelling lengthways along the surface it measures by either method. A directional microphone is effective therefore only in directions parallel to the wall. This suggests an end-fire cardioid microphone with its diaphragm normal to the surface, directed towards action at a distance. Again, a commercial version is available, but an alternative is to put a standard microphone inside a polyurethane pad: this is called a *microphone mouse*.

Commercial varieties can be engineered to be so flat that the greatest contribution to their thickness is the padding used to protect them from floor vibration. Several examples have pure pressure-operated capsules, so they are omnidirectional in principle, but have a hemispherical response in this housing. In all cases, the hard, reflective surface should, if possible, be clear for several feet around.

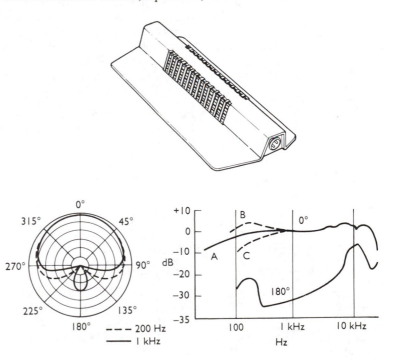

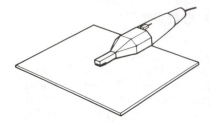

Pressure zone microphone on plate
This is easily fixed to a surface by tape (on top) or double-sided adhesive (below). The tiny capsule, very close to the surface, broadens out to allow for a conventional connector. The hemispherical polar response does not change with distance, a benefit when the distance of the source is unpredictable.

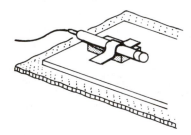

Reducing floor noise
To sample the boundary layer close to a floor, the microphone may be insulated by a spongy pad and taped either to the floor or, if this radiates too much noise, to a board which is itself insulated from the floor by a piece of carpet or underlay.

Lanyard microphone
An early, omnidirectional moving-coil design with a mechanically variable frequency response. A cylindrical clip could be raised to create a well over the diaphragm, in which the cavity resonance increased high frequencies. With the clip lowered (0° curve A) this was a general-purpose omnidirectional microphone. With the clip raised (0° curve B) the 7 kHz peak made it suitable for a lanyard position below the chin. With modern studio equipment, this is achieved more flexibly by electronic equalization.

Floor-mounted boundary microphones are in danger of picking up noises conducted through the floor. A simple way of minimizing directly conducted sound is to mount the microphone on a stand in the orchestra pit, with the capsule on a flexible swan-neck bending over the edge of the stage and angled slightly down, almost to touch the surface. As before, the directional response will be horizontal. The final limitation will be noise radiated from the stage surface and travelling through the air to the microphone.

Personal microphones

A *lavalier microphone* may be suspended round the neck in the manner of pendant jewellery – but rather less obtrusively. In this case it may also properly be described as a *lanyard* or *neck microphone*. Alternatively, it may simply be called a *personal microphone*, a term with more general application, because the small capsules used today are more often simply attached to clothing. For the best sound quality the microphone will be in vision; for neater appearance it can be concealed beneath a tie or some other garment.

The microphone should be omnidirectional because, although it is situated fairly close to the mouth, it is not at a predetermined distance. Commonly, a tiny electret is used. Such microphones work reasonably well because the pattern of radiation from the mouth is also essentially omnidirectional; though there is some loss of high frequencies reaching the microphone, particularly if it is hidden behind clothing. Early microphones with a strong, broad resonance between 2500 and 8000 Hz compensated for this but were unsuitable for speech only: this is better obtained by equalization. Other possible uses for a small unit of this type, e.g. held in the hand, or hidden in scenery to cover an otherwise inaccessible position in a set, require a level response.

Personal microphones must be very light in weight as well as being tiny, but they also need to be robust, as they are easily bumped or knocked about. Some electrostatics have a separate head amplifier,

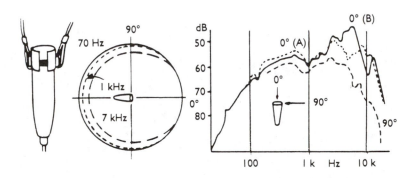

Small personal microphone
Size: 0.7×0.42 in (19×11 mm).
Weight of condenser capsule 0.16 oz
(4.5 g), plus cable permanently
attached to supply unit. The
microphone can be used with one of
several clip attachments or a
windshield, W, which very slightly
affects the high-frequency response.
A more recent version (far right) is
even smaller, and deterioration of
the electrostatic field with age has
been reduced; it is also more
resistant to temperature and
humidity impairment.

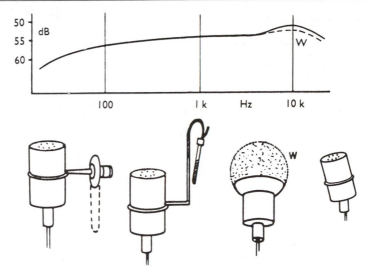

connected by a short cable: this is thin (and therefore relatively
unobtrusive) but inevitably it is vulnerable to damage, and is also a
possible source of conduction noise when rubbed against clothing.
Others are smaller still, and the cable is moulded to the housing. In
some, a tiny amplifier is included in the capsule itself.

In the studio and on location a personal microphone may be used
either with a cable (heavier beyond the head amplifier) or with a radio
link.

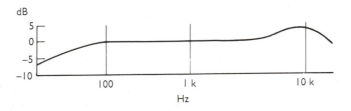

Radio microphones

In radio, television or film the user of a personal microphone may
need to be completely free of the cable because:

● in a long shot the cable may be seen to drag; this is unacceptable in
drama and distracting in other programmes
● in a sequence involving a number of moves, the cable may get
caught on the scenery or completely tangled up; or movement may
be unduly restricted to avoid such mishaps
● the performer may have to move about so much in the intervals
between the various rehearsals and takes that it is preferable to
allow full freedom of movement rather than keep plugging or
unplugging the microphone.

In these cases the cable may be replaced by a battery-powered radio
transmitter and aerial. The transmitter pack should be small enough

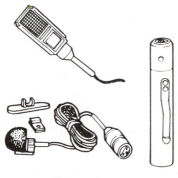

Versatile, tiny personal electret
This measures 4.4×7×13 mm, and
is supplied with a moulded
connector and a range of holders –
clip, tape down strip, tie-bar and tie-
pin – in black, grey, tan and white, or
with a small wind-shield or pressure-
zone adaptor. The response is level
in the mid-range, with a smooth,
extended presence-peak of 3–4 dB
in the 5–15 kHz range. Its relatively
large power supply can be clipped to
a waist-band or go in a pocket.

A radio microphone transmitter pair
Externally, the appearance is similar, although the controls on top are slightly different, and the connector pins or socket reflect the direction of the signal.

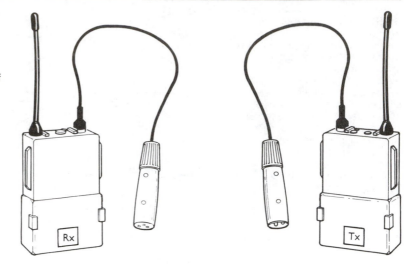

Radio microphone and dipole
The microphone contains its own UHF transmitter and has a stub aerial emerging from the base. Capsules are available for either omnidirectional or end-fire cardioid operation: the latter should be directed towards the mouth or other source accordingly. The base of the stick has switches including on/off, tone, and base roll-off (for cardioid close working) and LED condition and warning lights. An aerial can be mounted at a convenient nearby location and attached to a receiver.

to be slipped into a pocket, pinned to the inside of the clothing or hung at the waist beneath a jacket. If simpler answers fail, put it in the small of the back, using a cloth pouch of the exact size, with tapes to hold and support it. (A similar pouch might be required for a television presenter's radio communication equipment.) If an external VHF aerial (approximately a quarter wavelength long) is used, it may perhaps be allowed to hang free from the waist or, if it would then be in shot, drawn upward within outer clothing. For studio use the screening of the microphone lead can itself be made the right length and used as an aerial. A UHF aerial is shorter, in one example housed in a vertical stub below a hand-held microphone. A receiving antenna is usually a simple dipole placed so that there is an unobstructed electromagnetic path from the performer. The equipment can 'see' through sets made of wood and other non-conductive material, but is likely to be affected by metallic objects (including large camera dollies, which, of course, move around during the production).

When setting them up, i.e. before giving the transmitter to the performer, checks should be made that the antennae (often two for each transmitter) are suitably placed. This needs two people: one of them walks wherever the performer is expected to go, talking over the circuit to a second person who listens for dead spots. Several positions may have to be tried before a suitable one is found.

In a television studio, existing monitor circuits can be used to carry the signal from the dipoles to a radio receiver in the sound gallery. In order to smooth out unexpected signal variations, the receiver may possibly be set for a modest amount of automatic gain (but not in cases where the performer is going to speak loud and soft for dramatic intent). As the transmitter controls have to be preset, the signal can easily be overloaded, causing distortion, or undermodulated, reducing the signal-to-noise ratio. These can be avoided by introducing some degree of compression at the transmitter, which is compensated by expansion at the receiver: the whole acts as a *compander.*

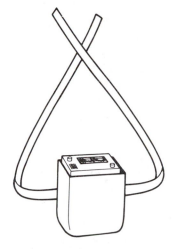

Communications pouch
A radio microphone transmitter or talkback receiver can be carried in a specially-made pouch held by a tape in the small of the back. For a television performer with tight clothing, matching or blending material is used, and the tapes are taken through splits in the side seams.

The transmitter should be free from frequency drift and its signal should not 'bump' if it is tapped or jolted. Battery voltage generally drops suddenly when the power source is exhausted, so batteries should be discarded well within their expected lifetime (which should itself be reasonably long). Most now have automatic early warning of failure; otherwise a log of hours of use can be used to schedule their replacement. The BBC uses a transmitter radiating a radio-frequency power of 120 mW (more in locations where greater than normal distances are involved); a practical minimum is 100 mW. Unfortunately, different wavebands are used in Europe and the USA, so that equipment carried by film crews must be reset (not always easy) after crossing the Atlantic. The band used in the UK is allocated to television broadcasting in the USA.

Occasionally, other equipment in a studio interferes with the radio link; for example, a signal operating radio-controlled special effects. It is safest to anticipate the unexpected by having stand-by transmitter packs with different frequencies. Obvious frequencies to avoid are those used for production talkback continuously broadcast to the studio floor manager and for his switched reverse talkback (where this is used). If it is of a suitable type, the floor manager's own radio microphone is available in an emergency when all others have been used: it is therefore helpful to standardize, using the same model for both purposes.

As a final safety measure, wherever possible have a cable version on standby. For crucial, live broadcasts this should be rigged and ready for use.

Radio microphones can be used in filming (particularly in documentary work or news reporting); for roving reporters in television outside broadcasts; by the hosts of studio programmes; and for many other purposes. But, although the quality of the radio link may be high, it may be compromised by the position of a personal microphone. So alternatives should be considered: in the studio a boom, and on location a gun microphone.

Radio links can also be used with hand-held microphones, though in this case the sight of a cable disappearing into the clothes and not coming out again may be distracting to the audience, particularly in the case of lady singers in tight clothing.

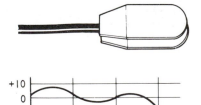

Contact microphone
This can be attached to the soundboard on a guitar or other stringed instrument. The principle is electrostatic but is unusual in having a rubber-coated diaphragm, in order to give it relatively large reactive mass. The polar diagram is figure-of-eight.

Contact and underwater microphones

Many solids and liquids are excellent conductors of sound – better than air. A device that is used to pick up sound vibrations from a solid material, such as the sound board of an electric guitar, is called a *contact microphone*. Since in this the whole microphone housing is moving with the vibration of the solid surface, one way of generating a signal is for a suspended 'active mass' to remain still, by its own inertia.

Sound from solids: active mass
1, Contact microphone. 2, Vibrating surface. 3, The mass of the suspended magnet holds it relatively still. The mass and suspension system can be arranged so that the resonant vibration of the magnet is at a very low frequency and heavily damped. For a reverberation plate or similar source in a fixed position, the microphone assembly can be mounted on a rigid arm, 4.

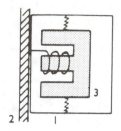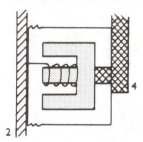

The mass of the outer capsule should be low, in order to avoid damping. A contact microphone that employs a piezoelectric transducer – bending, for example, a strip of barium titanate to generate an electric signal – emphasizes the mid-frequency range.

One electret has a broader response but still suffers a loss of high frequencies (above 4 kHz). This can be compensated by adding an acoustic electret to pick up high-frequency sound directly from strings: in the case of the guitar, it can be fixed underneath the strings, in the air hole, from which it will receive an additional component. Contact microphones can be attached with adhesive putty.

To pick up sound under water, a simple technique is to seal a conventional microphone in a rubber sheath (condom), but this means that the sound must cross the barrier from water to air, an acoustic impedance which reduces sensitivity and distorts the frequency response (totally apart from causing embarrassment, hilarity, or both).

It is better that the water is separated from the transducer only by a material such as chloroprene rubber, which is of matched impedence. One miniature hydrophone has a frequency range of 0.1 Hz to 100 kHz or more, which may seem excessive in terms of human hearing, but not for that of whales and dolphins. By recording and replay at changed speeds, or through a harmonizer to change frequency independently of time, the extended frequency-range of underwater communication and sound effects can be matched to our human perception.

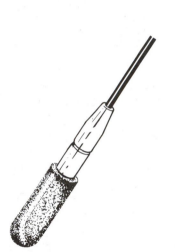

Miniature hydrophone
The ideal underwater microphone, 0.37 in, 9.5 mm diameter, with a rubber compound bonded directly onto a piezoelectric ceramic. This allows an excellent sound-transmission path from the water, and permits a frequency response which extends far above the range of human hearing, so that it can be used to capture extreme high-frequency signals used by some marine life (which we can hear by slowing the replay or other pitch transposition techniques). The response is flat over the normal audiofrequency range, in directions at right angles to the axis; there is a very slight reduction in some other directions.

Directional characteristics of A and B stereo pairs

For stereo, an important technique for giving positional information is the co-incident pair (see Chapter 3). If separate A (left) and B (right) microphones are used they should be matched by manufacture and model (at the very least), because even a minor difference in design may lead to big differences in the structure of the high-frequency response. An alternative to the matched pair is the double microphone, in which two capsules are placed in the same casing,

Stereo pair in one housing
For convenience the two diaphragms are usually placed one above the other, and the outer one can be swivelled to adjust the angle between them.

usually with one diaphragm vertically above the other, so that they can be rotated on a common axis. Sometimes (in either system) the polar responses may be switched between figure-of-eight, hypercardioid, supercardioid and cardioid conditions (as well as omnidirectional, which has no direct use in stereo). The angle at which the axes of the two elements is set is a matter of choice: it is part of the process of microphone balance. In practice, the angle may be as little as 60–70°, but for simplicity in describing how the signals combine and cancel, consider the case where they are set at a right angle.

A co-incident pair of figure-of-eight elements set at 90° to each other have their phases the same (positive or negative) in two adjacent quadrants. The overlap of these allows a useful pick-up angle that is also of 90°. Outside of this, the elements are out of phase, so that directional information is confused and meaningless to a listener, except at the back, where there is a further angle of 90° within which the elements are in phase once again. Sound sources at the rear of the pair are introduced in mirror image (but also reversed left-to-right) into the sound stage that the listener has in front of him, so that physically incompatible objects can appear to be occupying the same point in space.

In a hypercardioid pair (crossed cottage-loaf response) the frontal lobes must again be in phase. With the diaphragms still at 90° to each other, the useful pick-up angle is about 130° at the front. The corresponding 50° at the back is not much use for direct pick-up, as the frequency response of the individual microphones is degenerate in these directions. However, if a source that is being picked up on a separate microphone is placed in this angle, at least its position will not be distorted by the out-of-phase pick-up that would be found at the side.

Stereo pair on bar
This employs two directional monophonic microphones with their diaphragms close together. For nearby voices, it is usually better to have them side by side rather than one above the other.

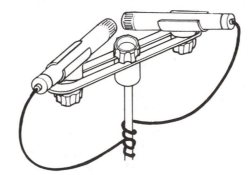

Crossed cardioids have a maximum usable angle of 270°, but for direct pick-up the useful angle is about 180°, as the frequency response of the microphone that is turned away from the sound source degenerates at angles that are farther round to the side.

There are two ways of looking at the polar response of a co-incident pair. One uses the normal response curves for the A and B microphones. The other considers their derived M signal (obtained by adding A and B) and the S signal (their difference).

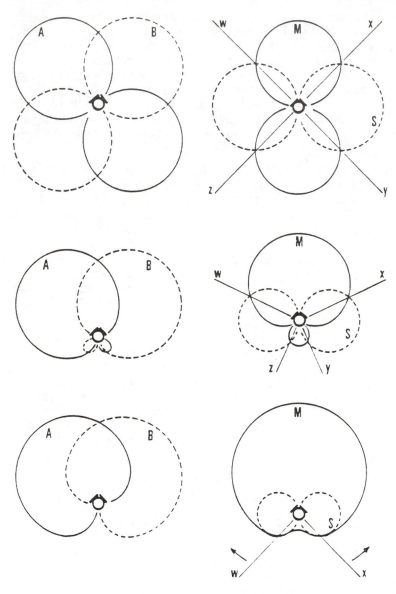

AB polar diagrams for co-incident pairs of microphones at 90°
Above: Crossed figure-of-eight.
Centre: Crossed supercardioid.
Below: Crossed cardioid. The MS diagrams corresponding to these are shown opposite. Note that although the subject of crossed microphones is introduced here in terms of elements set at 90°, a broader angle is often better for practical microphone balances. 120° is commonly used.

MS diagrams for the same co-incident pairs
Above: For crossed figure-of-eight the useful forward angle (W–X) is only 90°. Sound from outside this (i.e. in the angles W–Z and X–Y) will be out of phase. The useful angle at the rear (Y–Z) will (for music) be angled too high for practical benefit.
Centre: Crossed supercardioid microphones have a broader useful angle at the front (W–X). *Below:* The crossed cardioid has a 270° usable angle (between W and X, forwards) but only about 180° is really useful, as to the rear the response is erratic.

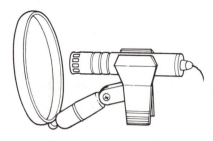

Pop-shield
A framework covered with gauze protects a microphone from gusts of air caused by plosives, etc, in speech and song. An improvised version can be made from a nylon stocking and clothes-hanger wire!

1

2

3

Windshields
These come in a variety of shapes for different sizes of microphones and may be made of 1, plastic foam; 2, foam-filled wire mesh; or 3, mesh with an outer protective cover and air space between the mesh and the microphone.

The M and S diagrams correspond to the polar responses of another pair of microphones that can be used to obtain them directly – a 'middle' (or main) one which has a range of different polar diagrams and a 'side' one which is figure-of-eight. Sometimes A + B and A − B stereo information has been taken directly from main and side microphones, thereby cutting out several items of electrical equipment. A potential disadvantage of this is that the combined frequency response is at its most erratic at the most important part of the sound stage: close to the centre – though the effect gets less as the S contribution gets smaller. Plainly, the quality of the S component must be high, and its polar response should not change with frequency. The 'M' capsule, which occupies the tip of the casing, operates as an end-fire microphone, while the 'S' element, behind it, is side-fire.

AB pairs are widely used in radio and recording studios; MS microphones may be more practical in television and video recording.

Windshields

The term *windshield* (*Am*: wind-screen) covers several distinct functions, for which different devices are used.

The smallest are used to reduce breath effects in close working, and are sometimes called *pop-shields* (*Am*: pop-screen). So long as they protect the diaphragm and its cover from the direct airflow from mouth and nose their shape does not matter. Some microphones have an integral windshield, typically two layers of acoustic mesh with a thin layer of a porous material between. Its effect on frequency response should have been calculated in its design. Many other microphones, designed primarily for use without one, have compact metal-gauze shields made to fit on to them; alternatively, shields made of foamed material can be stretched to fit over any of a number of shapes. To restore optimum frequency response, these should be taken off when the microphone is used at a greater distance.

For a talk studio, foamed shields may be supplied in a variety of colours, so that a speaker can be directed to sit at a clearly identified microphone, and will be reminded by its colour to work to that and no other.

Generally, however, the purpose of a shield is to cut down noise due to wind turbulence at sharp edges, or even corners that are too sharply curved. To reduce wind effects proper, a smooth airflow round the microphone is required. The ideal shape, if the direction of the wind were always known and the same, would be a teardrop, but in the absence of this information practical windshields are usually made either spherical or a mixture of spherical and cylindrical sections typically of 2 in (5 cm) radius. The framework is generally of metal or moulded plastic, with a fine and acoustically transparent mesh, often of wire and foamed plastic, covering it. This should reduce noise by more than 20 dB.

Stands and slings
Methods of mounting microphones (used in radio studios).
1, Suspension by adjustable wires, etc. 2, Boom (adjustable angle and length). 3, Floor stand (with telescopic column). 4, Table stand (but 1, 2 or 5 are better for table work if there is any risk of table-tapping by inexperienced speakers). 5, Floor stand with bent arm. Most methods of mounting have shock absorbers or straps.

Mountings

In addition to the equipment already described, there are many other standard ways of mounting microphones, including a variety of booms. In concert halls, most microphones may in principle be slung by their own cable from the roof of the hall, but for additional safety many halls insist on separate slings, and in any case guying (by separate tie-cords) may be necessary for the accurate final choice of position and angle. A radio-operated direction-control is also available.

Floor stands and table mountings should have provision for shock insulation, to protect the microphone from conducted sound, vibration or bumps. Where appearance does not matter, this may be a cradle of elastic supports; in all cases the microphone must be held securely in its set position and direction.

Studio boom
1, The microphone cradle pivots about its vertical axis, controlled by the angle of the crank (2). 3, The arm pivots and tilts, and is extended by winding the pulley (4). The whole dolly may be tracked, using the steering controls (5), and the platform (and arm fulcrum) height may be raised and lowered. 6, The seat is used only during long pauses in action. The operator has a script rack and talkback microphone (7). Some television organizations provide him with a personal monitor (8) attached to the boom arm: others use floor monitors. 9, The loudspeaker feeds foldback sound to the studio floor.

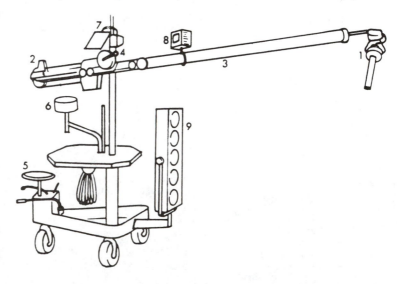

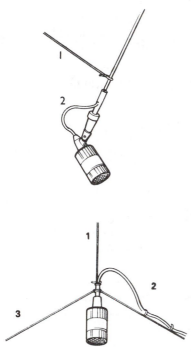

A typical boom used in television and film has a telescopic arm extending from about 10 to 20 ft (3–6 m). It can swing in any direction and tilt to 45°. The operator stands on the platform to the right of the boom, his right hand holding the winding handle that racks the counterbalanced arm in or out, and his left on a crank handle that rotates and tilts the microphone on the axis of its supporting cradle. With his left arm he also controls the angle of the boom. A seat is available but is used only for long pauses as it restricts mobility. Some television organizations provide him with a miniature monitor, which is fixed to the boom arm a little forward of the pivot. The platform and boom can be raised or lowered together: they are usually preset to a height such that the boom extends nearly horizontally, swinging freely above pedestal cameras. The whole carriage can be steered, tracked or crabbed (moved sideways) by a second operator.

A hand-held boom, which may be telescopic or have extension pieces, is called a *fishing rod* or *fishpole*. It takes up much less space and can in addition be brought in from directions that are inaccessible to the larger booms; for example, it may work close to the floor. In normal use, it is supported on a shoulder or braced under one arm for substantial periods. In exceptional circumstances it can be held high with both arms braced over the head, but directors should expect this only for short takes.

Control of microphone position
1, Tie-cord. 2, Microphone cable. 3, Guy-wire for time control of position.

Connectors

In the past a variety of connectors has been used. The most common types are XLR, DIN, Tuchel, Lemo and simple jackplugs. A minimum of three conductors is required for a balanced signal: two are for the signal circuit and a third, earthed (grounded), shields them from hum. If just two conductors are used, one inside the other ('unbalanced'), this may pick up mains hum, so can only be used where there is no danger of this.

XLR connectors are widely used in studios, and the number of pins is indicated by a suffix: XLR-3, for example, has three. In the most common arrangement, a cavity in the base of a microphone contains three projecting (i.e. male) pins: they point in the direction of signal flow. A cable therefore has a female socket at the microphone end and another male plug to go into the wall socket. Other arrangements are available, including right-angled plugs and other numbers of pins. Non-standard arrangements are sometimes encountered (more often in connectors to monitoring loudspeakers), so female-to-female cables may be required.

Shock-absorbing microphone cradle
This is bulky and unsightly but efficiently guards against all but very low-frequency vibration. A loop of the cable should be clipped to the stand.

All XLR connectors have a locating groove to ensure the correct orientation. Some are secured by a latch which closes automatically when a plug is pushed in, and released by pressing on the latch as it is pulled out. The way in which the pins are wired is also nearly always

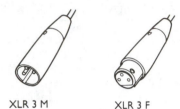

XLR 3 M   XLR 3 F

The figure is the number of pins.
'M' is 'male', with pins pointing in
the direction that the signal travels.
The output from a microphone would
normally have 'M' pins, the feed to
control desk would go into an 'F'
connector.

XLR 3 F

Wall mount with latch.

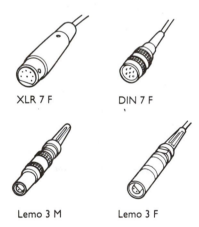

XLR 7 F   DIN 7 F

Lemo 3 M   Lemo 3 F

In lemo plugs and sockets the inner
connectors overlap: the 3M has two
pins and the 3F has one.

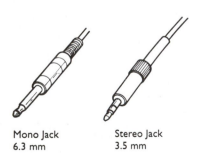

Mono Jack   Stereo Jack
6.3 mm   3.5 mm

standardized. If by any chance they are connected in a different order, this may lead to phase reversal, which therefore needs to be checked when installing unfamiliar equipment in a studio. Phase can usually be reversed in the desk, but to avoid a non-standard layout there, phase-reversal adaptors (and other accessories such as signal splitters) are also available.

The XLR system is robust and will withstand accidental bumps or rough use, but is also rather too massive for applications where low weight or small size are required.

Another common standard that is used for more compact and also much domestic equipment is DIN (the German Industry Standard). The pins are more delicate, and the plugs and sockets smaller. For connectors that must be secure, a screw collar may be provided.

Amongst others: Tuchel plugs have three flat pins angled on radii from the centre and are rather bulky; Lemo connectors (much smaller) overlap inside a sleeve; jackplugs have either 3.5 mm or 6.3 mm (mono or stereo) central pins – take care not to plug these into a headphone or loudspeaker socket, and replace them by XLR if possible.

Three wires (and therefore three pins), one being earthed, are sufficient for most purposes, including both the signal (mono or stereo) and the power plus polarizing voltage for electrostatic microphones. The same pins can be used for all of these by raising both of the live signal wires to a nominal 48 V above ground: this system is called *phantom power*, phantom being a term that had previously been used for grouped telephone circuits that shared a common return path, thereby saving wires. In practice, many modern electrostatic microphones will work with any phantom power from 9 to 52 V and some with batteries. The difference in voltage between this pair and the (grounded) third wire provides the polarization for electrostatic microphones that need it (any but electrets).

More complex systems have more wires (and pins) – for example, for switching polar response or bass-cut, and also for radio microphones. Old designs of electrostatic microphones containing valves (vacuum tubes) – collectors' items or working replicas that are treasured by some balancers for their distinctive sound quality – require an additional 12 V supply to the heater.

Trailing or hard-wired audio cables will usually have substantial but flexible, colour-coded conductors, with braided or lapped shielding to protect against electrical interference, and with a strong outer sheath. Somewhat more complex 'star-quad' cables have higher noise rejection: in a studio it will soon become apparent if these are needed. For longer distances, multichannel cables (with additional coding colours) will terminate in a junction box for the individual microphone cables that lead through a multipin connector to the control desk.

A radio studio may be hard-wired to offer a choice of two sockets for each channel, one on each side of the studio. The sockets may be numbered 1 and 1A, 2 and 2A, and so on, to make the choices clear.

In television studios, routings are often more complex, as they may have to go round or through the set (which may include extensive backings) at suitable points, and to avoid camera tracking lines and areas of the studio floor that will appear in vision. So here a very large number of points is provided: one BBC studio has over 100; several have over 50. Mostly they are distributed around the walls in groups, but some are in the studio ceiling, so that cables can be 'flown' or microphones suspended from above. Some terminate in the gallery area itself. It is convenient to have several multi-way single-socket outputs at the studio walls so that a single cable can be run out to a terminating box on the floor, to which, say, 10 microphones can be routed. In order to minimize induction, sound wiring should be kept away from power cables (to lighting or other equipment) and should cross them at right angles. This is necessary also for wiring built in to the studio structure.

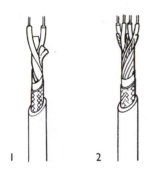

Cables
1, Screened pair. 2, Multichannel cable. 3, Star-quad cable.

Microphone checks

When a microphone has been plugged in – and where more than one is in use – make a few simple checks. First, it must be positively identified at the desk. Gently tap or scratch the casing while a second person listens, to be sure which channel carries the signal. If the desk is already in operation, it may have PFL (pre-fade-listen) facilities which can be used so that the main programme signal is not interrupted. The polar response is confirmed by speaking into the microphone from various sides, identifying them verbally in turn: a dead side will be obvious. For a stereo pair, these checks ensure that the 'A' channel is on the left, and that both are working. Such checks can also be carried out by one person (such as a recordist on location) by using headphones.

In addition, gently check cable connections and plugs by hand to see if these make any noise, and if necessary the microphone itself for handling noise. Check whether it is best to gather up a loop of cable, or if a pop-shield is needed for close speech.

A new or unfamiliar microphone may require more exhaustive tests. The manufacturer's literature gives some indication of what to expect, but is not always completely clear. It is best to start by making a direct comparison with some similar microphone if one is available; failing that, another by the same manufacturer (for a common standard of documentation). This 'control test' should show up any obvious deficiency in sensitivity or performance. Try a variety of sound sources: speech, a musical instrument and jingling keys (for high frequencies). Check the effect of any switches – to change polar response, roll off bass or reduce output.

A large organization using many microphones may check them more comprehensively, using specialized equipment in an anechoic chamber, thereby establishing its own standards, rather than relying on statements in different manufacturers' literature, which may not be directly comparable.

Lining up a stereo pair

If co-incident microphones do not have the same output level, the positional information is distorted. Here is a more detailed line-up procedure that can be carried out in a few minutes by two people, one at the microphones and one at the monitoring end:

1 Check that the loudspeakers are lined up reasonably well, e.g. by speaking on a mono microphone (or one element of the pair) that is fed equally to A and B loudspeakers. The sound should come from the centre.

2 If they are switchable, select identical polar diagrams for the co-incident pair. These may be the ones that will subsequently be used operationally, or if this is not known, figure-of-eight.

3 Visually set the capsules at about 90° to each other. Check and identify the left and right capsules in turn. Where microphones are mounted one above the other, there is no agreed standard on which capsule should feed which channel.

4 Visually set the pair at 0° to each other, ensuring with microphones of symmetrical appearance that they are not back-to-back (there is always some identifying point, e.g. a stud on the front, or the mesh at the back of the housing may be coloured black).

5 Listen to A − B only, or if there is no control to select this, reverse the phase of one microphone and listen to A + B.

6 While one person speaks into the front of the microphones, the other adjusts for minimum output. If possible, this should be on a preset control for one microphone channel or the other; most conveniently, this may be a separate microphone balance or 'channel balance' control.

7 Restore the loudspeakers to normal stereo output. As the person in the studio walks in a circle round the microphone, his reproduced voice should remain approximately midway between the speakers whatever the position of the speaker. Disregarding minor variations when the speech is from near the common dead axis or axes of the microphones, if there are any major shifts in image position as the person speaks from different places there is something wrong with the polar response of one of the microphones; the test must be started again when the microphone has been corrected. The earlier part of the line-up procedure is then invalid. As the person walks round, loudness should vary in a way that coincides with the polar characteristic chosen.

8 Restore the angle between the capsules to 90° (or whatever angle is to be used operationally) and identify left, centre and right once again.

Chapter 6

Microphone balance

This chapter outlines the principles governing the relationship between microphones, sound sources, ambient sound (reverberation) and unwanted sound (noise). Their application to speech and music are described in Chapters 7 and 8, and to effects in Chapter 15.

Its surroundings may indicate the way in which a microphone is used. In a very noisy location, for example close to an aircraft that is about to take off or among a vociferous crowd, the only place for a speech microphone is right by the mouth of the speaker. In these and many other cases the first criterion must be intelligibility. But if clarity is possible in more than one limiting position of the microphone (or with more than one type) there is a series of choices to be made. The exercise of these choices is called *microphone balance*. In America, the vaguer term *microphone placement* may be used.

The first object of balance is *to pick up the required sound at a level suitable to the microphone lines and audio system that it is fed to; and at the same time to discriminate against unwanted noises.*

In professional practice this means deciding whether one microphone is necessary to cover the sound source adequately, or several, and choosing the best type of microphone, with particular regard to its directional and frequency characteristics. In addition, it is often possible to choose the position of the source or degree of spread of several sources, and to select or modify their acoustic environment.

Where the natural acoustics are being used, the second aim of a good balance is *to place each microphone at a distance and angle that will produce a pleasing degree of reinforcement from the acoustics of the studio* – or, in other words, to achieve an appropriate ratio of direct to indirect sound.

Acoustic reinforcement may, however, be replaced by artificial reverberation which is mixed in to the final sound electrically. In this case each sound source can be individually treated. For this reason,

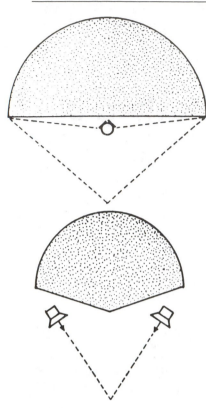

and because picking up a sound on a second more distant microphone will reintroduce the effect of the studio acoustics, we have as an alternative aim of microphone balance: *to separate the sound sources so that they can be treated individually.*

For many kinds of material (and especially for popular music) the term balance means a great deal more: it includes the treatment, mixing and control functions that are integral parts of the creation of this single composite sound.

Stereo microphone balance

In stereo, microphone balance has an additional objective: *to find a suitable scale of width.* Whereas indirect sound occupies the whole sound stage, it may be quite acceptable that direct sound should occupy only part of it.

A large sound source such as an orchestra should obviously occupy the whole stage and, to sound realistic, must be set well back by the use of adequate reverberation. On the other hand, a string quartet with the same reverberation should occupy only part of the stage, say the centre 20° within the available 60°; anything more would make the players appear giants. The same quartet can, however, be 'brought forward' to occupy almost all of the sound stage (provided that the balance itself is sufficiently dry). In either case, whatever reverberation there is would occupy the whole available width.

Compression of scale of width
A sound source that might be spread over 90° or more in real life (*above*) may be compressed on the audio stage to 60° or less.

With co-incident microphones, image width can be controlled in several ways:

● The microphones may be moved closer to or farther from the sound source. Moving farther back narrows the source and also (as with mono) increases reverberation.
● The polar response of the microphone elements may be changed. Broadening the response from figure-of-eight through hypercardioid out towards cardioid narrows the image. It also reduces the long-path reverberation that the bidirectional patterns pick up from behind and permits a closer balance.
● The ratio of A + B to A − B (M to S) may be changed electrically: if S (the A − B component) is reduced the width is reduced. If M (the A + B component) is reduced the width can be increased.
● The angle between microphone axes may be changed. Narrowing the angle also brings forward the centre at the expense of the wings, and discriminates against noise from the sides.

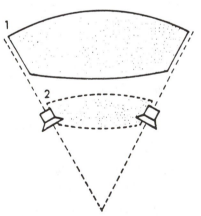

Orchestra in correct scale of width
A broad subject needs enough reverberation to make it appear a natural size (1); if the reverberation is low the subject appears close and under-size or cramped together (2).

With spaced microphones, their separation and the directions in which they point have the strongest effect on image width; and the M/S (A + B/A − B) ratio may again be manipulated. With both kinds of stereo balance, further reverberation may be added from additional distant microphones.

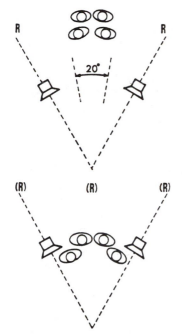

A quartet in stereo
Above: The quartet uses only a third of the audio stage, but reverberation fills the whole width. *Below:* The quartet occupies the whole of the space between the loudspeakers. Here the reverberation must be low in order to avoid the players appearing gross: such reverberation as there is appears from behind them.

Where individual sources for which no width is required are balanced on monophonic (or 'spotting') microphones, the balance must be close enough (or the studio sufficiently dead) to avoid appreciable reverberation, or there will be a 'tunnel' effect – extra reverberation beamed from the same direction.

A position on the sound stage may be established by steering (or *panning* – see later) portions of the signal to the A and B channels. If a second monophonic microphone picks up the same source unintentionally it joins with it to form a spaced pair, giving a false position; what is more, the degree of displacement also shifts if the level of one of the microphones is changed. Close balance, directional characteristics, layout of sources in the studio and screens may all be used to make sure this does not happen.

Monophonic spotting techniques cannot be used for subjects that have obvious physical size – such as a group of instruments or singers, a close piano or an organ. In this case, two monophonic microphones may be used to form a stereo pair, or an additional co-incident pair set in.

The balance test

A sound balance may be intended as a close representation of the original sound or an adaptation of it to suit different listening conditions – or, indeed, for any effect selected from a broad spectrum of modified sound qualities that might be offered for the listener's pleasure. It is never an exact copy, but always an interpretation: it is the balancer's job to exercise choice on behalf of the listener, selecting for him from a vast range of possibilities.

Good sound is not just a matter of 'going by the book'. It is a subjective quality, but the balancer's subjectivity must be allied (paradoxically) to a capacity for analytical detachment. And whenever possible, he judges his results in the best listening conditions that the intended audience is likely to use.

The surest way of getting a good balance – or at any rate the best for the available studio, microphones and control equipment – is to carry out a balance test. Such tests are not made on every occasion: with increasing experience of a particular studio and the various combinations of sources, a balancer spends less time on the basics and more on detail. A newcomer to microphone balance is likely to see a great deal of the latter, and this apparent concentration on barely perceptible changes may obscure the broader objective, and the way that is achieved.

Consider first the single microphone (or stereo pair) picking up a 'natural' acoustic balance. The preferred technique is to rig at least two microphones in positions that (according to the general principles outlined later) should provide a reasonable sound. Compare two balances by using the mixer desk and monitoring loudspeakers to listen to each in turn, switching back and forth at intervals of 10 seconds or so, considering one detail of the sound after another until a decision can be made. The poorer balance is then abandoned and its microphone moved to a position that seeks to improve on the better. The process is repeated until no further improvement is possible or time runs out. In practice, this means advance preparation, providing opportunities to move microphones with minimum disturbance and loss of rehearsal time, or to select from a wider range that has been pre-rigged.

For a multimicrophone balance the same process may be carried out for each microphone. Here, what might appear to be a lengthy process is simplified: there is no need for an acoustic balance between many sound sources with a single microphone, as this is achieved in the control desk; nor (in dead music studios) to balance direct with ambient sound.

Where multitrack recording is used (as in many music recording studios today) some of the balance decisions can be delayed until the recordings themselves are complete. Mixing down – still using comparison tests – is a part of balance that may be continued long after the musicians themselves have gone.

For speech, in those cases where the speaker works to the microphone (which includes most radio and some television) a similar method may be used: place two microphones before the speaker, replace that which gives the poorer balance, and so on.

A television balance in which the microphone is permitted to appear in vision may be similar to those already described. Static microphones are set to musical instruments or to speakers in a discussion; or a singer may perform to a stand microphone. But where there is movement and the microphone itself should not be seen, the process is reversed and the microphone is made mobile (on a boom), working to the performers. The balance changes from moment to moment – ideally in harmony with the picture.

In either case the visual element increases the complexity of the balance problems, to which the simplest answer may be additional microphones.

In any kind of balance here are the main things to listen for:

- The clarity and quality of direct sound. In the case of a musical instrument this radiates from part of the instrument where the vibrating element is coupled to the air. It should comprise a full range of tones from bass to treble, including harmonics, plus any transients that characterize and define the instrument.
- Unwanted direct sounds, e.g. hammer (action) noises on keyboard instruments, pedal operation noises, piano-stool squeaks, floor thumps, page turns, etc. Can these be prevented or discriminated against?
- Wanted indirect sound, reverberation. Does this add body, reduce harshness and blend the sound agreeably without obscuring detail or, in the case of speech, intelligibility?
- Unwanted indirect sound, severe colorations.
- Unwanted extraneous noises.

Balance test
Compare the sound from microphones A and B by switching between the two. The microphones lie in a complex sound field comprising many individual components. For simplicity, the illustration shows monophonic microphones. In stereo, the principles are the same, but the sound paths are more complex.

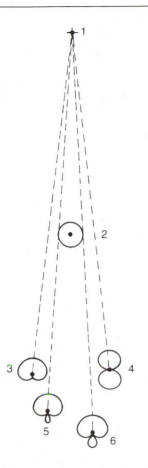

Distance from source
Their directivity (whereby they pick up less reflected sound) allows some microphones to be placed further away from the source. 1, Compared with an omnidirectional microphone, 2, the cardioid or bidirectional response, 3 and 4, can be set back to 1.7 times that distance; a supercardioid to 1.9 and hypercardioid twice that distance.

When trying out a balance, listen to check that bass, middle, and top are all present in their correct proportions, and sound in the same perspective. If they do not, it may be the fault of the studio, in that certain frequencies are emphasized more than others, so changing both proportion and apparent perspective; or it may be more directly the fault of the balance in that a large instrument is partly off-microphone (this can happen with a close piano balance). In a live acoustic, a balance that works on one day may not on the next, when the air humidity is different.

Here are some of the variables that may be changed during a balance test:

- distance between microphone and sound source
- height and lateral position
- angle (in the case of directional microphones)
- directional characteristics (by switching, or by exchanging the microphone for a different type)
- frequency response (by using a different microphone if an extended response is required, or otherwise by equalization).

For greater control over these variables, as few should be changed between successive comparisons as is feasible in the time available. Another possibility for improving a poor balance is to set in additional microphones to cover any parts of the original sound that are insufficiently represented.

The balancer working in a studio, having a separate room for monitoring, is able to use high-quality loudspeakers to judge the results. But where these ideal conditions are not available, and artistic quality is both important and possible, balance tests can take the form of trial recordings – if necessary, a whole series of them – so that comparisons can be made on playback. Since the recordist uses these trial runs to judge how he can use the acoustics of the studio to the greatest advantage (or to the least disadvantage, if the reverberation is excessive or highly coloured), he must use reasonably high-quality equipment for playback. Headphones can be far better than small loudspeakers.

A recordist working on location may be at a particular disadvantage here; one that will not be viewed sympathetically back in the studio. So skill – or in the absence of experience, very great care – is needed.

For reportage, an unchecked balance generally yields results that have an adequate documentary value. Here, experience is invaluable.

Visual conventions for microphones

In film and television the normal criteria for the placing of microphones for good balance still apply, but in addition a second set, based on what is visually acceptable, must also be considered.

One important convention concerns the match of sound and picture perspectives for speech. Obviously, a close picture should have close sound. Also, if possible, a wide shot should have sound at a more distant acoustic perspective. However, the need for intelligibility may override this: a group walking and talking across a crowded room may be clearly audible in a relatively close perspective, whereas nearer figures may be virtually inaudible; again, a reporter seen in long shot may be audible in close-up. In neither case will the audience complain – because they want to hear, and they can.

Other conventions concern whether microphones may appear in vision or not.

For drama, the microphone should always be unseen, as its presence would destroy the illusion of reality.

For television presentations where such total suspension of disbelief is not required of the audience, microphones may be out of vision or discreetly in vision. In-vision microphones are used where better or much more convenient sound coverage can be obtained in this way; but because much of television consists of programmes in which microphones *may* appear in vision they may be seen a great deal without being noticed. However, their very ubiquity is a point against them: there is visual relief in getting rid of them if the opportunity allows.

For film, including television films, where the illusion of reality does not have to be fostered, microphones may appear in vision; but the reason needs to be stronger than that required in a television studio; and the use even more discreet. This is because most film is shot in places where microphones are not part of the normal furniture, so that to show a microphone may remind the audience of the presence of all of the technical equipment (including the camera). Also, in a film there may be greater variety of picture, or a more rapid rate of cutting than may be possible in the studio; in these circumstances a microphone that is constantly reappearing is more intrusive.

When it is decided that a microphone really has to be in vision, its choice depends on its appearance and also, perhaps, mobility as well as on the balance required: some of the microphone types that are available have been described in the preceding chapter, and their use is discussed in the next two. But when microphones must not be seen, the best balance will probably be obtained with studio booms. Here, too, the equipment has already been described; and as it is used mainly for speech balance, boom operation is dealt with primarily under that heading. Television musical performances present special problems: their balance should compare well with that on radio or record even though television studios are usually fairly dead. Unless a concert hall is used, this provides an extra reason for multimicrophone balance.

Multimicrophone balance

In a multimicrophone balance a broad range of techniques is drawn together, often including ways of handling reverberation and equalization, of building up a recording by adding track upon track, and of mixing down from 16, 24 or even more pre-recorded tracks to a mono, stereo or quadraphonic finished product. More detailed description will follow in later chapters, but here are brief notes on some of the techniques used:

Equalization. Nominally the word means restoring a level response to a microphone that has inherently or by its positioning an uneven response. But in addition it is a second stage in the creative distortion of sound – the response of the microphone in its chosen position being the first. Equalization controls, whether located in an integrated dynamics module or assigned to the microphone channel by computer, include midlift and bass and top cut or boost. Most commercial desks offer several bands of midlift or the corresponding dip or 'shelf' in response.

Artificial reverberation. This augments or replaces the natural reverberation given by conventional acoustics. Greater control is possible, as feeds taken from individual microphones can be set at different levels. On commercial desks, an auxiliary feed which can be taken from one or more channels is sent to an external unit and then returned to the desk as a separate and usually stereo source. Sometimes several artificial reverberation devices with different settings are used. Digital reverberation equipment may include a range of other effects.

Compression. Reduction in dynamic range may be used to maintain a high overall level (in order to ensure the strongest possible recorded or broadcast signal), or it may be applied to single parts of a backing sound which without it would have to be held low in order to avoid sudden peaks blotting out the melody, or to the melody itself, which might otherwise have to be raised overall, and as a result become too forward. Limiters as well as compressors are used for this.

Overdubbing (track-laying). This is building up a complete sound by additive recording. For example, the instrumental backing may be recorded first; then the soloist listens to a replay on headphones and adds his contribution, the mixed sound being recorded on another track or machine. Or, for a particular item, an extra instrumental track may be required from a musician who has already played in the original recording.

Post balancing and mixing down. On many commercial desks each channel has an array of selector buttons, each taking a feed from it to a particular numbered track head on a multitrack recorder.

Music recording studios lay down many tracks side by side on broad tape, with 16–48 stacked heads to record and replay them. Most instruments are recorded in mono, but some, such as piano or a vocal backing group, can be in stereo and require two tracks; some (notably drums) may have many. With 8-track recordings it is likely that some preliminary mixing of groups of instruments will be necessary, with the stereo elements remaining split across pairs of tracks. A reference balance is obtained during the session and may be recorded. The levels chosen for all channels are noted, or stored by computer. After the departure of the musicians the tapes are replayed so that the original individual tracks, perhaps with further treatment, can be mixed down (*reduced*).

For the preparation of tapes as the first stage in record production a mixing-down session is often arranged on a later day, in order to allow fresh judgement to be applied to experiments in combining and placing the sounds. Panpots can be used to position the individual tracks on the stereo stage, with components that are already in stereo being either compressed and placed within the picture or retained at their full width. Some of the techniques used for popular music are considered further in Chapter 8.

In radio and television, where time is more limited, the multitrack is often used only for minor additions, improvements or second thoughts. When combined with subsequent material (e.g. in a film) the original mix may not work, but a remix can easily be made.

Computer control. There are advantages to systems that 'remember' mixing instructions and can reproduce them on demand – on another day if necessary. Subsequent experiments then involve only trying out and then feeding changes to the store of mixing instructions. Internal variations in level are reliably reproduced; tracks with nothing on them for the time being are consistently faded out or muted, and forgettable operations, like briefly taking out a channel to eliminate a cough, are handled automatically.

Such a desk has 'write', 'read' and 'update' modes and may also have additional 'write' and 'update' buttons associated with each fader (moving fader systems normally have touch-sensitive fader knobs). The initial balance is made with the desk in the 'write' mode. When the tape is replayed with the desk in the 'read' condition, all the earlier fader settings and changes are reproduced (irrespective of the current fader positions). If while the mix is being read a 'hold' button is pressed, the replay continues but with the controls unchanged from that point on: this could be used (for example) to get rid of a fade-out that had been put on the end of the initial mix.

Except in the case of moving fader system, to update the mix, all of the faders are set to zero or to some nominal setting which becomes the 'null' level. After the relevant channels have been set to 'update', the tape is replayed and any changes from the null position (and their variation in time) are fed to the data store, changing the selected components of the mix by the amounts indicated. Alternatively the whole of the previous control sequence for a channel can be rejected and written afresh by using the 'write' button instead of 'update' and operating the fader normally. Also associated with each fader is a grouping switch. If several faders are dialled to the same number to form a subgroup, the 'write' and 'update' instructions for any one of them operate on the control data for all members of the subgroup.

Early computer storage of settings was used for the faders only; later versions include equalization, compression, auxiliary feeds (including that to reverberation equipment) and so on.

Artistic compatibility: mono and stereo balances

If an item recorded or broadcast in stereo is likely to be heard by many people in mono it is obviously desirable that the mono signal should be artistically satisfying. This may mean modifying the stereo version somewhat. Several problems may occur:

Loudness. Whereas a sound in the centre appears equally loud in both channels, those at the side are split between the M and S channels. As the S information is missing in mono, such subjects may be lower in volume than is desirable.

Reverberation. The S signal has a higher proportion than does the M; the mono version may therefore by too dry. A possible way round this is to make the main stereo balance drier still, and then add more reverberation on the M channel (if this format is employed). In crossed figure-of-eight balances, in particular, there is heavy loss of reverberation due to cancellation of out-of-phase components, and this is one reason for preferring crossed hypercardioids (in which the effect is less) or crossed cardioids.

Aural selectivity. Individual important sound subjects may not be clearly heard in the mono version when in stereo they are distinct because of spatial separation. It may be necessary to boost such subjects or so place them (towards the centre of the stage) that in the mono version they are automatically brought forward.

Phase distortion. Spaced microphone techniques may not adapt well to mono, as the subjects to the side have phase-distortion effects: systematic partial addition and cancellation of individual frequencies which no longer vary with position. To check on these, feed A + B to one loudspeaker.

In addition, there are things that are less of a nuisance in mono than stereo. For example:

Extraneous noises. Noises that are hardly noticed in mono may be clearer in stereo because of the spatial separation. Moving noises (e.g. footsteps) may be particularly distracting unless they are part of the action and balanced and controlled to the proper volume, distance and position. Noises from the rear of bidirectional pairs will appear to come from the front, and movement associated with them will be directionally reversed.

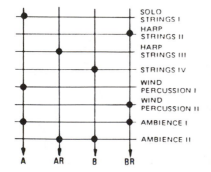

Two kinds of quadraphony
Mix-down from 8 tracks.
Blobs indicate a feed through the mixer. *Above:* In this example the A and B channels carry conventional stereo; reverberation only is fed to the rear channels. *Below:* Direct sound is fed to all four channels, with the two reverberation tracks split diagonally, and the wind and percussion also divided diagonally. In this example a premix appropriate to the music has been planned. In practice, modern multitrack techniques will involve many more tracks than this and the mix-down will be more complex.

Quadraphonic microphone balance

Quadraphonic sound had a short commercial life, during which techniques were developed for broadcasting and the record industry

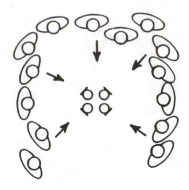

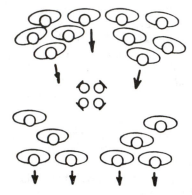

Quadraphonic microphone cluster

Above: All performers working to the centre. *Below:* All performers working in the same direction. The cluster technique works well with discrete quadraphony; less so with matrixed systems.

which are now of little more than academic interest (though in the cinema a wider spread of sound is often employed). Since records and recordings may still be encountered, here is a short account of how some quadraphonic balances were achieved:

● From a conventional stereo balance, direct sound (with some reverberation) is fed to the A and B channels and reverberation only to AR and BR (where R indicates rear). Direct sounds are steered to the rear channels for special effects only.

● A multimicrophone balance is employed, using mainly mono microphone techniques but with stereo pairs for broad sources such as piano or organ. The images are steered to their desired position around the periphery. Control of depth may be achieved by splitting individual direct sounds between diagonally opposed channels (to bring the image forward) or by adding reverberation (to carry it back, as in two-channel stereo).

● A cluster of four hypercardioid microphones is set in the middle and directed outward to sound sources arrayed around them. Generally these sources are all pointing inward, but a directional sense can be preserved by having the sources all directed towards the same wall of the studio.

Chapter 7

Speech balance

Speech balance depends on four factors: the voice itself, the microphone, and the surrounding acoustic conditions and competing sounds. In a good balance the voice should be clear and sound natural, perhaps reinforced and brightened a little by the acoustics of the room that is being used as a studio. There should be no distracting noise, no sound that is not an appropriate and integral part of the acoustic scenery.

A studio designed for speech and discussion programmes may be about the size of a fairly large living-room, with a carpet on the floor, and perhaps treatment on the walls and ceiling to control reverberation. The nature, as well as the duration of reverberation, should be considered: a large, open studio, even if dead enough, sounds wrong, and a speaker needs screens around him or her, not to reduce the studio's reverberation, but to cut down its apparent size by introducing more short path reflections. A warm acoustic might match that of a domestic sitting-room; but for a film commentary an audible studio ambience might conflict with picture, so for this and certain other purposes there is more absorber, and a lower reverberation time.

Microphones for speech

Microphone and acoustics are interrelated. For a given working distance an omnidirectional microphone requires far deader acoustics than a cardioid or bidirectional microphone, which by their directivity discriminate against the studio reverberation and any coloration that goes with it. A hypercardioid, with its greater directivity, reduces these still further. On the other hand, the omnidirectional types (including hypercardioids or cardioids which become omnidirectional at low frequencies) permit closer working.

Microphone position
This shows a good position for speech, with head well up. The script is also held up and to side of microphone.

Consider first a microphone for which, by this reckoning, the studio acoustics are important: the ribbon – which despite its archaic image is still preferred in many studios on account of its excellent figure-of-eight polar response. At its normal working distances, 2–3 ft (60–90 cm), it has a reasonably flat response throughout the frequency range of speech. There is some roll-off above and below that range, but that is no disadvantage, because these limits discriminate against low- and high-frequency noise. To study the interplay of voice and unfamiliar acoustics, experiment with balance tests as described in Chapter 6. Try several voices: some benefit more by a close balance than others. Do this well ahead of any important recording or broadcast, as a last-minute series of complicated microphone adjustments can be disconcerting.

The best way of getting a good signal-to-noise ratio is to speak clearly and not too softly. Two speakers may be balanced on opposite sides of a bidirectional microphone – easy enough unless they have radically different voices. An interviewer can sometimes help by adopting a voice level that acoustically balances voices. Since the primary aim of most interviews – as generally expected by most of the audience – is to bring out the views and personality of the interviewee, it has sometimes been suggested that an interviewer's voice should be slightly lower in volume.

If the relative loudness of two voices does not sound right on the balance test it may help if the microphone is moved slightly to favour one or the other. But watch out for any tendency for the more distant voice to sound acoustically much brighter than the other; rather than allow this to happen it is necessary to introduce some control (i.e. adjustment of the gain) – just a slight dip in level as the louder speaker opens his mouth to speak – or to balance the speakers separately, perhaps on hypercardioid mircophones. Random variations in loudness are much less tolerable when heard over a loudspeaker than when listening to a person in the same room, so if nothing can be done to stop people rocking backward and forward, or producing considerable changes of voice level (without, for example, breaking the flow of a good spontaneous discussion), some degree of control is necessary.

It may help if the microphone is less obtrusive, e.g. mounted in a well in a table between interviewer and interviewee. Sight lines are better, but paper noise from notes lying on the table may be worse.

An omnidirectional microphone at a similar distance would give far too full an account of the acoustics, so for this pattern a closer balance is employed, and the effects of movement in and out relative to a fixed microphone become more pronounced. However, with a very close balance the importance of the acoustics is diminished, so compensation by fader alone becomes less objectionable.

In practice, omnidirectional microphones lie somewhat uncomfortably between those designed to use acoustics and those that eliminate them. As a result, microphones that unequivocally de-emphasize acoustics are often preferred. This is where hypercardioids come into their own. They pick up less ambient sound and allow closer balance – partly because of the reduced pressure

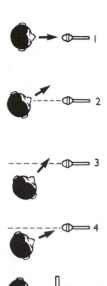

Popping
1, Speaking directly into a microphone beyond its normal range. 2, Speaking on-axis but to the side of the microphone may cure this without changing frequency response. 3, Speaking across and past the microphone may cure popping and also change frequency response. 4, Speaking across the microphone but still towards it may still cause popping. 5, A simple screen of nylon mesh over a wire obstructs the puff of air.

gradient component (compared with bidirectional types) but rather more from their carefully cultivated tendency towards an omnidirectional response at low frequencies. This last may, however, begin to bring a boomy low-frequency component of the room acoustic back in to the balance which, in turn, is de-emphasized by any bass roll-off in the microphone itself. In hypercardioids there is scope for a wide range of design characteristics. Variables include angle of acceptance, the way in which the polar response changes at low frequencies, the degree of intrinsic bass roll-off, and whether additional switched bass cut is provided. At high frequencies the differentially changing response at angles of around 30° or 45° to the axis may be used for fine tuning the response to particular voices. For increased intelligibility, especially against competing background noise, use a microphone (or equalization) with a response that is enhanced somewhere between 3–10 kHz, but check for sibilance.

With early omnidirectional designs the erratic high-frequency response made it desirable to add acoustic resonance to the voice, if only to disguise the deficiencies of the microphone itself. Modern hypercardioids are more satisfactory, so many voices can be close-balanced and acoustically dead without sounding harsh. This style is suited not only to voices over film but also to radio documentary narrative, news reporting and announcements between segments with different qualities and acoustics. But it does also emphasize the mechanical noises from the lips and teeth that sometimes accompany speech, and increases the danger of blasting, particularly on the plosive 'b' and 'p' sounds.

There are several ways of dealing with popping. Experiment by projecting 'p' sounds at your own hand: you will feel the blast of air which would push a microphone diaphragm beyond its normal limits. Then move the hand a little to one side. This will demonstrate one solution: to speak past the microphone, not directly into it. A windshield (or pop-shield) may be needed, if not already included in the microphone design.

Some disc jockeys prefer a close, intimate style, with very little of the studio acoustic in the sound. It is possible that the wish to work very close to the microphone is partly also psychological: the performer is helped if he can feel that he is speaking softly in the audience's ear. A disc jockey with a soft, low voice does not improve matters by gradually creeping in to an uncorrected directional microphone and accentuating the bass elements in his voice, particularly when he is presenting pop records, which have what in other fields of recording would be regarded as an excess of top. Once a good balance has been found, microphone discipline is necessary.

Interviewing with a hand-held microphone

For an interview using a hand-held omnidirectional, cardioid or hypercardioid microphone, one of four basic balances may be adopted:

Interviewing techniques
Position 1: Microphone at waist or chest height. Sound quality poor. Position 2: At shoulder or mouth level. Sound quality fair, or good if interviewer and inverviewee are close together in acoustically favourable and quiet surroundings. Position 3: Moving the microphone from one speaker to the other. Sound quality good. In a fourth variant, the microphone favours the interviewee most of the time.

Round table discussion
Showing six speakers working to a cardioid microphone (1) suspended above them, or (2) sunk into a well.

● With the microphone at or just above waist level, so that its presence is unobtrusive. This balance is sometimes attempted for television work or where it is felt that the interviewee will take fright at a closer technique. It gives poor quality (very poor, if omnidirectional) with either too much background sound or too reverberant an acoustic. In television, however, the picture distracts conscious attention from the sound quality, so that provided speech is clearly intelligible it may sometimes be accepted. The cone of acceptance should be as narrow as possible.

● Between the two speakers at chest or neck level, the face of the microphone pointing upwards. With cardioids to hypercardioids both speakers must be within the angle of acceptance, which may be adjusted during the dialogue to favour each voice in turn. This is preferable for television.

● By moving the microphone to each speaker in turn, especially necessary with omnidirectional microphones, the distance being judged according to the acoustics and noise (unwanted) or 'atmosphere' (wanted) in the background. The interviewee may find the repeated thrust of the microphone distracting, but if the interviewer times the moves well the overall sound quality will be good.

● Close to and favouring the interviewee, with only slight angling away during questions, the interviewer relying more on clarity of speech, voice quality and well-judged relative volume.

Where a monophonic balance is heard as a component of stereo, discriminate against background sound or reverberation, so that it does not all seem to come from the same point. With a stereo end-fire interview microphone, check the directional response, which may be clearly indicated. If not, as a reminder, you might mark it with arrows on tape on the side of the microphone.

The interviewer should be close to the subject. This may best be arranged by standing at about 90° or sitting side by side. If the loudest extraneous noises come from the side they are facing, the microphone will have to be angled to discriminate against it.

Three or more voices

When three or four people are taking part in a radio interview or discussion in the studio, the ribbon microphone has been widely used in the past, and still has its advocates. Although a round-table layout is not possible – four speakers must be paired off two-a-side, and three divided up two and one – good sound is consistent with comfort for the speakers. The microphone can be moved a few inches, or angled slightly, to favour lighter voices; and two speakers on the same side of the microphone should be encouraged to sit shoulder to shoulder so that there is no tendency for anyone to get off-microphone. The seating should be arranged so that those sitting side by side are more likely to want to talk to the speakers across the table. The change in

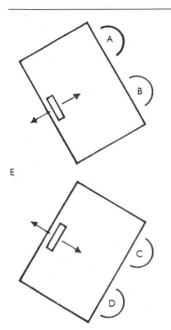

quality that occurs when one of a pair turns to his neighbour can be marked, and if he leans back at the same time the effect is magnified.

For a discussion (in mono) where there are more than four people, a cardioid microphone may be suspended above the group or lowered into a well cut into the centre of a round table between them. The acceptance of ambient sound is the same as for a ribbon, but its proportion may be greater, because for the same working distance (which must be maintained in order to avoid crowding the speakers too closely together) the direct sound is reduced by several decibels by being some 45–60° off-axis. The result is more reverberant unless the acoustic itself is deadened to restore the balance.

An alternative is to provide separate microphones (which can be hypercardioids) for each speaker. The ambient sound pick-up is that of all the open microphones combined, and in total is broader than that of a single cardioid or bidirectional microphone, but the much closer balance can more than compensate for this. If the order of speakers is known (or can be anticipated by watching the discussion) some faders can be held down a little, favouring the speakers who already have the floor.

Any nearby reflecting surface affects a balance: the reflected wave interferes with the direct sound, producing phase cancellation effects and making the quality harsher. To reduce this, tables in radio studios (and for some television programmes, too) can be constructed with a surface of acoustically transparent mesh. If a microphone well is used, it should be open at the sides. The best arrangement is a small platform suspended by springs or rubber straps. For insulation against bumps or table taps, rubber in tension is more effective than rubber in compression.

In a sufficiently dead acoustic there is one further possibility: an omnidirectional pressure-zone microphone on a hard-topped table.

Two bidirectional microphones in mono
Voices A and B are on the dead side of microphone 2 and vice versa: this is a useful balance for quiz games, etc., recorded before an audience. Any voice at E would be picked up on both – provided the two were in phase.

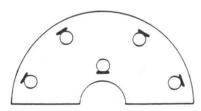

Host and guests
Layout for radio show, allowing individual control of each contributor's voice.
Guests take their places at the table while records or pre-recordings are played.

Acoustic table with well
The table top has three layers.
1, Perforated steel sheet. 2, Felt.
3, Woven acoustic covering (of the kind that is used on the front of loudspeakers). These are not bonded together, but held at the edges between wooden battens.
4, Wooden microphone platform, suspended at the corners by rubber straps (5).

Studio noise

For many purposes (e.g. narration, plays, etc.) background sounds are undesirable. They are reduced to a minimum by good studio discipline, the elimination of potential problems during rehearsal,

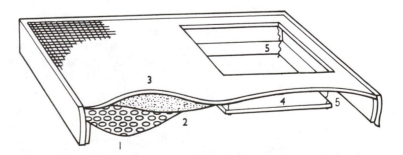

and critical monitoring. A check at the time of the initial balance will show whether *room atmosphere* itself is obtrusive. 'Atmosphere' means the ambient noise that pervades even quiet rooms; and some check is needed to see whether voice level or balance have to be changed to reduce it. But the sort of voice quality that stands out really effectively against background noise is more suitable for a hard-sell commercial than ordinary talk.

Even in the studios of radio stations built to discriminate against external noise, difficulties can occur: ventilation systems may carry noise from other parts of the building, or may themselves sound noisy when quiet speech is being recorded – simply due to the flow of air. Structure-borne noises are almost impossible to eliminate, and rebuilding work is complicated by the need to avoid all noisy work when nearby studios are recording or on the air.

Noise may come from the speaker himself. A lightly tapping pencil may resound like rifle-fire. A slightly creaky chair or table may sound as if it is about to fall to pieces. Lighting a cigarette may be alarming; lighting several, a pyrotechnic display. Tracking fainter sounds to their source may prove difficult – embarrassing, even. A loose denture may click, a flexed shoe may creak. Many people develop small nervous habits before the microphone: for example, retractable ball-point pens can be a menace – people will fiddle with them, and produce a sharp, unidentified click every 10 or 15 seconds.

There are several things that make these noises so irritating. The first is that many people listen to broadcast or recorded speech at a level that is much louder than real life; and a quiet talker is allowed much the same volume at the loudspeaker as one who shouts. So the extraneous noises become louder, too – and, worse, if heard monophonically they appear to come from the same point in space as the speech. Also, whereas a visually identified sound slips immediately into place in the mind and is accepted, the unidentified one does not. Many practised broadcasters realize this and take it into account. A studio host will notice that a guest's movements are not absolutely silent and say, 'And now, just coming to join me, is . . .' and the noise is given a reason and thereby reduced to its proper proportions – so that it is not consciously noticed by the listener.

Handling a script

A frequent source of extraneous noise at the microphone – and a particularly irritating one – is paper rustle. A few comments by the producer on the handling of scripts or notes is common practice at the start of any recording or broadcast where these are used, and where the speaker is not experienced. Here are a few general rules:

● Use a fairly stiff paper (duplicating paper is quite good). Do not use a copy that has been folded or bent in the pocket. Avoid thin paper (e.g. airmail paper or flimsies).

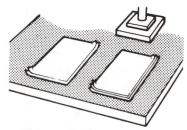

Avoiding script noise
Corners of script turned up to make it easier to lift noiselessly to one side.

Script-rack
No reflections from the script can reach the microphone, but the speaker may still drop his head at the foot of the page. The rack may be made of acoustically transparent mesh.

● If you are sitting at a table, take the script apart by removing its paper clip or staple. It is difficult to turn over pages quietly, so it is a good idea to turn up two corners of the script, and then each page can be lifted gently to one side as it is finished. Any sliding of one page over another will be audible. Better still, hold the script up, and to the side of the microphone.

● Hold or place the script so that it is not necessary to turn or drop the head in order to read.

● Standing at a microphone (necessary for actors with moves to make) the script can be left clipped together. The pages should be turned well off-microphone – or in the case of a bidirectional microphone, on the dead side of it.

● When there are several people in a studio (as in a play) they should not all turn the pages together. If actors remember to stagger the turnover, they will also remember to turn quietly.

● If there are only a few sheets, they may perhaps be slipped inside clear plastic envelopes.

One device sometimes used is a script-rack. Sloping up toward the microphone, it avoids awkward reflections but does not prevent a persistent offender from dropping his head. Indeed, as he will be speaking more directly into the microphone at the top of the page, results may be worse than without the rack.

Speech in stereo

Much speech for stereo radio transmissions is, in fact, balanced on monophonic microphones. In particular, announcements are normally taken mono, and are generally placed in the centre, so that listeners can use this to balance the two loudspeakers. If the announcement is in a concert hall a monophonic microphone is again used with a relatively close and dead balance, perhaps with a stereo pair left open at a distance to add stereophonic reverberation. The announcer and microphone are so placed that little direct sound is picked up in stereo; if it were (and were not central) the position of the announcer would be offset. In discussions, contributors are panned to a range of positions along an arc.

One reason for avoiding a stereo pair for speech balances where movement is not expected is that any slight lateral movements close to the microphone are emphasized and become gross on the sound stage.

Boom operation

In radio the microphone is usually static, and the performer works to it. In television the performer's position is determined by the scenery and lighting, and the microphone must work to him. Radio and television differ most strongly when, in addition, the microphone must be out of the television picture, and the performer moves freely. The 'microphone with intelligence' that moves and turns with him, ever seeking the best practicable balance, is the boom microphone. Its operation is central to the subject of speech balance in television, and film, too. For stereo, MS microphones are often preferred.

From a given position the boom can cover a substantial area, and the platform itself can be tracked to follow wider action, or repositioned between scenes. Its main position is chosen bearing in mind the lighting and the fact that it is quicker to swing the arm than to rack in and out when moving from one subject to another. In addition, the operator can turn the microphone to favour sound from any one direction, or (almost as important) to discriminate against any other. Characteristically, in mono, or for the middle (M) signal in stereo a cardioid or near-cardioid is used, and its diaphragm must be affected little by rapid movement and insensitive to or protected from the airflow this causes, so that the operator is not restricted in the speed of his response to action. Note that an effective windshield may be bulky, and can make movement clumsy. In practice it is best to use a directional electrostatic or perhaps moving-coil microphone, with shock-mounting and windshield. In the ideal balance for the best match of sound to picture, the microphone is above the line between subject and camera, but because of the need for clarity and internal balance between different voices (and, in addition, frequent changes in the picture itself), no single compromise position is likely to last for long.

The boom operator has a script (where one is available) and generally moves his microphone to present its axis to favour whoever is speaking or expected to speak. It may be asked why, when he is using a cardioid with a good response over a wide range of directions, he needs to move the microphone at all for most work, instead of splitting the difference between two people and leaving it. The answer is that there may be a 2 or 3 dB high-frequency loss at about 45° to the axis, increasing to perhaps 5 dB at 60°. The microphone will be dropped as close to the performers as the picture allows and this closeness accentuates any angle there may be between speakers. This means that by proper angling there can be not only the best signal-to-noise ratio for the person speaking but also automatic discrimination against the person who is not. If both speak it may be decided to let this balance ride, to favour the person who has the most important lines, or perhaps to cover both equally. With actors of different head heights the microphone must be placed either well in front of them or to the side that favours the shorter.

The position for the microphone must be a compromise between the positions of the various speakers, and if they become widely separated, or one speaker turns away, the split may become impossible to cover by a single boom. In this case a second microphone may be placed on a stand out of vision, or suspended over the awkward area, or concealed within the picture. Sometimes a second boom is used, but this may cause lighting problems.

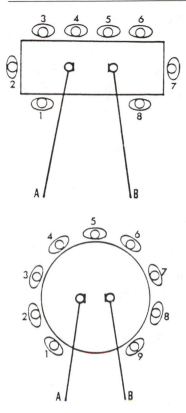

Boom coverage of large group on TV
This might be suitable for a meeting in a play.
Above: Boom A covers speakers 4–8; B covers 1–5. *Below:* Boom A covers speakers 5–9 (with 1 and 4 at low level); B covers 1–5 (with 6 and 9 at low level). Cardioid microphones are used.

For 'exterior' scenes in plays produced in the studio, the microphone must be kept as close as the limits of the picture allow. To provide a contrast with this, interiors may have a slightly more distant balance or a little added reverberation.

Because a number of different types of microphone may be used for a single voice, the control desk in a studio must have equalization associated with each channel, so that it can be used to make microphones match as closely as possible. If correction is needed, the boom microphone is usually taken as the standard and the other microphones matched to it.

Another justification for the boom is that given suitable (rather dead) acoustics, it automatically gives a perspective appropriate to the picture: close shots have close sound; group shots have more distant sound. The boom operator uses vision monitors in the studio to check the edge of frame, dropping the microphone into the picture from time to time during rehearsal. Then by taking a sight line against the far backing he sets the microphone for recording or transmission.

In television the sound supervisor can call the boom operator (and other staff on the studio floor) over his own circuit, which cuts in on the director's circuit to the operator. (This may be automatically preceded by a tone-pip to indicate that the message is from the sound supervisor.) During rehearsal and any time when studio sound is faded out, the boom operator can reply or call his supervisor on a separate sound reverse talkback circuit. A boom therefore has a multicore cable, permitting communication as well as transmitting programme sound, and returning the mixed sound to his headphones.

In certain types of television discussion programme the frame size is unpredictable, but this is partly compensated by the influence of an audience, which limits how far performers will turn. For this, the boom may be equipped with a highly directional microphone. Although the inertia of a gun that is nearly 2 ft long makes it clumsy to handle for work where gross and rapid movements are necessary, it can sometimes be placed farther back.

Where two people are having a 'face-to-face' discussion, but are far enough apart for a boom operator to have to turn a cardioid microphone sharply to cover each in turn, it may be better to use two microphones pointing outwards at about 90° to each other. The overall effect is a very broad cardioid which is suitable for low close working. In a play such an arrangement would restrict the use of the boom for other scenes, so in a similar situation – or, for example, for a number of people sitting round a large table – two booms might be brought in.

A boom can cover action taking place on rostra built up to a height of about 6 ft (1.8 m) provided that the performers keep within about 10 ft (3 m) of the front of the high area. The boom dolly should be positioned as far back as possible so that the angle of the arm is kept reasonably low. For scenes on acting areas above this height it may be necessary to hoist the boom up on to rostra at a similar level (as a result of which it will, of course, be lost for scenes in the more distant parts of the studio to which it could normally be taken). Another possibility is to de-rig the microphone boom arm and attach it to fixed scaffolding. The arm will require rebalancing, and the operator will need more time in rehearsal.

The boom operator should occasionally take off headphones to check for talkback breaking through or for other sound problems in the studio and report them back to the supervisor in the gallery. In addition, being visible to artists, the operator should react positively to their performance.

Slung and hidden microphones

In television balance one aim must be to keep the number of microphones to a practicable minimum, and the use of booms helps towards this. But unless set design and action are to be restricted there may be places that are inaccessible to booms. For these, additional microphones may be suspended above the action – ideally, just out of frame, and downstage of the speaker's position, so that sound and picture match. If a slung microphone has two or three suspension cables, its position may be adjusted in rehearsal, and if there is any danger of its shadow falling in shot, provision should be made for it to be flown when not in use.

Microphones are sometimes concealed in special props that are constructed of wire gauze, then painted or otherwise disguised, though an electret is small enough to be hidden more easily. Placed on or very close to a reflecting surface, it will act as a pressure-zone microphone, but check for the effect of hard surfaces further away that may modify the top response. Sometimes equalization can help by reducing the high-frequency peaks so that the quality is not so obviously different from that of the boom, but it cannot restore cancellation.

Microphones in vision

The presence of a microphone in vision must be natural and undistracting. A boom microphone which is visible in the top of the picture is very disturbing, particularly if it is moved to follow dialogue.

Microphones used in vision are either *personal*, fixed below the neck or held in the hand, in which case they are moved around by the performer, or they are *static*, on table or floor stands.

For a speaker who is not going to move, a fixed microphone in the picture gives better sound. It will generally be angled to favour the speaker (or may be split between two): if this deliberately chosen angle conflicts with the designer's choice of 'line' within the picture the microphone may be more noticeable, as it will also be in a simple uncluttered design.

The base, for stability and to cushion shocks, needs to be reasonably large and can therefore sometimes appear more obtrusive than the microphone itself. Sometimes it may be recessed into a desk, but such a cavity needs to be designed with care: if fully boxed-in, it would modify the local acoustics, so one or more sides of the recess should be left open. The microphone capsule itself should stand clear of the surface, so in a shot showing the table it is in vision, unless hidden behind something standing on or in front of the table – which in turn may be more obtrusive than the microphone itself. Alternatively, a specially constructed desk might have its front continue upward a little beyond the level of the desk top. An acoustically transparent top can be included in the design.

A microphone placed to one side may be neater than one immediately in front; it should go on the side to which the speaker is more likely to turn. However, any turn the other way may go right off-microphone, so it is therefore useful if the next person's microphone can cover this.

Where many people are spread over a large area of studio, a microphone is sometimes used for every individual – but good quality demands that as few as possible be open at any one time. So that the right ones can be faded up in time, it must always be made clear, either from a script or by introduction by name or visual indication, who is going to speak next.

The use of hand microphones is a widely accepted convention of television (and television film) reporting. It is convenient for the reporter plus cameraman combination, and also because the microphone can be moved closer if the surroundings become noisier. In extreme circumstances lip microphones have been used in vision. But hand-held microphones can easily be overused, particularly where they make an interviewer look unduly dominating and aggressive.

Alternative techniques, which for most purposes should be preferred, are the use of personal (lapel) microphones or, for exterior work, gun microphones.

In interviews or discussions, participants using personal microphones can be steered by the use of panpots to suitable positions on the left or right of frame, roughly matching those in the widest shot. In an unscripted discussion it is better to make no attempt to match position in frame: this makes editing difficult and in a television picture a small discrepancy is not likely to be noticed. The pictures establish a 'line' which the camera should not cross, and sound can follow the same convention. Two speakers (as in an interview) can be placed roughly one-third and two-thirds across the stereo image. But note that as two omnidirectional microphones are moved towards each other, they will 'pull' closer still, as each begins to pick up the other subject.

The main benefit from stereo here is that separation of voices from each other and from sound effects and reverberation (which should be full width), as well as from any noise, aids intelligibility and adds realism. A separate stereo pair can be directed into the scene and faded up just enough to establish a stereo background effect.

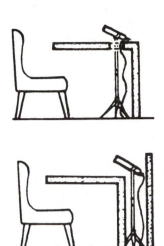

Television desk and microphone
To avoid bumps, the microphone is on a floor stand: *above*, emerging through hole; *below*, behind false front.

Using gun microphones

The gun microphone is most useful for filming outdoors, held just outside the edge of frame and pointed directly at each speaker in turn. Co-operation between camera and sound is required to avoid the microphone getting into picture: its windshield is obtrusive even when still, and more so if it moves.

An advantage of this microphone is that it allows great mobility – though not of course for the subject to turn right away from the microphone while speaking. Accidentally off-microphone sound is at low level in exterior work, and bad quality in interiors. The gun microphone can, however, be used in the relatively dead surroundings of a television studio (in an area well away from strongly reflecting surfaces) for picking up voices at a distance. But in general its directional qualities are not trustworthy in normal interiors or even reverberant exteriors.

The short gun microphone can also be attached to a video camera, pointing along its line. It picks up sound effects adequately within the field of the average wide-angle lens, but good voice balance only when the camera is close in. The principal application is to one-man news coverage.

When a highly directional microphone is used to cover effects such as tap dance, or to follow the action in sports (to which it adds great realism), it is often best to filter frequencies below about 300 Hz. A gun microphone with bass cut has sometimes been used in conjunction with other microphones to cover large press conferences, supplying presence that other distant microphones lack. For this, a parabolic reflector could also be considered: it would be out of frame, or possibly hidden behind a gauze that is lit to look solid.

Video camera with short gun microphone
This ensures that loud effects in the field of view are recorded. For television sports, a full-scale gun microphone attached to an outside-broadcast camera is used to capture more distant effects.

Using lapel microphones

For location work, an electret is often fixed to the clothing of each speaker – but take care if two are brought close together, as this can affect the balance. (Older, heavier personal microphones were hung around the neck.)

The microphone itself can often be hidden behind a tie, though some synthetics cause crackling, and a thick tie can produce a correspondingly woolly response. Whether it is concealed or not it is best to ensure that the means of suspension is also invisible – for example, by using a clip or cord of the same tone or colour as the clothing against which it may otherwise be seen. Some capsules are available in a range of colours, such as black, white, silver or tan (flesh tone). Otherwise a slip-on cover of thin, matching material can help.

The user may have to avoid large or sudden movements, especially those that disturb the hang of clothes (as sitting down or standing up does with a jacket). However, normal movement should not be discouraged unless necessary. A speaker who turns right away from the camera may still be heard clearly – though turning the head to look over the wrong shoulder may cause speech to be off-microphone.

The cable may be hidden beneath a jacket for a tight mid-shot (which, on its own, may not warrant the use of a personal microphone) but in the more general case it will go to a hidden radio transmitter or continue on its path beneath the clothing, down a trouser leg, or inside a skirt – to a connector not too far from the feet, so that the wearer can easily be disconnected during breaks. If the subject is going to walk it is best to attach the cable (and connector) to one foot, perhaps tying it in to a shoelace. To walk dragging a considerable length of cable is awkward, and may be visibly so. It may also cause cable noise – reinforcing the case for a radio link.

Because of its position, a lapel microphone may produce a characteristic quality that is difficult to match, so once started, it is best to persist with it until there is either a marked change of location (e.g. exterior to interior) or a long pause before the same voice recurs. This is true also for many other microphones, though to a lesser degree. The effect of disregarding this is obvious on many television news broadcasts.

A disadvantage of personal microphones when compared with booms is that there is no change of aural with visual perspective. This is more important in plays than in other programmes. But they do usually offer reasonable intelligibility in fairly noisy surroundings, or in acoustics that are brighter than usual for television. The quality gains a little from good reflected sound.

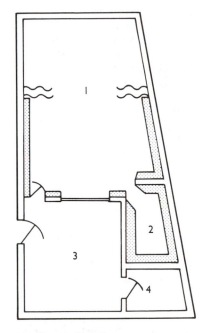

Drama studio suite
This is typical of the specialized studio layouts adopted for broadcast drama in the days when radio was at its peak of popularity. Many such studios still exist (as at the BBC) and are used for their original purpose. The principles involved in the layout remain valid for many other purposes. The areas shown are:
1, Main acting area – the 'live' and 'dead' ends of the studio can be partially isolated from each other (and their acoustics modified) by drawing double curtains across.
2, 'Dead' room, with thick absorbers on walls. 3, Control cubicle.
4, Machine room. In this example a virtue has been made of the irregular shape of the site. The use of non-parallel walls avoids standing wave coloration.

Radio drama

In many countries radio drama is still vigorously alive. In drama the acoustics of a studio are used creatively, the 'scenery' being changed by altering the acoustic furniture around the microphone. Specialized studio suites are divided into areas with different amounts of sound-absorbent or reflecting treatment; each area provides a recognizably different quality of sound. The live areas offer a realistic simulation of the actual conditions they are supposed to represent; the dead areas, less so. In most cases the same microphone type is employed in most areas: this means that the voice quality itself does not change but is matched from scene to scene.

One factor in the aural perception of studio size is the timing of the first major reflection – so that delay may sometimes be used to increase apparent size without creating excessive reverberation. But the quality of the reverberation itself also plays an important part. A small room sounds small because certain frequencies are picked out and emphasized by its dimensions. A combination of reverberation time and quality suggests both size and the acoustic furniture of a room: a sound with a reverberation time of 1 second may originate from a small room with little furniture (and therefore rather bright acoustically) or from a very large hall with a rather dead acoustic. The two would be readily distinguishable.

For example, a 'bathroom' acoustic would be provided by a small room with strongly sound-reflecting walls, and can be used for any other location that might realistically have the same acoustic qualities. Similarly, a part of the main studio (or a larger room) with a live acoustic can be used to represent a public hall, a courtroom, a small concert-hall, or anything of this sort – and indeed, when not needed for drama, can be used as a small music studio. When going from one type of acoustic to another, balance (in particular, distance from the microphone) can be used to help to differentiate between them; for example, the first lines of a scene set in a large reverberant entrance hall can be played a little farther from the microphone than the main body of the action which follows – it would be hard on the ears if the whole scene were played with heavy reverberation.

For monophonic drama, ribbon microphones have been widely used: their frequency range is well adapted to that of the voice, and with working distances of about 3 ft (1 m) give a generous account of the studio itself. Actors can work on either live side of the microphone, standing two or three abreast, and can move 'off' either by increasing their distance or, more often, by easing round to the dead side (or by a combination of the two). The conditions are comfortable, flexible and easily understood. The actor is working *with* the acoustics.

For a large, open acoustic with many actors, such as a court room, public meeting, or council chamber scene, an alternative to the ribbon is a cardioid suspended above the action. Working distances and speed of approach or departure are set by listening to the acoustic balance. An additional hypercardioid microphone can be set in for quiet asides that require little reverberation.

If there is a narrator as well as the 'living-room' action the two must be recognizably different. There will, of course, be differences in the style of acting, and the use of the fader at the start and end of each scene will encapsulate it, and so further distinguish between the two. But acoustics can also be used to emphasize the differences: if the scene is played well out into the open studio the narrator can be placed nearer to one of the walls and at a closer working distance to his microphone, which may be of a different type – often a hypercardioid. The distance between the narrator's microphone and that for dramatic action must allow for any rapid turn from one to the other (though more often they will be recorded separately). It may help to provide a satisfying contrast if there is a screen near the narration microphone. Even for stereo drama, a narrator will normally be mono.

It is sometimes necessary to mix acoustics – to have two voices in acoustically different parts of the same studio, but taking care to avoid spill. There is more danger of spill where the voice in the brighter acoustic is louder. Suitable placing of directional microphones will help, but may not prevent the spill of reverberation from the live

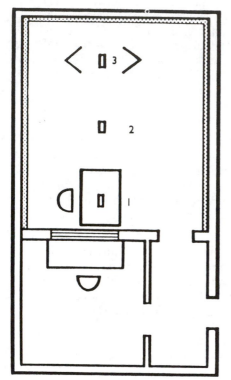

Simple layout for drama (mono)
A simple set-up using three ribbon microphones for dramatic work.
1, Narrator (close to window). 2, 'Open microphone' – for normal indoor acoustics. 3, 'Tent' – to represent outdoor quality. This is a much less versatile layout than can be obtained in a multiple-acoustic studio, and the results sound more stylized. This might be used for broadcasting directed to schools, because: (a) school audiences often have poor listening conditions; (b) casts for a variety of reasons are small; and (c) educational broadcasting generally has a low budget.

acoustic to the dead one. Thick curtains running out from the walls may help to trap the sound – or screens may be used for the same purpose – but in extreme cases use a separate studio (or the studio lobby) or pre-recording.

Screens can be used to add realistic touches. For example, the interior of a car consists of a mixture of padded and reflecting surfaces: a 'box' of screens, some bright, some padded, gives a very similar quality of sound.

Open-air acoustics

Open-air acoustics, frequently required in dramatic work, are characterized in nature by an almost complete lack of reverberation. Even when there are walls, etc., to reflect some of it, the total reflected sound at the source is low, and is likely to produce only a small change in quality.

It is, of course, possible to simulate this realistically by building a section of the studio with almost completely dead acoustics, i.e. a dead-room. There are clear advantages:

- it provides the best possible contrast to other acoustics in use, thus making a wider range of sound quality possible
- the muffling effect of the treatment causes the performer to lift his voice and use the same amount of edge on it as he would in the open air
- effects recorded outdoors blend in easily
- off-microphone voices and spot effects can be blended with on-microphone voices (in mono by using positions round towards the dead side of a directional microphone)
- the amount of space round the microphone is greater than that within a tent of screens.

Unfortunately, there are also disadvantages:

- such an acoustic can soon feel oppressive and unreal
- the transient sounds in speech tend to overmodulate, even if the general level is kept low
- it is less pleasant to listen to completely dead sound for any length of time.

In fact, the true dead-room is not a practicable proposition, and some sort of a compromise solution has to be found. Where a script is set almost entirely out of doors, it is more convenient to go for a formalized style of production with an acoustic that is hardly less reverberant than a normal speech balance. On the other hand, if there are only a few isolated lines set outdoors, something fairly close to the dead acoustic is more appropriate. In general, try to set the average acoustic demanded by the script as one that is fairly close to a normal speech balance, and arrange the others relative to this.

Of the various possible sound-deadening systems, it is best to try to avoid those that give a padded-cell effect. This is, literally, what a

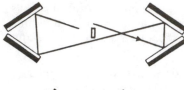

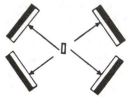

Tent of screens with bidirectional microphone
In a double-V of screens it is better that sound should be reflected twice on each side (*above*). Standing waves cannot form unless there are parallel surfaces directly opposite each other (*below*).

Use of screens
Double reflection of sound in a V of screens set at an acute angle. For any particular path some frequencies are poorly absorbed, but they are different for the various possible paths. Sound that is reflected back to the region of the microphone is less coloured with this arrangement than with a broad V.

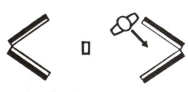

Speaking 'off' bidirectional microphone
No actor should go further off than this, or turn further into the screens. More 'distant' effects must be achieved by changes of voice quality or by adjusting the microphone fader.

thinly padded dead-room sounds like, and the same sort of sound can all too easily be obtained by laying out a cocoon of screens. For monophonic drama, try two Vs, one on each side of a bidirectional microphone.

Here are some points to remember with this layout:

● Keep the screens as close to the microphone as is conveniently possible; the actors should accept some restriction on their movements. This keeps the sound path lengths short and damps out reverberation (and coloration) as quickly as possible.
● Avoid setting the screens in parallel pairs: it is better to keep the V of the screens fairly acute. This helps to reduce bass as well as top. Also, actors should avoid standing too far back into the apex of the V.
● Exit and entry speeches can be awkward. In nature, distant speech is characterized by a higher voice level arriving at a lower final volume. In the studio, a voice moving off is likely to be characterized by a sudden increase in studio acoustic. So stay inside the screens at all times, and avoid directing the voice out into the open studio.
● Off-microphone (i.e. 'distant') speeches may be spoken from the screens, speaking across the V. If this is still too loud to mix with other voices, the actor should change his voice quality, giving it less volume and more edge. He should not try to deaden his voice by turning further into the angle of the screens: this produces a cotton-wool muffled quality.

All in all, the effect of screens is not that of the open air. But it is not a normal live acoustic, either. It is different, but acceptable as a convention.

Stereo drama

For stereo, co-incident pairs are commonly used for each main acting area. Crossed cardioids or hypercardioids give the best use of space. Where actors are unfamiliar with the technique, the studio floor may be marked up with strips of white adhesive tape on the carpet showing the main limits of the acting area, together with any special positions or lines of movement. Movement in a straight line on the sound stage is simulated by movement along an arc of the studio: the acting area can therefore be defined by an arc with the microphone at its centre, or by arcs roughly limiting it. Such arcs can be drawn either as having constant radius – for convenience – or alternatively at distances that give constant A + B pick-up. The latter is better for more complex situations, such as group or crowd scenes in which the relative distance of different voices is important. From the microphone polar diagrams it can be seen that for a crossed cardioid this arc is farther away at the centre line and loops in closer at the sides. In addition, the studio itself has an effect: unless the acoustic treatment is evenly balanced, the line may 'pull' closer to the microphone on one side

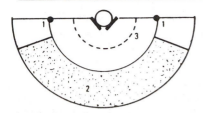

Stage for stereo speech
1, If crossed hypercardioids are used, voices in these two positions, roughly opposite to each other in the studio, appear to come from the loudspeakers. 2, Most of the action takes place in this area, at a distance of about 2–3 m from the microphones. 3, More intimate speech may be closer.

than on the other. Indeed, sound stage positions may also be offset laterally from studio positions: the whole studio may seem to pull to the left or right.

Positions 'in the loudspeaker' on the two sides may also be set: for crossed cardioids, these would in principle be at the 270° limits of the theoretically available stage. In fact, the cardioid response of practical microphones is imperfect at the rear, so the best 'in speaker' positions would usually be found at about 200–220° separation. In practice, it is often convenient to reduce this further, to 180°, by setting microphones with variable polar response one step towards hypercardioid.

In mono, a long approach from a distance can be simulated by a relatively short movement by the performer, who simply walks slowly round from the dead to the live side of a directional microphone. In stereo, long approaches really have to be long, and must be planned in terms of what space is available: a decision on this in advance will probably determine studio layout.

If a microphone with capsules one above the other in the same housing is used, the difference in the height of the two capsules may cause problems in close working: a tall actor is nearer the top capsule and 'pulls' one way; a short actor is nearer the low one and 'pulls' in the other direction. Working side by side, a short and tall actor may seem to be either separated more or shifted to the same position. The use of a pair of microphones which are set on a bar with the capsules almost touching, avoids this.

Balances more distant than those that are normal for mono may be necessary in order to avoid movements being exaggerated: for safety the nearest part of the real stage in which much lateral movement will occur should be at least as broad as the sound stage. In consequence, the studio must be somewhat deader than for a comparable scene in mono, and reverberation added if brighter-than-average sound is required.

For this, either stereo reverberation or a second stereo pair within the studio may be used. If a co-incident pair is chosen, it should be placed,

Stereo drama in practice
This layout emphasizes extreme side positions and allows long approaches from left (1) and right (2). In many studios (as here) there may be asymmetry in the acoustic treatment, distorting the sound-stage, and minimum working distance around the microphone pair. The positions may be judged by ear and marked on the floor accordingly. 3, Drapes screening live studio area beyond. 4, Announcer, centred.

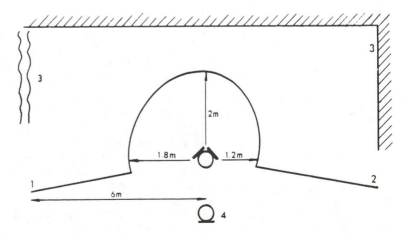

say, 4 ft (1.2 m) above the first pair. This provides some degree of automatic compensation as a speaker moves in close. An alternative arrangement is to use a spaced pair pointing back to the end of the studio that is away from the stage.

Whereas a small studio for drama in mono may have several acting areas, the same studio used for stereo may have room for only a single stage (this allows for a reduction in reverberation and an occasional need for deep perspective). Therefore there may be good reason to record all scenes that have one basic acoustic before resetting the studio for the next, eventually editing the programme to the correct running order. If very different acoustics are required in the same scene – e.g. a distant open-air crowd scene including studio 'open-air' voices, mixed with close 'interior' conversion – a pre-recording has to be made, the studio reset, and the second part dubbed in to a playback of the first.

To simulate exterior scenes in a studio it may be unavoidable to use the same basic studio acoustic. Cutting bass on the microphone output or on the desk may help, but the result depends on the particular coloration of the studio acoustic. For the effect of a more distant voice in the open air, the actor may help by staying well forward and projecting and thinning the tone of his voice, or, alternatively, by turning away from the microphone and speaking into a thick screen. Ideally it is bass that needs to be absorbed more.

Stereo drama problems

Dramatic conventions for stereo are well established: one is that movement should occur during speech from the same source, or while other sounds – e.g. footsteps – are being made; certainly, moves should never be made in a short pause between two speeches, except for blind man's buff comic or dramatic effects – easily overused.

Problems that are met in mono drama may be greater in stereo. For example, extraneous noises are worse because:

● there is no dead side of the microphone
● with a more distant balance of voices, the signal at the microphone is at a lower level; channel gains are higher and therefore noise is amplified more
● noises may be spatially separated from other action, and therefore more distant.

Particular difficulties are raised by a need for artistic compatibility between stereo and mono. Consider, for example, a conversation between two people trying to escape unnoticed through a crowd: they talk unconcernedly as they move gradually across the stage until eventually they are free of the other actors and can exit at the side. The mono listener to the stereo play needs as many positional dialogue pointers as he would be given if the production were not in stereo. Indeed, in mono the microphone would probably stay with the main subjects and other near voices would drift in and out of a clear hearing distance; in stereo this convention, too, could perhaps be established – though with greater difficulty. In such a case, imagine or listen to both stereo and mono versions of the chosen convention and decide whether changes could or should be made for the benefit of one group or the other; or whether to disregard potential mono listeners (who may be an important part of a radio audience).

A more extreme case might be a telephone conversation: a theoretical possibility would be to split the stage, putting one actor in to each loudspeaker. This might work well in stereo, but for mono would be meaningless. In this case artistic compatibility requires the voice of one of the actors must be filtered. The filtered voice could still be put in one of the loudspeakers, but the other, 'near' voice would probably be closer to the centre of the sound stage and in a normal stereo acoustic ambience, but offset to balance the other.

Audience reaction

Audience reaction is a vital element of many comedy programmes, while for music broadcasts it provides a sense of occasion that the home audience can share. The two differ in that for the first the performers' and audience microphones must be live at the same time, while for music performances they are sequential. For all types of programme, applause (and other forms of audience reaction) must be clear, without being unbalanced – that is, individual pairs of clapping hands should not be loud or assertive.

To achieve this, two conflicting requirements must be met: the audience must be reasonably large (ideally, at least several hundred strong); and the microphone coverage must not be so distant as to lose clarity. In addition, there must be adequate separation between audience and other live microphones. Two or more audience microphones will often be used, and for musical performances these may double as 'space' microphones. They are premixed on the control desk and fed to a single group fader.

It may help to have a compressor or limiter before the main audience fader. A further limiter or compressor can be inserted after mixing audience and dialogue, and used in such a way that speech automatically pushes down the audience reaction as the voice itself is compressed.

In stereo, applause is best spread across the full sound stage. It must be picked up on the main and any announcer's microphone only in such a way that controlling the level of the applause does not pull it to one side.

The audience reaction to a comedy show will obviously be poor if the words are not clearly audible. This requires p.a. loudspeakers above the heads of the audience, fed by one of the auxiliary send channels available on studio desks. A line-source loudspeaker radiates most

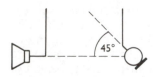

Public address and audience microphone

Even when the microphone is nominally dead at the rear it is usually safer to set loudspeakers at a rear angle of 45°, where the frequency response is cleaner.

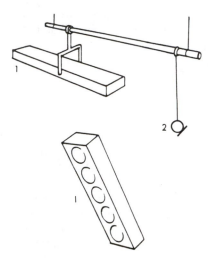

Audience loudspeaker and microphone

1, Line loudspeaker. Except at low frequencies the elongated source radiates at right angles to its length, and low frequencies are reduced by bass roll-off. The microphone, 2, on the dead side of the loudspeaker, is itself dead-side on to it. Both are attached to the same bar, so that their relative position is fixed.

strongly in a plane at right angles to its length, and a microphone placed out of this plane will receive its middle, and high-frequency sound only by reflection. Unless they are well separated from the loudspeakers, the audience microphones will be directional and dead-side on to them. The rear axis of a unidirectional microphone is generally avoided, as in many cases the frequency response is most erratic along that line. It is safer to place the microphone so that the nearest loudspeaker is at 45° to its rear axis. The p.a. send is in this case a specially tailored mix of only those channels required for intelligibility. Bass can be cut or rolled off, and equalization introduced to increase clarity and avoid coloration.

In television, for which the performers' microphones may be set well back from the action, the p.a. loudspeakers must be brought as close over the audience as is practicable for adequate coverage. The loudspeaker volume can be raised by 3–4 dB without increasing the danger of howlround if the frequency of the public address sound is shifted (usually downward) by a few Hertz, using a harmonizer.

If the response from a particular individual in the audience becomes obtrusive, fade down the microphones gently, in turn, until that nearest to him is located, then reset it accordingly.

The volume of applause at the end of a performance must be held back a little: prolonged clapping at high level is irritating to the listener, who may be impelled to turn it down. On the other hand, laughter within a programme may have to be lifted a little: it should be held down between laughs in order to avoid picking up coloration, raised quickly as the laugh begins, and then eased back under subsequent dialogue just enough to avoid significant loss of clarity in speech. The performer relates his own timing to the audience's response, and sensitive control anticipates the progress of this interaction.

Chapter 8

Music balance

There are two basic types of music balance. One is the 'natural' balance which uses the studio acoustics for reverberation and is generally preferred for classical music. For a single instrument or for a group having perfect internal balance, one microphone (or a stereo pair) may be sufficient. However, such perfection is rare: a singer may be overwhelmed by all but the quietest of pianos; an orchestra may be dominated by its brass. So even 'natural' balance often requires additional microphones.

Most music is balanced in stereo, for which the techniques may be further subdivided into those using co-incident pairs and others with various spaced microphone arrangements. Mono is still used in parts of the world where for economic or language reasons stereo broadcasting is restricted. (In television, if several language versions are required, this may take up audio channels that could otherwise be allocated to stereo.) But mono is also used for individual instruments or sections within a stereo balance. Any further treatment of individual components of the whole sound is normally only for purposes of correction.

An extreme version of the multimicrophone balance is used for popular and most light music. The many close microphones, predominantly mono, are steered to some appropriate place within a stereo image. Natural acoustics are all but eliminated, to be replaced by stereo artificial reverberation. Treatment of the individual components is an expected part of the process, which may be extended over time by multitrack recording before the final mixdown.

Each of these types of balance will be considered first in general terms and then by taking individual instruments and sections in turn.

'Natural' balance

In a balance that is taken on a single microphone (or co-incident stereo pair), its distance depends on both the polar response of the microphone chosen and the acoustics of the studio. The most flexible approach is to use an electrostatic microphone with a polar diagram that can be switched by remote control. For each, try various distances and listen to the combinations of clarity (of individual components and melodic line) and blending, or warmth of tone. If possible, set several microphones (or pairs) of the same type at different distances, in order to make comparison tests. If all else fails, consider moving the sound source.

A close balance sounds more dramatic, but may be difficult to live with. Where the acoustics are difficult, it may be necessary to accept the brilliant close balance, with artificial reverberation if desired: this can always be added, whereas a bad acoustic cannot be subtracted.

Frequency ranges
A microphone with a frequency response which matches that of the sound source will not pick up noise and spill from other instruments unnecessarily. The lowest frequency for most instruments is clearly defined, but the highest significant overtones are not. At high frequencies, an even response is usually considered more valuable than a matched one.

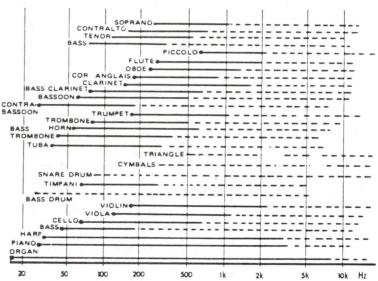

The spread of orchestra or ensemble must be matched to polar response and distance. With a sufficiently distant balance a bidirectional response may be possible, but as the microphone is moved a progressively broader pick-up is required. In many cases the best mono is achieved by a cardioid with a clean, even response (in which the polar response is independent of frequency). Omnidirectional microphones will have to be nearer still, often by enough to be noticeably closer to some instruments than others. They are therefore more suited to spaced microphone techniques.

As spotting microphones, smooth cardioids are again favoured. Be cautious of hypercardioid microphones in which the directional response is engineered by means of a phase-shifting network, as some are a little erratic at high frequencies. Ribbons are undervalued today, but note that they can provide an excellent, smooth response in all but the extreme top – which in any case lies beyond the range of the human voice and many instruments.

When they have high-quality loudspeakers many listeners prefer a more distant and therefore more blended sound than that which some balancers enjoy. But for listening on poorer equipment, greater clarity compensates for some of the deficiencies: at least you can hear the main elements of the music. For this reason (and also because it helps to motivate close-ups), a drier balance has been preferred in television.

The arrival of an audience can change the balance dramatically. Experience will show how much too live the balance must be in rehearsal for it to be correct with a full hall. It will usually be safer to have some separate means of controlling this, either by combinations of microphones or by adding artificial reverberation.

Music studio problems

In a 'natural' balance we record the characteristics of the studio just as much as those of the player.

It is rather as though the studio were an extra member of the orchestra. With stringed instruments, very little of the sound we hear is directly from the strings; almost all is from a sounding board to which the strings are coupled and the character of the instrument depends on the shape and size of that radiator. In the same way, the studio acts as a sounding board to the instruments: its shape and size gives character to the music. But there is an important difference. By uniformity of design, the character of all instruments of one particular type is roughly the same. But music rooms, studios and halls differ widely in character: no two are the same.

The sound will be right if directional microphones are angled to pick up reverberation from all three dimensions. Theory also suggests that a central position should be avoided, because in the middle of the room the even harmonics of the eigentones – the basic resonances of the room – are missing, thereby thinning out its acoustic response. Fortunately, in practice, the number of studios where there are serious eigentone problems in music balance is small. Stereo microphones are usually placed on the centre line of concert halls and rarely with ill effect (but note that concert-hall eigentones are very low pitched).

In a studio that is comparatively dead for its size, omnidirectional microphones may help. But there are times when even a distant balance with an omnidirectional microphone does not do enough. Some percussive instruments, such as timpani, continue to sound for some time, and have decay characteristics somewhat similar to those of reverberation. If they tend to stand out in too dead a hall, artificial reverberation should be added.

The average television or film studio is the dead studio on the grand scale. But the same problems may also arise in microcosm when one or two singers or instruments are being recorded on location – perhaps in a domestic living-room, which is not designed for music-making. Here the answer is different, for taking the microphone to the other side of a small room with heavily damped acoustics only emphasizes their inadequacy. In this case the only effective solution may be furniture removal: if it is possible to get rid of heavily padded furniture – armchairs and so on – take these out first; but, in any case, rugs should be rolled up and soft drapes pulled aside. And avoid having too many people in the room.

For monophonic balance (which may be a component of stereo), a directional microphone gives the greatest control in cases where acoustics cause difficulty, as it allows an extra dimension for experiment: the angle can be varied, as can the distance from the source and position within the room.

Stereo music balances that are based primarily on co-incident pairs of microphones with cardioid elements angled at 90° may be modified in practice by changing the polar diagram, or by changing the relative angle of the elements, as described earlier.

In particular, for two reasons the double figure-of-eight is rarely used. The most important is the poor compatibility with mono: there are bound to be some listeners to a stereo broadcast or recording who will hear it in mono, and if the A − B signal is very large they lose a significant part of the information. What happens, in fact, is that they lose much of the reverberation, which is very strongly represented in the A − B component of a double figure-of-eight. The second problem is that reverberation is picked up on both back and front of the microphones in roughly equal proportions, so that at low frequencies there is a phase-cancellation effect that takes some of the body out of the stereo reverberation, making it thin and harsh. This is noticeable in a concert hall and very marked in the more reverberant surroundings of, say, a cathedral. Switching the polar diagram even a quarter of the way towards cardioid reduces the effect significantly.

One microphone or many?

The extreme alternative to studio acoustics and the natural blending of sounds is the multimicrophone balance in which an essential, but not necessarily conventionally natural, quality is sought from each instrument or group and blended together electrically. Neither technique is any more 'right' than the other, but opinion has tended to polarize to the view that most classical music is best balanced on relatively few microphones, and that popular music demands many.

There is a rightness about this: the 'simpler' microphone technique for classical music reflects the internal balance of the orchestra and existing acoustics that the composer wrote for, whereas the top-twenty hit of the moment is more likely to reflect the search for novelty and opportunities for experiment. But the question, 'one microphone or many?' is still open in a number of areas lying between the extremes: somewhere in the region between light orchestra and showband there is room for this to be a practical question.

Before choosing a 'natural' balance, consider how the sound source is distributed. Even if the acoustics are suitable, can it be arranged that the sound source falls within the pick-up field of a single microphone position? Given a satisfactory instrument layout, careful microphone placing can ensure that the distance between the microphone and each source is exactly right.

Instruments of orchestras and pop groups

Orchestral players and pop musicians play instruments that have a great deal more in common than in difference. All orchestral instruments are fair game for popular music; but there are others that rarely make the journey in the opposite direction, notably the electric guitar and synthesizer but also some of the older sounds such as the piano accordion or banjo. Nevertheless, for all types of music balance we are dealing with the same basic sound sources, which can be treated in a range of different ways, but for which the various possible treatments required in different types of music overlap. An orchestral balance may require reinforcement of a particular instrument; a stereo balance may require individual mono spotting: both of these demand close microphone techniques which approach those used in multimicrophone balance. In classical music a distance of several feet is 'close' or even 'very close', depending on the instrument; for pop music, closeness may be measured in inches (centimetres) or replaced by contact. There may also be slight differences in terminology.

But there is no difference in the basic problems and possibilities. In the following sections these are explored in terms of individual instruments, sections and combinations; and full orchestras and bands. The instruments to be examined are divided into the following groups:

- *Strings:* plucked, bowed or struck, and coupled to a sounding board acting as a resonator (or from which vibrations are derived electrically and fed to a loudspeaker).
- *Wind:* where the resonator is an air column excited by edge tone, vibrating reed, etc.
- *Percussion:* where a resonator (often with an inharmonic pattern of overtones) is struck.

This far from exhausts all possibilities: for any instruments that have not previously been met it is useful to determine (or guess) the radiation pattern for different frequencies and try microphones in

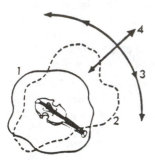

Radiation pattern and microphone position for violin
1, Low-frequency radiation. 2, High frequencies. 3, Move along this arc for more or less high frequencies. 4, Move closer for clarity; more distant for greater blending.

Close violin balance in pop music
The player has encircled the microphone stand with his bowing arm. A directional microphone is not subject to much bass tip-up (proximity effect), as the lowest frequency present is a weak 196 Hz. A filter at 220 Hz may be used to discriminate against noise and rumble from other sources.

different directions and at different distances accordingly, using (wherever possible) comparison tests to find the preferred balance for a particular purpose.

It is well also to remember that such a balance is for a particular player and a particular example of the instrument: some players and some instruments produce very different sounds from others of the same apparent style or type (this is notably true of double basses and their players). It is a balance also for the particular microphone or microphones used, and in all but the closest balance, for a particular studio – and perhaps even for particular weather conditions. If it is very dry, a good natural balance is difficult, because high frequencies are absorbed by dry air.

Violin, viola

A violin radiates its strongest concentration of upper harmonics in a markedly directional pattern, and it is to be expected that, for all but the lower frequencies produced, much of the sound goes into a lobe directed over the heads of the audience. With it goes a great deal of harsh quality, squeak and scrape that even good players produce. So it must be recognized that an audience normally hears little of the high-frequency sound that is produced, as even in the reverberation it may be heavily attenuated in the air. At its lowest frequencies the radiation pattern is omnidirectional, but the violin is an inefficient radiator of sound at low frequency owing to its size: the lowest notes are heard mainly in their harmonics.

For a concert balance the microphone is placed well back from the instrument to add studio reverberation; and the frequency content can be controlled by movement in an arc over the instrument. The upper harmonics are heard at their strongest a little off the axis, towards the E string (the top string on the violin).

The lowest string of the violin is G of which the fundamental (196 Hz) is weak, so bass tip-up is not a serious problem with directional microphones; indeed, a bass (i.e. high pass) filter at 220 Hz may be used to discriminate against low-frequency sound from other sources. A ribbon microphone can give a satisfactory response (though possibly requiring some reduction in top).

To make a *close* violin sound pleasant requires more than the art of the balancer; it demands a modified style of playing in which the unpleasant but normally acceptable components are much reduced. But very close working for pop music work may be limited by the athleticism of the player. In ordinary circumstances 3 ft (0.9 m) is the minimum, for during pizzicato passages there is some danger of the bow hitting the microphone, though some players are capable of encircling it with the right (bowing) arm.

The pick-up from a contact microphone near the bridge may lack high frequencies: experiment with the position and judge whether or

Close balance for three desks of violins
As an alternative to this, for very clean separation it is possible to work with a separate microphone for each desk, but that produces a very strong high-frequency response which must be discriminated against or corrected for.

not this is an advantage. Alternatively, try an electret pinned to the violinist's lapel: this position will discriminate against the stridency of the instrument.

For a close balance on an orchestral violin section, the players sit in their normal arrangement of two at a desk, with the desks one behind another and the microphone is slung directly over and slightly behind the first two players. This should favour the leader sufficiently for solos, but should not lose the rear players as a close balance farther forward would. If this balance is still too distant for adequate separation in, say, a dance band balance in live acoustics, a microphone may be placed over each desk. Some roll-off of high frequencies will be required to counter the harsher quality.

Even if used as part of a stereo balance, a mono microphone must be used for close working, as the slightest movement near a co-incident pair becomes gross. For stereo, the instrument or group is steered to its appropriate position on the sound stage (for a soloist on his own, in the centre) and full width reverberation added as required.

For a balance in which stereo microphones are used, they must be more distant. Slight movements of body and instruments then produce an effect that is pleasant, unless they substantially alter the proportions of sound reaching the microphones. This could happen very easily with spaced microphones, but may still do so with a co-incident pair: a change in reflected sound may produce spurious movement. Once again, reverberation must be adequate in quality and distributed throughout the sound stage.

Violas may be treated in the same way as violins, except that very close working on directional microphones may produce bass tip-up, and that a filter at 220 Hz cannot be used, as this is well inside the frequency range of the instrument.

Cello, bass

Cellos and basses differ from violins and violas in that they are much more efficient as radiators in the lower register, because they have a much larger area of resonator to drive the air. The high, extended spectrum of resonance which makes the violin shrill and 'edgy' if heard directly above the instrument is scaled down to the upper middle frequencies in the case of the cello. In order to get the full rich sonority of this instrument's upper harmonics, it becomes necessary to place the microphone more directly in line with its main radiating lobe. Again, we may use this as an aid to balance. If we merely want to add depth to a fuller orchestral sound we may be content to balance near-sideways-on to the instrument, but if we want richness we turn the instrument to the microphone, as a cello is turned to the audience in a concerto.

If cellos or basses are mounted on light rostra, these may act as sounding boards, introducing unwanted coloration. Rostra should therefore be solidly built or heavily damped. Strategically placed, the mass of the player himself may help. If microphone stands are used, avoid placing them on the same rostra or pay particular attention to their acoustic insulation. In an orchestra, the basses in particular may be helped by a spotting microphone, placed about 3 ft (1 m) from the front desk and angled towards the bridges.

For popular music the bass, or double bass – plucked rather than bowed – is one of the three basic instruments of the rhythm group; the others are piano and drums. Of the three, the bass is the most difficult to separate: a very close balance is essential, as the volume is relatively low. Pop music balancers are disinclined to recommend any single balance that will always work, pointing out that there are great differences between one bass or bass-player and another. The following have been suggested:

- a cardioid, hypercardioid or figure-of-eight microphone 1 ft (30 cm) from the bridge, looking down at the strings
- a similar microphone directed towards the f-hole at the side of the upper strings
- a miniature ('personal') electret suspended from the bridge, with the diaphragm pointing directly upwards to face the bridge
- a miniature electret wrapped in a layer of foam plastic, suspended by its cable *inside* the upper f-hole; this gives virtually total separation
- so too does a contact microphone somewhere near the bridge

Most directional microphones used for the first or second of the above suggestions require equalization to roll-off bass aided by proximity effects. In general, the first has the most to recommend it, because it hears the percussive attack quality from the string itself as it is plucked, and there can be fine control over the ratio of this to the resonance of the body of the instrument. In contrast, the fourth suggestion gives a heavy, hanging quality which has to be controlled to a lower level than the other balances. In stereo pop balances the bass is always taken mono and steered to its appropriate position.

More strings: acoustic balances

Other stringed instruments, such as the classical guitar and the banjo, the violin's predecessors (the viol family), and so on, all radiate their sound in ways similar to the instruments already discussed, and the control that the balancer has over the ratio of top to bass is much the same. Groups based on stringed instruments are also balanced in a similar way.

For a string quartet, if the players sit in two rows of two, the microphone can be placed somewhere along a line extending forward and upward at about 30° from the middle of the group. Check that the distance, together with the angle and position within the studio are right for taking full advantage of the acoustics. Listen for each

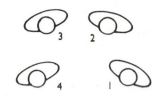

String quartet
1, Viola. 2, Cello. 3 and 4, Violins. Similar arrangements may be made for wind and larger chamber groups.

instrument in turn as though it were a soloist, and make sure that the cello can 'see' the microphone. Try to combine clarity and brilliance on individual instruments with a resonant well-blended sonority for the group as a whole. For radio or recording, the scale of width should be about one-third of the sound stage; for television it would be wider. As usual, reverberation is full width.

As more instruments are added, a similar 'natural' arrangement may be retained, moving the microphone back if necessary in order to keep all the instruments within its field. In this way we progress through the various types of chamber group or wind band to the small orchestra.

Electric guitar family

The electric guitar has steel strings and a pick-up, or possibly a contact microphone near the bridge of the conventional instrument which feeds an 'amp' (amplifier, loudspeaker system). Where a variety of effects is added by the performer, it is necessary to include their output, too, taking it either directly by splitting the signal or acoustically from the loudspeaker.

If the 'amp' is considered to be an essential part of the musical quality, then a microphone may be placed in front of the loudspeaker, perhaps 4 in (10 cm) away and at a slight angle to the axis of a cone. Directly along the axis, the high-frequency content may be too harsh, so experiment for the best position. A hypercardioid response is appropriate. If the volume of sound results in distortion, check first whether there is a switch on the microphone itself to reduce signal level. If the loudspeaker system itself is overloaded, reduce the acoustic volume (or the bass), restoring it in the mix.

In some cases the guitar serves double purpose as an acoustic and electric instrument. In the simplest balance the loudspeaker is placed on a box or chair opposite the player, with the microphone between loudspeaker and instrument. The player himself can now adjust the overall amplifier level, comparing it with the direct, acoustic sound. He has control of the loudspeaker volume from moment to moment by means of a foot control. A directional microphone is used for this balance, with any bass equalization that may be necessary at the distances involved.

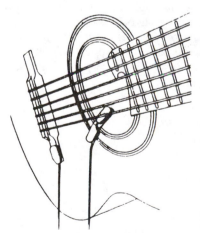

Guitar balance
For popular music, use contact microphone at bridge and acoustic microphone clipped over air-hole.

Other balances include taking a feed from the contact microphone and combining it with the high-frequency response from an acoustic microphone clipped to the edge of the sound hole, hanging inside it directly below the strings and angled to discriminate against noise from the fingerboard. These may be mixed with the loudspeaker balance, taking different frequency ranges, equalization and reverberation from each. Double check the combination by monitoring it on a small loudspeaker (via AFL).

Where a feed of any electrical signal is taken to the desk (via an isolating transformer, for safety) a radio-frequency rejection filter should eliminate unwanted police and taxi messages and, in TV studios, radio talkbacks.

To avoid spill to other acoustic microphones, some studios have a separate room for this instrument, with communication by window and headphones only.

The combination of contact microphone near the bridge, mixed with acoustic microphone clipped to the rim of the air hole, may also be used for other instruments that have this layout – such as dulcimer or zither. On a banjo, try the contact microphone at the foot of the strings, and clip an acoustic microphone to the bridge itself to pick up the string sound.

The piano

The piano radiates in a similar manner to the smaller stringed instruments: the vibration is fed to the soundboard, from which almost the entire wanted sound is radiated, although there is some contribution from the strings themselves at the higher frequencies. If the microphone were to be placed beneath the piano this higher component would be lost.

The radiation pattern from the soundboard allows a reasonable degree of treble and bass control: in particular, the bass is at its strongest if the microphone is placed at right angles to the length of the piano. In an arc extending from top to tail of the piano, the balance with the most powerful bass is that at the end of this arc closest to the top of the keyboard. As the microphone is moved round, the bass is progressively reduced until it is at its minimum at the tail. For very powerful concert pianos the tail position may be best, but there is a slight disadvantage in that there is a loss of definition in the bass, as well as reduced volume. For most purposes, a point somewhere in the middle of this arc is likely to give a reasonable balance – and this is as good a starting point as any for studio tests.

Piano balance
The best balance for a grand piano is usually somewhere along the arc from the top strings to the tail. A close balance gives greater clarity; a distant balance gives better blending. Of the positions shown:
1, Often gives a good balance.
2, Discriminates against the powerful bass of certain concert pianos. 3, Picks up strong crisp bass. A mix of 2 and 3 can be very effective. 4, Close balance (with lid off) for mixing into multimicrophone dance band balances (pointing down toward upper strings).
5, Discriminates against piano for pianist/singer. 6, (Angled down towards pianist) as 5. 7, One of a variety of other positions that are also possible: experiment rather than rule-of-thumb indicates best balance. 8, Concert balance 'seeing the strings'. 9, By reflection from lid. 10, By reflection from the floor: microphones set for other instruments may inadvertently pick up the piano in this way.

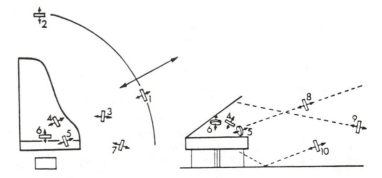

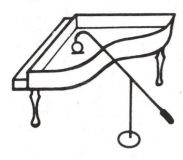

Close piano balance
Directional microphone suspended over the top strings – but not too close to the action.

The piano is the first individual instrument that we have considered that may benefit by stereo spread of the direct sound. But even for a solo the spread should not be too wide: half the stereo stage is enough, with reverberation occupying the full width.

Distance is governed largely by the desired ratio of direct to ambient sounds. A fast piece may need clarity of detail; a gentler romantic style benefits from greater blending. A closer balance is also required for brilliant or percussive effects, and to some extent these qualities may be varied independently of reverberation by changes of microphone field patterns as well as distance. A piano used for percussive effect in an orchestra might require a spotting microphone. Pianos themselves vary considerably: some are 'harder' (more brilliant) than others, so that for a similar result greater distance is required. Conversely, a muddy tone quality may be improved by bringing the microphone closer. A new piano takes time to mature and is better reserved as a practice instrument until the tone settles down.

In a dead acoustic such as that of a television studio try using two microphones – one close to control brilliance, the other more distant and with added reverberation. To give a concert-hall scale to the sound, ensure that the reverberation has suitable delay.

The height of the microphone should allow it to 'see' the strings (or rather, the greater part of the soundboard) which means that the farther away it is, the higher it should be. However, if this is inconvenient, other balances are often possible: for instance, use reflections from the lid. This is the balance that an audience generally hears at a concert. For the very lowest notes, the pattern of radiation tends to the omnidirectional; but for the middle and upper register, and the higher harmonics in particular, the lid ensures clarity of sound.

Depending on the acoustics, cardioid or bidirectional microphones may be used. Although out of favour with some balancers, a bidirectional ribbon often gives excellent results: its range is well matched to that of the instrument.

The closer one gets to an open piano, the more the transients associated with the strike tone become apparent; at their strongest and closest they may be difficult to control without reducing the overall level or risking momentary distortion on the peaks. Action noise – the tiny click and thud as the keys are lifted and fall back – may be audible, and in balance tests this and the noises from pedal action, etc., should be listened for. When a close balance is employed for pop music the lid is best removed, unless this makes separation problems more severe.

A balance suitable for a piano in a rhythm group may be as close as 6 in (15 cm) above the top strings. Surprisingly, a bidirectional ribbon microphone can still be used, as the bass tip-up at this distance actually helps to re-balance the sound. The exact position finally chosen varies with the melodic content of the music being played: one criterion is that the notes played should all sound in the same perspective, which can also be affected by the manner of playing. For

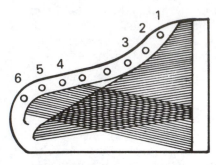

Piano holes
The focusing effect of holes in the frames of some pianos can be used for a pop music balance. A microphone 2 or 3 in (50–75 mm) over hole 2 probably gives the best overall balance, but those on either side may also be used, as also can mixtures of the sound at hole 1 with 4, 5 or 6.

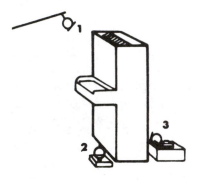

Upright piano
Position 1, microphone above and behind pianist's right shoulder. Position 2, to right of pianist below keyboard. Position 3, at rear of soundboard.

Piano and soloist on one microphone
Balance for the piano first, and then balance the soloist to the same microphone. The piano lid may have to be set in its lower position. For stereo, a coincident pair is used.

a percussive effect the microphone could be slung fairly close to the strikers. A baffle, perhaps a piece of cardboard, fastened to the upper surface of the outer casing of a ribbon microphone further emphasizes and 'hardens' the higher frequencies, and especially the transients.

Working at such distances, it may be better to use two microphones, one near the hammers in the middle top and the other over the bass strings – kept separate to give appropriate breadth to the piano in the stereo picture. If the lid has to be kept low to improve separation, the separate treatment of equalization and echo makes it a little easier to cope with the resulting less-than-ideal conditions.

For a fully controlled mono contribution to a group, wrap a small electret in foam, place it on the frame at the farthest point and close the lid. Another possibility is to use the lid as a baffle for a pressure-zone microphone: with the lid open, it has to go near the middle, in order to retain some bass.

An alternative is to use the focusing effect of the holes in the iron frame of some pianos: the different sounds can easily be heard by listening close to them. A microphone two or three inches above the second hole from the top may give the best overall balance. With two microphones, try the top hole in combination with the fourth, fifth or sixth hole.

Distortion (particularly of the percussive transients) due to overloading has to be avoided, and may require attenuation at the microphone itself.

With an upright piano, lift the lid and try the microphone somewhere on a line diagonally up from the pianist's right shoulder. But remembering that it is the soundboard that is radiating, an alternative is to move the piano well away from any wall and stand the microphone at the back (for a close balance), or diagonally up from the soundboard (for a more distant one). Behind the soundboard there is less pedal action noise, but also, of course, less brilliance and clarity.

One or other of these methods is also suitable for balancing a jangle-box – the type of piano that is specially treated, with leaves of metal between hammer and strings to give a tinny strike action, and with the two or three strings for each note of the middle and top slightly out of tune with each other.

An upright piano used for pop music may have two very close microphones, one for the bass, clamped to the frame near the top of G2, the other somewhere between one or two octaves above middle C (experiment for the best position). Where separation is a problem use a contact microphone on the soundboard for the bass. The output may be taken to the mixer (with a feed back to the two loudspeakers), or direct to the loudspeakers which are balanced as for the guitar. Equalization may be used to emphasize high-frequency percussive transients, or to add weight to the bass.

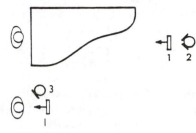

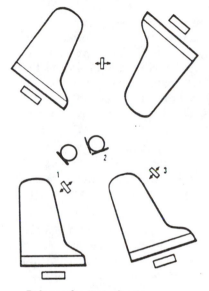

Piano and soloist side by side
Balance using two ribbons (1) or stereo pair (2) and hypercardioid (3).

Balance for two pianos
Above: Mono – two pianos on a single microphone. *Below:* Two pianos on individual microphones (1 and 3) or on a stereo pair (2).

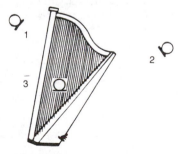

Harp
1, Frontal coverage, looking down to soundbox. 2, Rear position for microphone favours strings and discriminates against pedal noise, but also loses reinforcement from the soundbox. 3, Close on soundbox, can be used for light or pop music.

Piano and soloist, two pianos

In a studio with suitable acoustics it is possible to get good separation by turning the soloist's microphone, which may be bidirectional or hypercardioid, dead-side-on to the piano and mixing in the output of a second and possibly third microphone, arranged to pick up a little of the piano. This time start by getting a good sound from the soloist, and then gradually fade in the piano just enough to bring it into the same perspective.

A common technique in pop music recording is to employ a backing track. The soloist listens to the pre-recorded piano on headphones, and the levels can be adjusted at will at the mixer.

A pianist who also speaks or sings may wish to use less voice than would be possible without an additional vocal microphone: this achieves a more intimate effect at the expense of internal balance. For the voice, a hypercardioid can be placed well forward, above or slightly to the side, so that its dead angle (beginning 135° back from the forward axis) centres on the piano. However, any residual piano pick-up may have an erratic frequency response, so it is worth comparing this with an older style of balance which uses the smooth bidirectional response of the ribbon. This is suspended about 20 in (50 cm) from the singer (as for speech balance), again with its dead side toward the soundboard of the piano. It may be in front of the singer (tilted down) or to the singer's right. A wider (stereo) pair is faded up just enough to establish the piano stereo effect and natural reverberation; alternatively or additionally, differential artificial reverberation is added. Note that once the faders are set, if either is moved on its own it can change the balance radically.

The best way of balancing two pianos on a single microphone depends on studio layout, on which the wishes of the pianists should be observed if possible. With the pianos placed side by side, use a single ribbon or stereo pair at the tail. The equivalent two-microphone balance would use two ribbons, each in the curve of one of the pianos: these can be a shade closer than the single microphone, and give a marginally greater degree of control over the relative volumes and frequency pick-up.

Harp, harpsichord, celesta

The harp has a soundbox below the strings: in performance it is cradled by the player who reaches round on both sides of it to pluck (and otherwise excite) the strings. Owing to the relative inefficiency of the soundbox, the vibration of the strings themselves contributes relatively more than in a piano, and ideally any balance should cover both. The key of the instrument is controlled by pedals at the foot, and as these may be operated during the performance of quiet passages, their mechanical sounds may be audible. In an orchestra, the harp or

Celesta
A microphone near the soundboard behind it avoids mechanical noise which can be loud compared with the music in this relatively quiet instrument. An alternative position is in front of the instrument, below the keyboard, on the treble side of the player's feet at the pedals, but keeping well away from them.

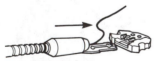

Piano accordion
A popular music balance can be achieved using close microphones at the bass and treble sides. These may both be attached to the body of the instrument or just one (over the bass), with a stand microphone for the treble. In this example, a special plate and clip are shown, which can also be used for guitar and bass 'amps' (loudspeakers).

harps generally benefit by spotting for presence and to locate them more clearly, usually towards the side of the stereo picture.

A microphone, about 4–5 ft (1.5 m) diagonally upward from the soundbox, may be directed to cover both that and the strings. A ribbon is excellent, giving tight coverage of one or both harps on its relatively narrow forward lobe, while the rear lobe points upward into the hall. An electrostatic cardioid will also cover the harps well, while a hypercardioid can be placed closer in to the strings, favouring the middle and upper parts of the soundbox, with some discrimination against pedals.

A harpsichord has many of the qualities of the piano, including mechanical action noise, plus the characteristic attack on each note as the string is plucked by its quill. Where the instrument is featured, it is useful to have control over the relative values of transients and resonance. For this, set out two microphones, one above and one below the level of the soundboard. The lower microphone takes the body of the sound, perhaps by reflection from the floor; the other, angled to 'see' down into the action, picks up the full sound but with the transients and upper harmonics somewhat over-represented. Mix to taste.

In a baroque orchestra (covered by a co-incident pair) harpsichord continuo needs to be located well back, to avoid its penetrating sound appearing too dominant, and also near the basses, with which it must closely cooperate. A spotting microphone can then be used to fine-tune the contribution of continuo.

In other instruments of the same general class, such as virginals, clavichord or spinet, apply the same principles, first identifying the soundbox or board, the sources of action noise, attack transients and higher frequency components and placing the microphone or microphones accordingly.

The celesta is a rather quiet keyboard instrument. In an orchestra it benefits by spotting; the needs of separation then generally require that a close balance be employed in order to avoid bringing up neighbouring instruments with it. This has the disadvantage of accentuating the already potentially intrusive action noise. A layout in which a balance at 6–8 ft (about 2 m) is possible may avoid the worst of this. If a closer balance is unavoidable try a microphone either in the middle at the back or below the keyboard, on the treble side of the player's feet – and keeping well away from them.

Piano accordion

In a close balance the piano accordion may require two microphones, one directed towards the treble (keyboard) end, the other favouring the bass. A closer balance on the air-pumping bass end can be achieved by clipping a goose-neck electret to the instrument itself, so that the microphone hangs over some of the sound holes. For greater mobility, attach two, one at each end.

Woodwind

Flute
A microphone behind the player's head discriminates against the windy edge-tone, but can 'see' the pipe and finger holes.

'Electric' flute
Pick-up from the end of the head discriminates very strongly against breath noise, and offers a pure, mellow tone.

'Electric' penny whistle
Pick-up can only be from the side of the body of the instrument.

Clarinet
Popular music balance using two microphones, including a miniature gooseneck for the bell.

Each of the wind instruments has its own directional pattern of radiation. In contrast to brass, for the woodwind with fingered or keyed holes the bell may make little difference to the sound: the main radiation is through the first few open holes. It follows that, unless we go closer we have considerable freedom in balancing woodwind; in general, a microphone placed somewhere in front of and above the player will prove satisfactory.

For wind, as with other sections of the orchestra, the transient, the way in which a note starts, is of considerable importance in defining the instrument for the listener and giving it character: as also is the breathy edge tone that we hear with the flute. These sounds must be clear but not exaggerated. The closer the balance, the stronger are the transients in comparison with the main body of the sound. Listening at an ordinary concert distance, the harsh edges of the notes are considerably rounded off. Listening to a flute at two feet, it can sound like a wind instrument with a vengeance! In close balances it is occasionally advisable to resort to devices such as placing the microphone *behind* a flautist's head, so reducing the high-frequency mouth noise and edge tones without affecting the sound radiating from the other end of the instrument.

At the other acoustic extreme is a headset carrying a miniature microphone with its axis directed past the embouchure towards the upper holes. Taking this a stage further, what might be termed an 'electric flute' is used to obtain an unnaturally pure tonal quality for popular music, as in the 'Pink Panther' theme. The microphone is attached rigidly in line with the end of the head joint, so that it picks up the wave from inside within the length of the flute, but no mouth noise or sound radiated from the holes. Radiated sound can be added to taste, from a stand microphone in front of the player.

An electric recorder or penny whistle cannot have such an arrangement, as the mouth covers the top end, but a microphone can be fixed in to the side of the body of the instrument above the finger holes.

When used in popular music the clarinet has been balanced using two close microphones, one on a stand, pointing down towards the keys, the other a lightweight electret on a spring clip attached to the bell, and angled by means of a short goose-neck to point back into it.

In a close orchestral balance the woodwind may sound relatively far back, so sling an additional co-incident pair or set in three separate microphones on stands at a height of 6 ft (about 2 m). The extra microphones are for presence, perhaps emphasized by midlift in the 2–4 kHz range. They are not intended to provide the main balance, and their signal must be added to it with discretion. But they do also allow precise control of the stereo spread in the central region, perhaps broadening it a little. The width required from a stereo pair on woodwind would still, of course, be much narrower than that of the main pair covering a whole orchestra.

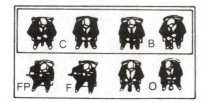

Woodwind orchestral layout
F, Flute. P, Piccolo. O, Oboe. C,
Clarinet. B, Bassoon. If the main
orchestral microphone is set for a
close balance, it may tend to favour
the strings at the expense of
woodwind, and an additional
microphone may be required.

Saxophones in dance band
In a studio layout up to five may be
grouped around a central
microphone. *Above:* a bi-directional
microphone set at the height of the
bells of the instruments. *Below:* C,
Two cottage-loaf microphones set
above bell level and angled
downwards to reduce pick-up from
outside this area. S, Solo
microphone for player doubling on a
second instrument.

Saxophones

In big-band music a saxophone section has perhaps five musicians.
An early balance with a bidirectional ribbon had three players on one
side and two on the other, with the microphone at a height such that it
is in line with the bells of the instruments. The lowest saxophone
(usually a baritone) was at one end of the row of three. Players in this
section may double on flute or clarinet, which was also played to the
same microphone.

However, for this balance, separation is less than it could be, as there
has to be compensation in volume for the angle of the bell of the
instrument; besides which the microphone is all too well placed to
pick up sound reflected from the walls of the studio. Another
arrangement has the same grouping of the players, but with two
hypercardioids angled down toward the centre of each side.

An alternative that some balancers prefer is to place this section in a
line or arc, with one microphone to each pair of musicians. This helps
particularly where saxophone players doubling woodwind fail to
move in for their second, quieter instrument, and also allows the
height of each microphone to be adjusted more precisely to the needs
of the second instrument. For a close balance on a clarinet, for
example, it should be lower than for a flute.

For greater separation of individual saxophones, try a miniature
electret, clamped (by a spring-loaded, bulldog-type clip) to the edge
of the bell, with a goose-neck angling it toward the rim. Try other
angles, too – including the central axis of the bell for breathy subtone
playing.

Saxophones may benefit from midlift at about 2.5–3 kHz.

Brass

In brass, the axis of the bell of each instrument carries the main stream
of high harmonics. Even when the music requires that this be directed
at the audience, only a relatively small group receives the full blast at
any one moment, the rest hearing the sound off-axis. In consequence,
for all but the most exaggeratedly brilliant effects, the main
microphone for an orchestral balance is usually quite satisfactory for
the brass. That its distance from the microphone is greater than that
for some other sections is no disadvantage, for the added
reverberation conventionally suggests the sheer volume produced.
Orchestral brass that is balanced well back in a spacious hall is usually
easier to control (requiring relatively little manipulation of level) than
when, by choice or necessity, closer microphones are used.

Spotting microphones may have to be set for an unusual reason: in
this case because the volume of sound from brass can cause the whole

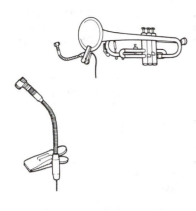

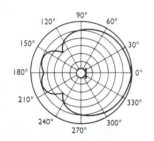

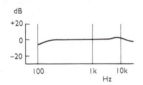

Miniature gooseneck
Electrostatic cardioid for brass (trumpet, trombone, tuba, and also saxophone) in popular music balances, allowing freedom of movement.

balance to be taken back, making this section seem even more distant unless there is separate control of presence – to which additional reverberation must be added to maintain perspective.

In a close balance, the volume from brass can reach 10–15 dB higher than the maximum recommended for electrostatic microphones, unless the signal can be reduced by switching at the microphone itself. In addition, the balancer should not rely on rehearsal levels of big band brass players he does not know, but should allow for as much as a 10 dB rise in level on the take, which for a live broadcast could be disastrous. It risks distortion, reduction of the working range on the faders and loss of separation. It also helps if the players themselves are prepared to cooperate by leaning in to the microphone on quiet passages with mutes, and sitting back for the louder passages. The microphone must be near enough for them to reach for close work but far enough back to pick up several players equally. Players are also well aware of the directional properties of their instruments and will play directly towards the microphone, or not, as the music requires; though some bands which have done little work in studios and more in dance halls may take this to extremes – perhaps by playing soft passages in to the music stand. Here the balancer needs to remind them that a cleaner sound can be obtained if some of the control is left to him.

The height of the microphone should be such that it may be on the axis of the instruments as they can be comfortably played – though with the trombones this is a little awkward, as the music stands get in the way of the ideal position. However, the leader of the section will say whether he prefers the microphone above or below the stand. The answer probably depends on how well the players know the music. In a band where the music played is part of a limited and well-rehearsed repertoire the trombones are often held up; but musicians faced with unfamiliar music will, understandably, wish to be able to look down to see it.

Miniature goose-neck electrets can also be clipped to the rim of the bell of a brass instrument, a technique which gives the player freedom of movement, but loses the effect of movement on performance. Again, try different microphone positions by bending the goose-neck.

Brass may benefit from midlift of 5 dB or even more at about 6–9 kHz. But some microphones have peaks in their response in this region and produce the effect without much need for equalization.

In the orchestra, horns can often be placed where their sound reflects from a wall or from other surfaces behind them. Such an arrangement – with a microphone looking down from above at the front – is suitable in almost any circumstances; though exceptionally, for complete separation and a strong, rather uncharacteristic quality, a microphone can be placed behind the player and in line with the bell.

Several layouts are possible for brass or military bands. In one, the various groups are laid out in an arc round the conductor in the same way as a string orchestra; in another, the performers form three sides of a hollow square, with the cornets and trombones facing each other on opposite sides. In this way the tone of the horns and euphoniums at the back is not blotted out by the full blast of more penetrating instruments. The microphone pair may be placed a little way back from the square, as for an orchestra.

The main body of euphoniums and bass tubas radiate upward and can sound woolly on the main microphones unless there is a reflecting surface above them to carry their sound forward. Failing this, an additional microphone above them will add presence.

A solo euphonium player should come forward to another microphone that is set high up and angled to separate him from other instruments of similar sound quality. A solo cornet may also have its own microphone, an electrostatic cardioid.

The military band behaves like a small orchestra from which the strings have been omitted. There is usually plenty of percussion, including such melodic instruments as the xylophone and glockenspiel. Care must be taken that these are given full value in the balance.

A military band in marching order is balanced for the benefit of the troops behind them; and in the open air, woodwind, at the back, will not be reinforced as it would indoors. A microphone balance from the rear helps to correct this, even though for television or film, pictures are more likely to be from the front. A band on the move is even trickier, as the best balance will rarely match preferred wide-angle camera positions. A marching band in an arena can be followed well with a tetrahedral array (Soundfield) microphone: suspended over the action, it can be steered to give an angle and polar response that covers the band, while discriminating against excessive reverberation or crowd noise.

Percussion, drums

Most percussion instruments can be recorded with ease. However, some percussion sounds register more easily than others, and on all but the most distant microphone set-up it is as well to check the balance of each separate item with care. Any rostra used need to be reinforced to prevent coloration. In orchestral percussion a microphone with an extended, smooth high-frequency response is required mainly for triangle, cymbals and gong. Spotting

Typical drum-kit for popular music
1. Bass drum. 2, Cymbals. 3, Tom-toms. 4, Snare drum. 5, Hi-hat.
6, Microphones with good high-frequency response used in two-microphone balance. 7, Additional microphone (moving-coil) for more percussive bass drum effect. Further microphones may be added, to give each drum and the hi-hat its own close pick-up, plus two above for cymbals and overall stereo.

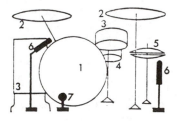
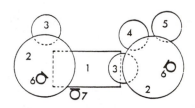

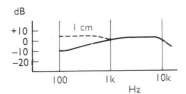

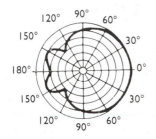

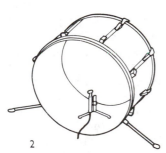

Miniature microphone for drums
1, Electrostatic hypercardioid, clamped on top hoop or tensioner. Tone varies with drum balance.
2, Stand the microphone inside bass drum shell fairly close to skin.

microphones may be set for many of the individual components: for the quieter ones, so that detail is not lost; and for the louder elements (as for brass), so that presence can be retained when overall volume is reduced.

Whereas instruments such as the glockenspiel and xylophone require no special balance, the characteristic sound of the vibraphone requires a close microphone beneath the bars, angled upward to the tops of the tubular resonators (the bottoms are closed).

In popular music, percussion shows itself mainly in the form of the drum-kit, including a snare drum, bass drum (foot-pedal operated), a cymbal (or several), a hi-hat (a foot-operated double cymbal), a large tom-tom, and perhaps a small one as well.

In the simplest balance, a single microphone (which also requires a good high-frequency response) points down at the snare drum. But any attempt to move in for a close balance from the same angle increases the dominance of the nearer instruments over the bass drum and the hi-hat. The hi-hat radiates most strongly in a horizontal plane, in comparison with the up-and-down figure-of-eight pattern of the top cymbals. So for a somewhat closer balance place a wide-frequency-range cardioid microphone at the front of the kit and at about the level of the hi-hat.

For a tighter sound move the main microphone closer to the hi-hat, also favouring the left-hand top cymbal or cymbals and some of the snare and small tom-tom, together inevitably with some bass drum; a second cardioid microphone is then moved in over the other side of the bass drum to find the other cymbal or cymbals, the big tom-tom, and again, perhaps, the snare drum. Changes of position, angle, polar diagram or relative level give a reasonable degree of control over nearly every element except the bass-drum sound. This may also be required to fill in the texture of some pop music and can be obtained by facing the diaphragm of a moving-coil microphone toward the front skin and close to its edge, where a wider and more balanced range of overtones is present than at the centre. This is one place where one of the older, more solid-looking microphones with a limited top-response is as good as a modern high-quality microphone (which might in any case be overloaded). For the dry thud required in pop music take off the front skin (remembering to tighten up the loose lugs to avoid their rattling) and put a blanket inside.

For greater control still, use separate microphones for each component of the kit. For some of the drums (snare, toms, congas, bongos) they may be clipped to the top edge (tensioning hoop). The same type of relatively insensitive electret can also be used inside a bass drum, but this needs a solid mounting. With a low stand microphone for the hi-hat, an overhead pair for the cymbals and basic stereo coverage the drum-kit alone can use up far more tracks in a multitrack recording than any other part of a group. Many pop balancers work with the drummer to tailor the sound, and in particular modify the snare drum by tightening the snare tension (perhaps also putting some masking tape on it), and may tape a light duster over part of the skin (away from where it is struck).

Midlift (at 2.8 kHz) plus additional bass and extreme top may be added. The tom-toms are less likely to require attention, but the quality of cymbals varies and the application of masking tape may again help to reduce the blemishes of poorer instruments. With such extreme close balancing it is wise to check the maximum playing volume in advance, and also to recheck the quality of the sound from each drum from time to time: it may change as a session progresses.

Stereo width is introduced by steering the various microphones to slightly separate positions.

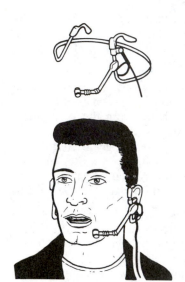

Headset (without headphones)
For singers, designed for light weight and minimal disturbance of hairstyle. The electret microphone is in a fixed position, and at this distance must be beyond the corner of the mouth in order to avoid popping.

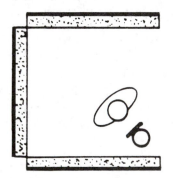

Vocalist
When the volume of surrounding instruments is high, the singer is screened. The microphone's live side is towards absorbers and its dead side towards the open studio.

Singers: solo and chorus

What counts as a close balance for a singer depends on the type of music. For a pop singer it may be close to the lips: for an operatic aria, at arm's length. For classical music a clean, flat response is all that is necessary: try a hypercardioid electrostatic microphone at 3–4 ft (1 m), angled upward enough to point over the top of orchestral instruments behind the singer. High-quality ribbons at the same distance can also be good – though with care for what might lie in the opposite lobe.

For lighter styles of music using a close balance, compensation for bass tip-up (proximity effect) is needed when working with a directional microphone. A hypercardioid discriminates well against the background. One double-ribbon hypercardioid has a reduced bass response such that a singer can work at a matter of inches without noticeable lift to the bass (indeed, if the singer moves back, bass may have to be added). A singer working very close does in any case produce less of the chest tone which in opera is used to deepen and strengthen the sound.

A skilled practitioner may use the angle and distance of a directional microphone to precise effect; less reliable singers are safer with an omnidirectional microphone, for which only the volume has to be adjusted (though even this may have some variation of frequency response with angle). For freedom of movement (and free hands), but constant-distance and angle, try a headset (with headband below the hair at the back, for those whose hairstyle is part of the act): the tiny electret must be a little away from the corner of the mouth in order to escape 'popping'. For other microphones this is minimized by directing the voice (as for speech) across the microphone – that is, by missing it with the air blast. Windshields, built-in or added, also help.

To improve separation in a studio, the singer is often given a tent of screens, open (for visual contact with any musical director) only on the dead side of the microphone. In many pop balances a vocal guide track is recorded (on a spare track) at the same time as the main instrumental sound, and the lead and backing vocals recorded to replay.

Some singers seem subject to a superstitious belief that close balance gives them presence. In fact if separation (the real reason for close balance) can be achieved in other ways, try setting in a second microphone a foot or two (30–60 cm) from the singer's mouth. Not only will popping no longer be a problem; it may also be found that the more distant balance is actually better, as well as easier to control. Presence can be added electronically, with midlift at about 1500–3000 Hz or more, depending on the voice. Sibilance can be cured by some combination of singing across the diaphragm, careful enunciation, high-frequency roll-off, and choice of a microphone with a smooth response in the 3–12 kHz region. Electrostatic cardioids are preferred for this quality rather than for their extended high-frequency response, which is not needed. On the other hand, some balancers are happy to use microphones with double high-frequency peaks.

A folk-singer with guitar can in principle be balanced on a single microphone, perhaps favouring the voice, but two rather closer microphones mounted on the same stand give greater control of relative level, together with the capacity for separate equalization and artificial reverberation.

Individual singers are rarely balanced in stereo (except where action is taken from the stage). Mono microphones are used and steered to a suitable position in an overall balance. But a co-incident pair is well suited to a pop music backing group, which can then be given any desired width and position in the final mix.

With choral groups, aim (generally) for clarity of diction. Set the microphone far enough back to get a well-blended sound, but without losing intelligibility. Spread bidirectional or cardioid microphones may cover a large chorus. This is also used for a chorus with orchestra (see below). Where a small group can conveniently work to a single microphone position use a cardioid or (for stereo width) a co-incident pair, arranging the front line of singers to match the microphone field pattern.

There is no special virtue in clear separation of sopranos, altos, tenors and basses into separate blocks: the stereo effect may be good, but the musical quality of a layout that blends all sections may be better. As an alternative to the block layout a chorus may be arranged in rows (here even more than in the other cases the microphones need to be high up in order to balance all of the voices). Occasionally, the singers may all be mixed up together. Opera singers must be able to do this, and so can some choirs. A high order of musicianship is required: when a group is able to and wishes to sing like this, the stereo balancer should not discourage them.

Church music, and particularly antiphonal music with spaced choirs of voices, benefits enormously by being brought round to the full spread of stereo: it was here that quadraphony came into its own without the need for special composition.

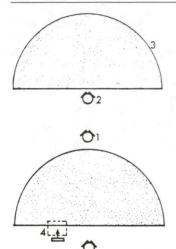

Orchestral balance
1 and 2, Two stereo pairs with different polar diagrams but giving the same stage width to the orchestra (3). Microphone 1 is double figure-of-eight; microphone 2 is double cardioid: a mixture of the two can be used to control and change reverberation where there is such a wide dynamic range that a great deal of volume compression is necessary. 4, For a soloist a spotting monophonic microphone is used, steering this to the same position where it appears on the audio stage, as established by the stereo pair.

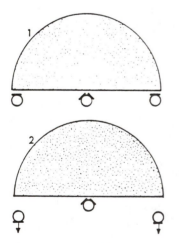

Stereo orchestral layouts
1, With a spaced pair to 'pin down' the corners of the orchestra (e.g. harp on left and basses on right). 2, With a spaced pair used to increase reverberation.

The orchestra

Our next individual 'instrument' is the full orchestra. And, indeed, an orchestra may rightly be considered to be an individual instrument if it is well balanced internally.

The detail of orchestral layout is far from standardized, but the strings are always spread across the full width of the stage, grouped in sections around the conductor, and with the woodwind behind them in the centre. In the older form of string layout the violins were divided, with the first violins on the left and second on the right, and with the violas and cellos between them and the double basses also in the centre. For stereo this has the advantage that the violin sections are clearly separated, and the weight is in the middle. But that the upright instruments (cellos and basses) face forward is unnecessary for microphone or audience, as they are virtually omnidirectional radiators anyway; and there is also the slight disadvantage that the high-frequency radiation from the second violins is directed backward away from both audience and microphone. The more modern layout has the sections placed, from the left: first violins, second violins, violas, cellos; and raised behind the cellos on the right, the basses. Apart from the woodwind, which are always across the centre, there are further variations in the rest of the placings although brass is often at the back towards the right, and percussion and timpani to the left. That leaves horns, harps, piano or harpsichord (as orchestral instruments, not soloists); all of these may appear in different places.

While the balancer's job is generally to reflect what is offered, in the studio a well-established and experienced balancer may also offer suggestions for improvement – usually through a producer. However, a conductor who is thorough may consult the balancer directly, not only on matters of detail in placing but will also ask how the internal balance sounds. Many conductors, however, concentrate solely on musical qualities, so that if, for example, there is too much brass it will be left to the balancer to sort it out, if he can.

In an orchestral balance that is based on co-incident cardioids, the microphones are placed to control width rather than reverberation: for a subject that is to occupy the major part of the audio stage the microphones look down from one edge of the pick-up area. For an orchestra that is to occupy about two-thirds to three-quarters of the sound stage, they are high over and set back somewhat from the conductor.

Reverberation may be added by a variety of means. One is to have a second co-incident pair set back from the first and facing toward the empty part of the hall or studio. Another has a spaced pair of cardioids at the sides of the hall, well forward again, but looking back along it. In either of these cases the two microphones must be steered to their correct sides of the studio, as some direct sound will arrive at these microphones, and even more by short-path first reflections.

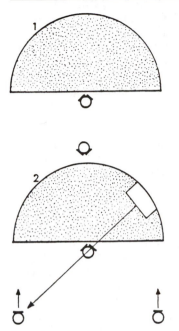

Reverberation in stereo orchestral layouts

1, A second co-incident pair with reversed output used to control reverberation separately: can also be used for four-channel stereo. 2, Wrong: a forward-facing spaced pair set back into the hall may give false information from the brass, which, directed towards the microphone on the opposite side, may suddenly appear to be in the wrong place.

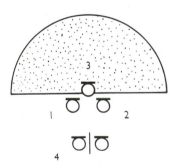

Microphone 'tree'

1–3, A spaced pair of omnidirectional microphones, with a third to fill in the middle, has been used for many years by a recording company. Alternatively, 4, a dummy head layout, with omnidirectional microphones on either side of a disc, has been favoured for its spaced effect. Discs that are head-sized and larger up to about 3 ft (1 m) have been used to separate the pair.

Another layout has a second co-incident pair of microphones, but this time figure-of-eights looking forward and set back to such a distance that the positions of the various parts of the image roughly coincide. For an orchestra, the separation may be about 10 ft (3 m). This arrangement has the additional advantages that the balancer can work between the two pairs to handle extreme changes of volume, and that the second pair gives an automatic stand-by facility in the event of failure or loss of quality of the first pair during a live transmission. Such failure is very rare, but the facility is nevertheless desirable; two pairs of microphones are in any case needed for balance comparison tests. Phase cancellation on the back of the second pair does remain a problem.

Any microphone that is set up in addition to the main pair and facing toward the orchestra (or a 'bright' reflecting surface) may give false information about the position of the brass, if significant direct or strong first-reflection sound can reach it. The trumpets are likely to be directed over the heads of the strings and may be picked up on any microphone in line of fire on the *opposite* side of the studio. There is then an extra trumpet signal, pulling the section across the sound stage, perhaps to a position occupied by other instruments.

Further mono microphones are generally added. Two, one on each side of the orchestra, may be placed to cover strings. They are faded up just sufficiently to lend presence to their tone, and panned to their proper places in the stereo balance, which is thereby more clearly defined. Woodwind, too, is often covered: where the main pair give good basic stereo coverage a mono microphone may be sufficient, though a stereo pair is often used here too. It lends presence to woodwind solo passages. Harps and some other instruments may be given spotting microphones, but again must be used with subtlety to add clarity and positional information. Spotting microphones can also be used to a limited extent to restore the balance of sections that are weak in numbers, but that implies lifting volume more than is needed solely for increase in presence. The result may sound like a small group given solo status.

Orchestra: spaced microphones

Previous editions of this book have firmly recommended co-incident pairs, but it must be recognized that spaced microphone techniques, too, have always had strong advocates, and that the availability of omnidirectional microphones with an exceptionally smooth response reinforces their case.

A 'bar' with two omnidirectional microphones separated by nearly $1\frac{1}{2}$ ft (about 40 cm) gives a rich, spacious sound. Their faders must be locked rigidly together, as the slightest difference swings the orchestra sharply to one side. The axes are angled slightly outwards to allow for high-frequency reinforcement normal to the diaphragm. In a concert hall, their distance may not be easy to change, so a curtain of some five

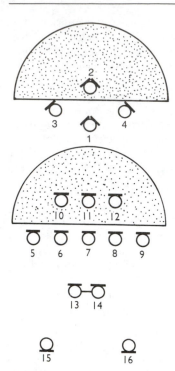

Orchestral multimicrophones
1, Main stereo pair. 2, Woodwind pair 7–10 ft (2–3 m) forward of this section. 3, 4, Spotting microphones for string sections, about 6 ft (2 m) from them, add presence and strengthen positional information. 5–9, A curtain of omnidirectional microphones. Phase and control problems are reduced by adopting a sufficiently close balance. Once the faders are set, these are controlled as a group, because changes in the relative level of adjacent microphones can affect the apparent position of instruments equidistant from them, including loud, distant sources such as brass. 10–12, Omnidirectional 'drop' microphones, mainly for woodwind. 13, 14, Spaced bar of omnidirectional microphones used as an alternative to the curtain. 15, 16, Spaced reverberation microphones. Although these are nominally omnidirectional, the main axes are directed away from possible sources.

extra microphones may be added, close to the front line of the orchestra, with three more set deeper to favour the woodwind. Their outputs are all panned to their appropriate positions within the stereo picture, and after their relative levels have been set by listening to the effect of each in turn, are controlled as a group, adding just enough from them for presence and positional information (rather weak from the space-bar) to be strengthened. As usual, fade up until 'too much' is heard, then set them down a little.

Spotting microphones (as well as those for the woodwind) are added, and considered as candidates for 'equalization' (or rather, selective frequency reinforcement) and artificial reverberation. Digital delay (of 10–16 msec) can be used to avoid time discrepancies.

The virtues of this mass of microphones are still debated fiercely. Where a series of concerts is to be covered by several balancers it is best if the hall is rigged in advance with all the main systems currently in contention, so that direct comparison tests can be made. Critics of space-bar and curtain techniques describe them as 'muddy'; critics of co-incident pairs call them 'clinical'. Even 'dummy heads' have their advocates, but in direct comparison may be beaten by the space-bar.

Pipe organ with orchestra

If a *pipe organ* is used with orchestra there is a possibility (in stereo) that its distance will cause its image to be too narrow.

In fact, as a good organ is laid out in a hall, it should have some width (indeed it is bound to), but not so much that the spatial separation of the various voices are marked: like a choir, organ tone is better blended than broken up into obvious sections. However, in stereo, excessive width (such as that of the Royal Festival Hall in London) is narrowed by a distant co-incident pair and an average width may even appear narrow. A second pair at such a distance that they 'see' the organ as a whole can be used to broaden it on the sound stage. They should be set high and angled to pick up as little orchestra as possible; and the orchestral pair set to discriminate against direct sound from the organ.

Orchestra with soloists
or chorus

Concert works with solo instruments placed near the centre can in principle be taken using a standard orchestral balance: the main microphones favour the soloist. However, it is usual to provide solo spotting microphones for additional presence. For a piano concerto, however, it may be necessary to angle or place the main microphones to discriminate somewhat against the volume of the solo instrument.

The relative volumes in a curtain of microphones will be set to exclude the piano; any microphones hanging over the woodwind will be brought up to compensate, in effect taking the curtain in an arc behind the piano.

If there are several soloists, as in a double or triple concerto, or several solo singers, they need to be placed with a thought for layout on the sound stage.

In public performances a chorus is normally placed behind the orchestra. If it is already weaker than the orchestra this position does not help; and moving the microphone in only makes the orchestra stronger in comparison with chorus. Several spaced directional microphones are therefore added to give presence, and if necessary volume, to the choir. These are set on very high stands (or suspended) at the back of the orchestra. As they look down at the choir, giving it increased body, presence and intelligibility, their dead side is toward the orchestra. The stereo spread can be increased if on the main microphone they appear to be bunched too close together.

This arrangement gives more separation and control than would be obtained using a second stereo pair, which is also more likely to obstruct television coverage. To clear sight-lines, suspended omnidirectional microphones have been used, but require equalization, with sufficient lift at 3 kHz to restore intelligibility.

There are further problems in stereo when the chorus is supplying not words but orchestral texture, as in Ravel's *Daphnis and Chloë* (the difficulty here is to get sufficient spread when spotting microphones would bring the chorus too far forward), or in the last movement of Beethoven's Ninth Symphony (where the soloists are better placed for perspective at the back, but prefer to come and stand at the front).

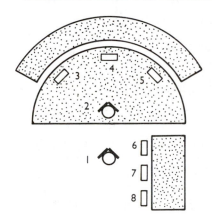

Orchestra with choir
1, Main pair. 2, Choir pair. 3–5, Directional spotting microphones. 6–8, Alternative for choir. Since any spill onto a central co-incident pair will be reversed, 6 should be steered to the right and 7 to the left.

Opera

Opera may be recorded or broadcast in concert form; or it may be taken from a public stage performance. A range of methods has been used for each. The simplest employs a standard layout for orchestra, with chorus behind, and soloists to the fore: this remains excellent for oratorio and for concert performances of opera where there is no pretence at dramatic presentation.

For opera on records, an advance was made by introducing stereo movement laterally and in perspective. Historically, a layout that produced excellent results had a line of five cardioid microphones beyond the orchestra. These covered an acting area marked in a chequered pattern. The square from which the player should sing was marked on his score; moves were similarly indicated. A disadvantage was that the conductor could not hear the solo singing as well as with

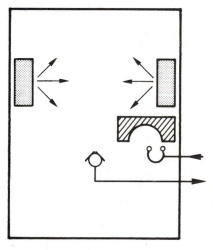

Organist on headphones
Where a keyboard is badly placed for the organist to hear a good balance, it may be fed back on headphones.

Stage stereo pair set in pit
With capsules on goose-necks angled slightly down, almost touching the stage. These act as boundary microphones, with directional responses that are parallel to the stage surface, which should be clear of absorbers for at least 3–4 ft (1 m). Additional spaced microphones will be required to cover action out to the sides, or which approaches too close to the central pair.

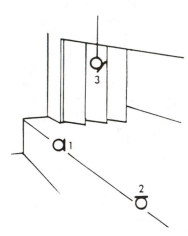

Microphone positions in the theatre (e.g. for opera)
1 and 2, A pair of (cardioid) microphones in the footlights. Though very close to the orchestra their response discriminates sufficiently against this. 3, A single microphone placed high in the auditorium – but well forward. Cardioid response again favours singers.

the oratorio layout. An alternative was to have the singers on either side of him, but facing the orchestra, and working to cardioids which therefore had their dead sides toward the orchestra. The singers were recorded on separate tracks and their perspective and position (with movement provided by panning) was arranged at the mixdown stage. For this, less preparation was required in advance.

For the best use of studio space, a standard orchestral layout may be combined with a laterally reversed balance of singers. With the orchestra at one end of the studio, the chorus is placed at roughly a right angle to it along one side of the studio, so that one corner of the orchestra adjoins one end of the chorus. The chorus, arranged in rows on ascending rostra, work to a line of microphones, as do the soloists in front of them, These microphones are 'folded' back across the orchestral sound image.

For opera from the stage, coverage is possible using a co-incident pair for the orchestra and a row of cardioids along the front of the stage for the singers. In its best position, the main pair may obscure sight-lines. If it is raised, close microphones are required in the orchestra pit.

For the stage itself there are several further possibilities. One has a co-incident pair of cardioids at the centre of the footlights, with a spaced pair to the sides, filling in the downstage corners. By mounting them on a bracket, so that the diaphragm is low and close to the edge of the stage, the effect of sound reflections from the stage itself is minimized. The main pair also picks up some (though relatively little) of the orchestra.

A similar arrangement employs directional boundary microphones or mice on the stage itself. Again, there may be a coincident pair at the centre and single microphones at the corner, but on a wide stage three separate pairs can produce good results. The width of each pair is reduced, with the inner microphone of each outer pair steered to the same position as the corresponding element of the central pair. In practice, this gives smooth coverage of movement across the stage. With a central prompt-box, two pairs might serve. Note, however, that if pressure-zone microphones or mice are placed on a flexible rather than solid surface, their directional response may change in ways not predicted by simple theory. In particular, they might pick up more orchestra, nominally in the dead sector of the cardioid, than expected.

A deficiency of any footlight microphones is that they favour footsteps on the stage. If the feet approach any particular microphone too heavily, take that one out and rebalance on a microphone to one side. Check the effect on the stereo picture: it may be acceptable, or it may seem better to pan the remaining microphone to restore the image to its former position.

Important action in an upstage area that is not covered adequately on the downstage microphones can perhaps be taken on a 'drop' microphone (that is, a microphone suspended by its own cable). Take care not to overdo its use, bringing intentionally distant scenes too far forward. But in general, balance with an ear to the clarity of vocal line – even when, on television, subtitles are added.

Studio opera layout with chorus and soloists reversed left to right and superimposed on stage

A converted theatre is used in this example. 1, Proscenium arch. The stage is extended into the main auditorium. 2, Soloists. 3, Chorus. 4, Conductor. 5, Announcer. This microphone can also be used for spotting effects (e.g. castanets, which can be steered to represent the movement of a dance). 6, An off-stage chorus can be placed underneath the balcony without a microphone. 7, Audience above. Orchestral layout: V1, 1st violins. V2, 2nd violins. VA, Violas. C, Cellos. D, Basses. H, Horns (with reflecting screens behind). F, Flutes. O, Oboes. C, Clarinets. B, Bassoons. T1, Trumpets. T2, Trombones. P1, Timpani. P2, Other percussion. Hp, Harp. Note that the harp is 'in' or near the right-hand speaker, and that the left-hand side of the orchestra and the side of the chorus and soloists nearest to it will be picked up on each others' microphones and *must* therefore be arranged to coincide in the audio stage. For the singers, examples of positioning for particular scenes using this layout are as follows: (i) Part of chorus left: principals centre; off-stage chorus, right. (ii) Duet at centre solo microphones; harp figures, right. (iii) Quartet at four centre solo microphones.

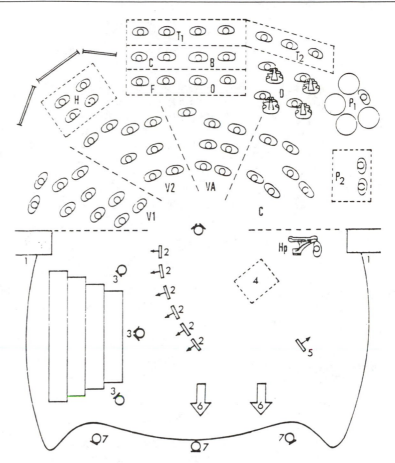

Classical music in vision

For classical music, a television picture makes only a little difference to the type of balance that is sought. But getting a good balance of any type is a little more difficult, because:

● television studios are generally too dead: concert halls (or even music sound studios) are therefore to be preferred in cases where they can conveniently be used
● microphones must not be obtrusive
● the layout must be visually satisfying, but a good rule is to adopt a layout that is as close as possible to that of a concert performance.

What is *not* needed is any attempt to make sound follow picture: where the camera zooms in to a close-up of the oboe, the sound remains at the concert perspective (note that this is the exact reverse of what may be done with speech). The big close-up in sound would be permitted only for a special effect; if, for instance, someone were talking about the role of the oboe in the ensemble, or alternatively, if

Ballet: playback and effects
Dancers need the music reasonably close, so a loudspeaker may be necessary. Here a series of loudspeakers set in a vertical line radiate high frequencies in a horizontal field; a microphone above them receives only low-pitched sound from this array. This can be filtered out, as the main component of the required effects is at higher frequencies.

this were a drama and we were hearing, as we were seeing, from a position within the orchestra. Even in these cases the overall orchestral sound would be distorted as little as possible.

One slight difference in balance is desirable for television: this is marginally greater clarity at the expense of the tonal blending that many people prefer for sound radio and records. The extra presence permits individual instruments or groups to be heard just clearly enough for close shots to be aurally justified. This does not necessarily mean shorter reverberation, but simply that the attack on each note must be heard with reasonable clarity: this is usually enough to identify individual instruments. A technique that can be used successfully to produce this effect is to mix the outputs of two microphones, one closer and one more distant than would be used for a single microphone balance.

A balance with this extra degree of clarity also takes into account the different listening conditions that may exist in the home: the sound quality of loudspeakers on the majority of television receivers is rather lower than that used for radio and records by many people who enjoy listening to music; a more transparent balance serves this better.

For *ballet* performed in the television studio it is often difficult to get both the dance settings and the orchestra into the same studio: this may be a blessing in disguise, however, as the solution is to take the orchestra away to a music studio (with their own remote cameras, if required). The music can then be replayed to the dancers over loudspeakers.

Limitations (due to cost) on orchestral rehearsal time means that orchestra and dancers really need to be rehearsed separately; if this also means that the orchestra can be pre-recorded the tempi are then set, and the dancers know from then on precisely what to expect. Ideally, of course, the principal dancers will have attended the recording. Piano transcriptions may have to be used for rehearsal, however, and it may be a responsibility of the sound department to make and replay a recording of this too. The dancers must have clear audio replay at all places where the action takes them, so the action studio must be liberally supplied with suitable loudspeakers. In some cases in ballet the dancers start before the music; where this happens it is up to the tape operator to be as precise and accurate as an orchestral conductor in timing the start of the music to match the action.

Effects microphones – usually gun microphones or hypercardioids angled to discriminate against loudspeakers – are needed in the studio to add just sufficient noise of movement (the dancers' steps, rustle of costume, etc.) to add conviction to the pictures of the dance.

For *singers*, the type of balance depends on whether a microphone may appear in vision; if not, a boom will probably be used. This may introduce separation problems: orchestral spill on to the singer's microphone can modify the balance to a noticeable degree. Electrostatic cardioid microphones are preferred.

One marginal advantage of the relative deadness of television general-purpose studios is that when artificial reverberation is used it can be added differentially – not simply by adding it in different proportions, but also by using different reverberation times in different parts of the sound: in general, singers require less than the orchestral music that may accompany them.

An important part of any singer's performance lies in his use of vocal dynamics. For broadcast sound some degree of compression may be necessary; in television the problem is made more complex by movement of both the singer and the microphone and by variation in the distance between singer and microphone to accommodate the size of shot. In these circumstances even more of the responsibility for interpretation of the artist's role may be given to the sound balancer, who must therefore think not only about problems of coverage but also – along with the director and the singer – about the overall artistic intention.

Where singers and accompaniment have to be physically separated, time-lag problems may hinder the performance, and foldback loudspeakers are needed to relay sound to the singer; the conductor, too, needs to hear how the singer's contribution fits into the ensemble.

For wide shots pre-recordings may have to be used: timing this to work in with live action requires careful planning and rehearsal. A spare track on the recorder can be used for cues that are fed to performers but not to line. Remember, always, that foldback loudspeakers may be safer when placed off the rear axis of a directional microphone.

Televised opera

There are several different ways of presenting opera on television:

● from the stage
● with actors miming to pre-recorded music (the actors may include some of the singers themselves)
● with singers in a television studio and orchestra in a sound studio
● with both singers and orchestra in some hall of suitable (large) size with good musical acoustics

For stage performances use the balance already described. If possible use a separate sound control vehicle which has good loudspeakers, a reasonable volume of space and suitable acoustic treatment.

Miming opera is a technique that has in the past been extensively used in Europe, but in Britain mime has been used less, in the belief that the singer's own performance has greater musical conviction. A successful arrangement employs two studios. That for the orchestra has good musical acoustics; the action studio has all of the normal facilities to cover sound and pictures – for which the original grand, and deliberately unrealistic, style of stage opera is generally modified

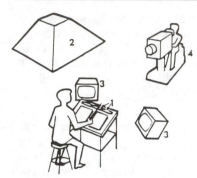

Opera or ballet in the television studio
Layout for conductor with orchestra in separate sound studio. 1, Line loudspeaker along top of music stand. 2, Hood loudspeaker over conductor's head. 3, Monitor for conductor to see action. 4, Camera to relay beat to repetiteur in television studio.

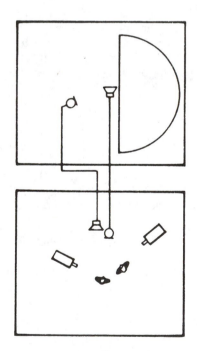

Studio communications
Links between a television studio and (in this case) orchestral sound. The singers' voices are fed back to the conductor.

to that of music drama. The singers are therefore covered by boom (and, where necessary, other microphones) in a way similar to that for a play. The conductor has a monitor for the action, and a camera relays his beat back to the action studio where a repetiteur takes it from another monitor and gives it to the singers. A similar arrangement is used in nearly every opera house.

Sound communication is a little more complicated. The orchestral sound is relayed to the action studio floor by using directional loudspeakers mounted on the front of the boom dollies. As the booms themselves are moved around to follow the action, so do the upright line-source loudspeakers that are used with them. As we have seen, a line source is directional at middle and high frequencies, radiating primarily in a plane intersecting the line at right-angles, so that good separation can be achieved. Relaying the voices of the singers to the conductor has been done in several ways: one employs a hood (a large pyramid-shaped baffle) with a loudspeaker in the apex and above the conductor's head. But if the conductor is to appear in vision this restricts the television lighting. The alternative is to give the conductor a line-source loudspeaker along the top edge of his music stand. Three very small elliptical speakers set in a line produce little bass and are directional at high frequencies, helping separation.

Even using this technique, it may be necessary to pre-record or film some scenes in advance. To employ an orchestra for such a short sequence may be disproportionately expensive, so a system of guide tracks may be used. Here the pre-recorded (or filmed) scenes are accompanied by a sound track consisting of bar counts or piano version, plus any verbal instructions such as warnings of impending changes of tempo. This guide is fed to the conductor only (by headphones in this case). Orchestral sound is then added at the main recording.

If singers' voices are included in the pre-recording this can itself be used as guide track or, better still, twin or multitrack sound can be used with a track for each purpose. On videotape, a separate audio track can be used. Where singers' voices are pre-recorded in sound only (so that miming can be used, perhaps in order to cover distant action), a guide track can again be accommodated on a multitrack recorder.

The fourth arrangement, with orchestra and singers at opposing ends of the same concert hall, employs many of the same communications systems: for example, the time lag between the ends of the hall still requires the use of the conductor's music-stand loudspeakers. Separate rehearsal, one of the advantages of twin studios, is lost; but in its place is an enhanced unity of purpose.

Popular music groups

Having worked up to some of the largest and most complex layouts for classical music, we can now take some of the same – or similar –

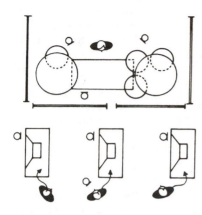

Pop group
Three electric guitars, drums and vocalist. For maximum separation the drums may have two low screens in front, high screens at the sides, and a further screen bridged across the top. For stereo, microphones are steered to suitable positions on the audio stage, and blended and spread artificial reverberation is added.

components and put them together again in a contrasting style: pop music.

A simple pop group may consist of drums, three guitars (lead, bass and acoustic), vocalist and perhaps a synthesizer. In the dead acoustics of a big recording studio it is usual to place the various components in booths around the walls, with tall sound-absorbent screens on either side of the drum-kit (unless it is in a corner) and half screens in front. Lead and bass guitar loudspeakers can be directed in to half-screened bays, and picked up on cardioid microphones, but an acoustic guitar may need a separate room. The synthesizer is angled to radiate any acoustic output into an opposite wall, and the amount of screening depends on whether there are any further quieter instruments to be balanced; if so, heavy screening may be required. The guide vocalist is kept physically well away from all of these but in good visual contact with the other players. Considerable use is made of tracking (sequential recording) for vocals, vocal backing, additional acoustic guitar and percussion and other effects.

Depending on the number of tracks available, there may be partial mixing as the session progresses, with some components 'bounced' from one track to another to leave space free for further material to be recorded. Some parts such as the main drum coverage, keyboard and vocal backing may appear on stereo pairs of tracks; the rest will be mono. The first and last tracks (those near the edges of the tape) are used for less important parts of the recording or for time-code.

The session may start by recording the basic rhythm plus guide vocal, followed by additional acoustic guitar and the synthesizer. Next the lead vocals are laid down, then the backing (as separate stages if complexity warrants it). Finally, the solo guitar, electric guitar riffs, etc., are overdubbed. Alternatively, all or most of the tracks may be recorded at the same time. The desk will have two faders for each channel, one of which may carry the full audio signal, which is then given nominal (but not irreversible) treatment on its way to the multitrack recorder via the buttons at the top of the strip. The second fader, below the first, controls a separate feed that is used to monitor a trial balance, and this may also be recorded for later reference. In a fully digital desk, the faders and all controls mimic their audio counterparts, but the signal is not fed directly through them.

Reduction, in a separate session, begins with the return of the tracks to the desk channels, checking and refining the treatment, starting with rhythm as before, and gradually building up a stereo composite from the separate components. A built-in computer may be used to record and update the control data, and then to make the final mix.

For radio, where less time and money are available, the process is often condensed, with the multitrack recordings used mainly for cosmetic changes, to correct errors or for the greater freedom to experiment that is allowed by the possibility of second thoughts.

The multitrack recorder can be used in this way for classical music too. To regain a sense of occasion and the feeling of a real, live performance, recordings may be made from concerts or road shows. A rock concert can itself involve a great deal of electronic rebalancing, even before any recording is contemplated. A group of three or four players may have some 30 channels linked to its own travelling mixer desk feeding the stage loudspeaker system: indeed 100 channels is not unknown, with a separate desk for the drummer alone. Studio-style main desks send feeds to audience loudspeakers while recording on multitrack tape.

Otherwise, to record from a concert, it is possible to take a feed from each channel by means of a passive splitter near the microphone. This is a low-level feed via a hybrid transformer. The increase in signal strength available in an active splitter (i.e. using an amplifier) is not needed and might even result in distortion, as the volume is so high to start with. Depending on the number of tracks available on the recording tape, some of the channels may have to be combined at the time of the performance – for example, the drums could be reduced to four channels, with all but the snare and bass mixed down to a stereo pair. The recordings are taken back to the studio and reduced in the normal way. Records produced from concerts are sometimes proudly marked 'no overdubbings'.

When recording on location, whether from a pop or classical concert, time is limited and it is sensible to establish a routine, setting microphones or taking split feeds as early as possible, then laying the cables back to the mixer. Microphones are identified by scratching them and indicating what instruments they cover. Radio microphones may be used for direct communication between stage and mixer. The recording arrangements are set up and checked out last. A mobile sound control room is sometimes used.

Light music and jazz

A basic component of much popular light music and jazz is the piano trio, i.e. piano, bass and drums, which forms the nucleus of most larger groups. To this may be added, stage by stage, guitar (acoustic/ electric), brass, woodwind (primarily saxophones), singer and other instruments for particular effects. These might include celesta, violin, flute, horn, harp, electric organ or any other instrument that interests the arranger. Once again, mono balances are normally used for most of them, but stereo pairs might be tried on trumpets, trombones and woodwind as well as piano and vocal groups. Here, too, tracking is often used, but economics may demand that a substantial amount of music be recorded in a single session. Or there may be limitations on the number of microphones, the mixing or recording equipment or in the studio acoustics. Despite such deficiencies it is often possible to achieve a balance that makes satisfying listening – even if not to the pop-music balancer's taste.

Normally the studio is dead, but sometimes only bright surroundings are available. In this case some spill may be inevitable, and it might be best to modify the ideal fully separated balance. Even in good

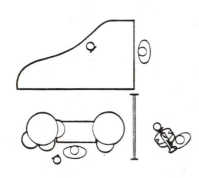

Rhythm group
The drum-kit and bass are separated by a low acoustic screen. A range of positions is possible for each microphone.

Small band
Rhythm group with: 1, Guitar (acoustic/electric). 2, Trumpet. 3, Trombone. 4, Woodwind (saxophones). Rhythm group microphones not shown.

conditions there may be enough spill in the studio to dictate some elements of the layout of the stereo sound picture. However, certain relationships will appear in both the real layout and its image: for example, the piano, bass and drums will always be close together. These might be steered to centre and left of centre with electric/acoustic guitar near the left speaker, singer and woodwind right centre, and brass on the right. The actual layout in the studio is governed not only by the need for separation (so that this or some other stereo picture can be achieved, and so that all the sounds can be individually treated), but also by a number of practical considerations – such as whether the players can hear or see each other.

Multimicrophone layout

When the musicians arrive in the studio they expect to sit at positions already arranged for them, with chairs, music stands, screens, piano, boxes for such things as guitar amplifiers and microphones all (at least roughly) laid out. They will have been set according to the balancer's instructions; and in the case of the microphones the balancer or an assistant will have actually placed them. Three different situations may occur:

● The group may already have a layout used for stage shows. The balancer finds out in advance what this is and varies it as little as possible (but as much as may be necessary for good separation) with the cooperation of the musical director (if there is one).
● The group may have made previous recordings or broadcasts with the same organization. The balancer finds out what layout was used before, and if it worked well uses it as a basis for his own. The group will not expect to be in different positions every time they come to the studio. However, minor changes for the purpose of experiment, or to include ideas suggested by a producer or which the balancer is convinced work better than those already used, may be tried – with the cooperation of the individual instrumentalist, leader of a section or musical director, as appropriate. Major changes may be dictated by the use of additional instruments (of which the balancer should have been informed), or new problems created by the music itself (which will become apparent in rehearsal).
● The group may be created for this session only, in which case the balancer – having found out the composition from the producer or musical director – can start from scratch.

Even starting afresh, the balancer can apply certain rules that make the musician's job easier. For example:

● The bass player may wish to see the pianist's left hand, to which his music directly relates. This is important if there may be improvisations.
● The drummer, bass player and pianist form a basic group (the rhythm group) who expect to be together.

Show band with strings

1, Rhythm group. 2, Electric/acoustic guitar. 3, Trumpets. 4, Trombones, 5, Woodwind. 6, 1st violins. 7, 2nd violins. 8, Violas. 9, Cellos. 10, Vocalist. 11, Musical director. Variations must be made for acoustic difficulties: in one regularly used BBC layout it was found necessary to bring the bass forward to the tail of the piano and separate him by acoustic screens from brass, drums and woodwind (the trumpets having been moved further away). Additional forces also complicate the layout still further. For example, the pianist may need an additional keyboard for organ or celesta: this could be placed to his right and the guitar moved forward. A wide range of different microphones is used: see the notes on individual instruments for details of type (including frequency and polar response) and position (distance, height and angle).

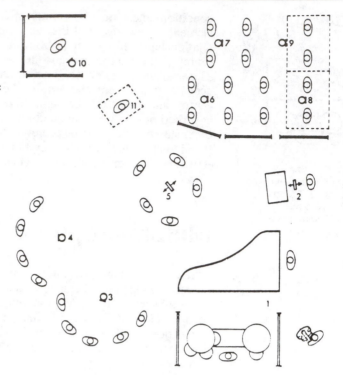

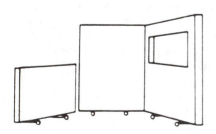

Studio screens

A dead music studio has a supply of full and low screens with absorbers about 4 in (10 cm) thick that are easily moved on rollers to provide acoustic separation without loss of visual contact. Vocalist's screen may have a window of stiff transparent plastics.

- In a piano quartet the fourth player can face the other three in the well of the piano; everybody can see everybody else. But a guitar player may be working particularly with the bass, so could be placed at the top end of the piano keyboard in order to see across the pianist to the bass player.
- When the bass player has fast rhythmic figures he needs to hear the drums, particularly the snare drum and hi-hat; these can be heard well on the drummer's left.
- Any other two players who are likely to be working together on a melodic line or other figures need to be close together.
- Where players double, e.g. piano with celesta, the two positions must be together.
- All players need to see the musical director or conductor.
- The musical director may also be playing an instrument.

This may appear to make the original idea of building up a layout according to good principles of separation less feasible. But although compromises may be necessary, the use of directional microphones and screens should make a good sound possible. A low screen is needed between bass and drums, and another to separate drums and brass. Other sections, and especially quiet instruments, such as strings, celesta, etc., will also require screens. A singer may be provided with a tent of screens well away from the players, but should still be able to see them.

Popular music in vision

Plainly the ideal recording-studio layout for pop music cannot be adopted in vision – though a band wholly or largely out of vision certainly can, and probably will use it. But apart from layout – which may produce some problems of separation – all other elements of the balance and treatment will be as close as possible to that of the recording studio, which after all, provides the standard by which television sound will be judged.

Show band in vision to support singer or featured acts
Typical layout: V, Violins. Va, Violas. C, Cellos. DB, Double bass. G, Acoustic guitar. T, Timpani. P, Percussion. D, Drum-kit. H, Horns. Tp, Trumpets. Tn, Trombones. Ac, Accordion. S, Saxophones, doubling woodwind. W, Overall wind microphone. All microphones may be electrostatic cardioids except for a moving-coil on bass drum.

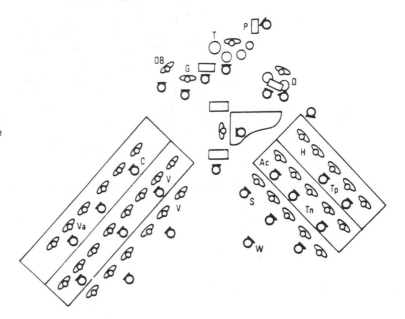

Low microphones in among the instruments are generally not obtrusive; indeed, they are readily accepted as part of the pop paraphernalia. Microphones at about or just above head height (particularly if they would be between the cameras and the front line of players) are avoided if possible; as are large microphone booms and stands. Indeed, there are more differences in mountings than in positioning.

The pop singer's hand microphone is accepted as a prop and not merely something to sing into – though some experienced singers also use it to balance themselves with skill. For the less mobile singer a stand microphone is an accepted convention.

For any popular music television programme the control desk is likely to have most of the equalization, compression and artificial reverberation facilities of the recording studio. It may be necessary to pre-record some tracks even in a live performance if a group is to match its expected sound. Miming to playback of records is cheap and easy – though some would call it cheap and nasty: the sense of occasion is prejudiced, and the audience may feel cheated. Such devices should be kept to a minimum.

Television has to a large extent adapted its pictures to the sound needs of popular music, rather than the reverse; nevertheless, occasions still occur where a singer must appear without a microphone in vision – so booms are still used, but perhaps with a high quality gun microphone to improve separation. Only where separation cannot be achieved by these means, or in complex song and dance acts should the music be pre-recorded and vision shot to playback of sound.

Most of what has gone before applies also to music on film (including film for television); though generally there is a somewhat greater effort to eliminate microphones from the picture.

Although multiple-camera techniques have, at times, been used, close-ups are normally shot separately, taking sound, but only as a guide to synchronization with a master recording. More is shot than is actually required, so that actions before and after the cuts will match. Occasionally film shot in sync for one sequence is eventually synchronized to the sound from another part of the work. Here the beat or action (e.g. drum beat) is marked on the track and matched to a close-up of the action.

Chapter 9

Monitoring and control

Monitoring means using ears, and therefore loudspeakers (or, at a pinch, headphones), to make judgements on sound; *control* is the use of hands to manipulate that sound, mainly in a mixer desk. This chapter describes loudspeaker layouts; the path of the signal through typical mixer desks; and touches on questions concerning sound quality, without which the more creative modification of sound (discussed in the following chapters) is wasted.

During rehearsal, recording or broadcast, the ear should be constantly searching out the faults that will mar a production. Here are just a few:

● *production faults*: e.g. miscast voices, stilted speaking of the lines, uncolloquial scripts, bad timing
● *faulty technique*: e.g. poor balance and control, untidy fades and mixes, misuse of acoustics
● *poor sound quality*: e.g. distortion, resonances, and other irregularities in the frequency response, lack of bass and top
● *faults of the equipment or recordings*: wow and flutter, hum, excessive tape hiss, noises due to loose connections, poor screening, or mis-matching.

Of these, the faults in the first group are certainly the most important. They are also, perhaps, the easiest to spot. As for techniques: the ability to see faults there comes with the practice of the techniques. But the various things that can be wrong with the quality of a sound – objective, measurable things – appear, somewhat surprisingly, to be the most difficult for the ear to judge in other than subjective 'I-know-what-I-like' terms.

Quality and the ear

The quality of sound that the ear will accept – and prefer – depends on what the listener is used to. Few whose ears are untrained can judge quality objectively. Given a choice between qualities, listeners are generally prepared to state a preference, but experiments in the USA have indicated that as many will prefer medium quality as prefer the best.

In a less systematic demonstration of this, a manufacturer of loudspeaker enclosures invited members of the public to judge and compare the quality of three different systems, without seeing which was playing or being told the prices. To a trained ear there could be little doubt about the order. A large number of people, including many audio enthusiasts, took part in the tests. Their votes split evenly between the three systems.

It seems that even among those who claim to like hi-fi (a term which seems to cover any sound equipment from medium quality upwards) the real preference is often for a box that sounds like a box; they find the sort of quality that turns their living-room into one corner of a great concert hall to be somewhat disturbing. Their preferences must be respected; but it must nevertheless be true that quality and enjoyment are often closely related. The high-quality performance of a musical work gives more *information* to a listener who is new to it; and it is often a technically superior recording that first 'sells' music content, and interests a listener in something new.

The difference between a good loudspeaker and a poor one boils down to this: that the good one allows more of the original conception of the sound to reach the listener. Plainly, for technical monitoring a high-quality speaker is essential.

High-quality sound is not – as with, say, wide-screen film – an advance that imposes a new set of limitations. Different degrees of quality may be acceptable to the same person when applied to different programme ends. For example, high quality may be best for brilliant modern music (or for any orchestral music heard as a concert); medium or low quality can be satisfactory for ordinary speech, where the primary reason for listening is usually to hear what is said; and 'mellow' quality is often felt to be less obtrusive when music is required for a companionable background noise. On high-quality equipment these different types of programme are all available at the appropriate levels of quality; on poor equipment they are all debased to the same indifferent standard.

So it is undoubtedly worth persevering with the education of one's ears, and allowing other people the opportunity of listening to sound that is of rather better quality than they are used to.

Nevertheless, any organization that is conscious of cost must decide what is to be the practicable economic upper limit to the audio-frequency range to be offered. Tests were carried out by the BBC in which the audio-frequency range was restricted to 7 kHz, 10 kHz and 12 kHz. Sounds heard within the test material included tubular bells,

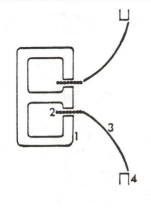

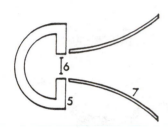

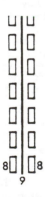

Loudspeaker types
Above: Moving-coil. The most commonly used for all purposes (bass, treble and wide range). 1, Permanent magnet. 2, Moving coil. 3, Cone. 4, Baffle or enclosure. *Centre:* Ribbon unit. Sometimes used for treble. 5, Permanent magnet. 6, Ribbon (seen in plan view). 7, Acoustic horn. *Below:* Electrostatic. 8, Perforated fixed electrodes. 9, Diaphragm.

brass, cymbals and triangle, snare and military drums, maraccas, handclaps, and female speech. The listeners were all under 40 years of age, both male and female, and included a number of people who were experienced in judging sound quality. The results of the tests suggested that:

- only a few critical observers hear restriction to 12 kHz
- many inexperienced listeners can detect restriction to 10 kHz; experienced listeners do so readily
- a filter at 7 kHz was detected even by the inexperienced.

In fact the BBC, like many other broadcasters, transmits up to 15 kHz on VHF/FM and NICAM television stereo. For digital audio used in broadcasting, the sampling rate (which must be more than twice the highest audio frequency) is 32 kHz. In the recording industry, which can offer products (CD and digital tapes) with a 20 kHz frequency range, the sampling rate is 44 or 48 kHz.

Loudspeakers

The main obstacle to good sound has nearly always been the loudspeaker. It is, in principle, fairly easy to get good middle range and high-frequency response, but efficient production of bass requires either a fairly large surface area to push the air, or a horn with an equally large aperture.

One apparent alternative to this is to have a relatively small area driving the air and to boost the bass so that the inefficiency of the coupling is compensated in advance. But there is a limit to this, as the excursions of the diaphragm get so big that it becomes a complication in itself, as does the extra power that is required from the amplifier.

Most loudspeakers use a stiff diaphragm working as a piston. It is operated by an electromechanical device that is usually very similar in principle to those used in microphones – except, of course, that it converts an electrical signal back into physical movement. The most common type is the moving-coil, but electrostatic, ribbon and other types are sometimes used.

It is not easy for a single diaphragm to respond equally well to all frequencies: usually the signal is split and fed to separate loudspeakers covering two or three different parts of the frequency range. When the sound is divided up in this way there is no need for a sharp cut-off as one loudspeaker takes over from another, so long as the acoustic output adds up to a full signal. But careful relative positioning of the two units is necessary, so that there is no phase cancellation at the switch-over frequencies.

For the low-frequency unit a cone of paper may be used as the diaphragm. This has to be corrugated concentrically in order to stop unwanted harmonics from being formed (these may include a subharmonic, below the main frequency at which the cone is being

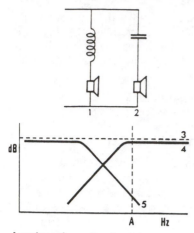

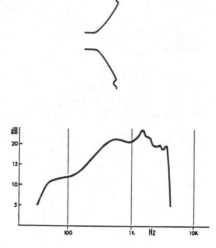

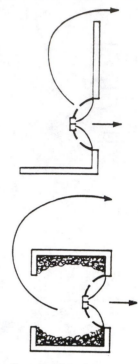

Loudspeaker: simple cross-over network

1, Bass unit. 2, Treble unit. 3, The total power is divided evenly between the two units 4 and 5. For each it falls off at 6 dB/octave beyond the cross-over point. This is a very satisfactory type of network provided that there are no severe irregularities in the response of the cone itself that are close enough to the cross-over point to make a significant contribution even though attenuated. For example, a very severe fluctuation in the bass cone output at A will still affect the overall response: the answer is to make the cross-over point lower (if the response of the high-frequency unit permits) or to use a more complex dividing network with a steeper cut-off at the cross-over point.

Loudspeaker cone, old BBC design

In its year (1967), this was claimed to be better than any other commercially available cone of equal size (12 in, 30 cm) or larger used in a bass unit (with cross-over at 1600 Hz). The material used was a polystyrene/synthetic rubber compound with a PVC surround.

Loudspeaker housings

Above: Baffle. *Below:* Enclosure with opening at rear. Both reduce the frequency at which sound radiated from the back of a cone can recombine with and cancel the direct sound. A box may have internal resonances which need to be damped by lining it with absorbent material.

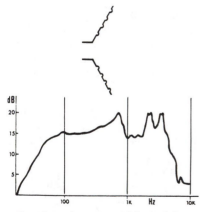

Loudspeaker cone with straight sides of corrugated paper felt

The performance diagram shows axial frequency response with output in dB (arbitrary scale), taken from a particular example.

driven). Cones or pistons of other materials are also used. They must be stiff but have adequate internal damping or they may ring like a wineglass at certain frequencies. The cone, of whatever construction, must also be of low mass or its inertia may affect the response.

A problem with a cone – or any other diaphragm – is that as it radiates the wanted signal from the front, a second signal, in the opposite phase, is being radiated from the back; and if there is nothing to stop it, some of this back-radiation leaks round the sides and joins the sound from the front. The long-wavelength low frequencies are most affected by this: as the two signals combine, they are still partly out of phase and therefore tend to cancel a signal that is already weaker than it should be because the cone is too small for effective low-frequency work.

There are several partial cures for this. One is to house the loudspeaker in a baffle of relatively large area. The size determines

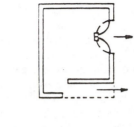

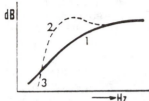

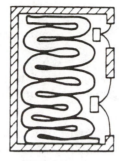

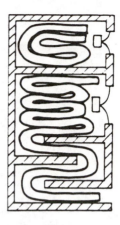

Bass reflex loudspeaker
This uses the internal cavity and a port at the front as a Helmholtz resonator with a low-frequency resonance peak. Added to the normal bass response of the loudspeaker (1) it can be used to lift the bass for about one octave (2). Below this (3), the cut-off is more severe.

Infinite baffle enclosure
The sound from the rear of the speaker is trapped and much is absorbed. In practice, the stiffness of the air within the box somewhat restricts the movement of the speaker. The cavity contains folded sound absorbent material.

Complex loudspeaker enclosure
The tweeter (h.f. unit) has an infinite baffle and the woofer (l.f. unit) backs on to an acoustic labyrinth (also called transmission line or pipe). This introduces both phase change and attenuation. Again, both cavities are padded.

the limiting wavelength, below which cancellation still occurs: a baffle some 30 in (75 cm) square is about the minimum for a moderately good low-frequency response. A second technique is to turn the space behind the cone into a low-frequency resonator with a port that radiates some of the power. It can be arranged that by phase change this boosts the response in the region in which it previously fell off. But this bass reflex loudspeaker (as it is called) is bulky, may be subject to cabinet coloration in the bass, and below its resonant frequency cuts off even more sharply than before. It has a characteristically recognizable quality – which is hardly a recommendation for any loudspeaker. Yet another idea is to try to lose the back radiation completely. Simply feeding it out into another room would be the best way, but as there are usually objections to this the next best answer is generally adopted: to stifle it within the box in such a way that the diaphragm is not affected by reflections or by the 'stiffness' of the enclosed air.

The design of high-quality loudspeakers is a problem to which there have been many ingenious part-solutions attempting, in particular, to get round the difficulty of the sheer size that is naturally associated with the longer wavelengths of sound. This is a big subject, and beyond the scope of this book.

Monitoring layout

The best distance to place a monitoring loudspeaker is probably 3–7 ft (1–2 m). Closer than this the sound has a somewhat unreal quality owing to standing waves close to the cabinet; at a greater distance the acoustics of the listening room begin to take over. It is desirable that these acoustics should be similar to those of a domestic living-room. But it is important to produce a well-blended signal and to know whether some minor fault is due to the studio or listening room; and the only way to be sure would be to sit rather closer to the loudspeaker than when listening purely for pleasure at home. This is, however, possible only for mono.

For monitoring stereo the listener should be at the apex of an equilateral triangle, or perhaps a little closer. The speakers will be about 8 ft (2.5 m) apart and the listener therefore 6 ft or more from them. The producer sits immediately behind the balancer. In a studio layout the ideal arrangement is one in which the balancer and producer do not see directly into the studio as they face the sound stage, but can do so, to sort out positioning or other problems, by simply turning round. Because the volume from these loudspeakers is usually set very high, there is an attenuator of some 20 dB which can be cut in (by a '*dim switch*') when using telephones, for example; when using talkback it automatically switches in.

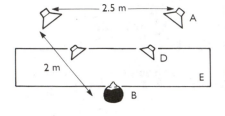

Monitoring stereo
A, High-quality loudspeakers.
B, Balancer. C, Producer, also equidistant from loudspeaker.
D, 'Mini-speakers' offering a distinctively different lower quality.
E, Desk.

An additional pair of small loudspeakers is mounted closer and in line with the main monitoring loudspeakers. These '*mini-speakers*' (acting also as *near-field loudspeakers*), set at lower volume, serve as a check on what many of the audience will actually hear: all vital elements of the high-quality balance must still be audible when output is switched to them. They are also used for talkback and for listening to channels that have not been faded up (*pre-hear* or *prefade-listen*, PFL). *Afterfade-listen* (AFL) takes its feed later, as its name implies. Switching over to these small speakers does not affect output. On some desks the PFL for any channel can be switched in by pressing down on the fader knob when it is fully faded out – this is called *overpressing*.

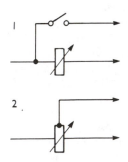

Prefade-listen (PFL, prehear)
1, Output is taken before channel fader. 2, Symbol for 'overpress' facility: when channel is faded out, pressing the fader knob routes signal to prefade-listen. This can then be heard on near-field or studio loudspeakers as selected, or on headphones. A variety of other options are often available: for simplicity, these are not shown in the above circuit diagrams.

Most desks have a wide range of further switches that are associated with the two loudspeaker systems. '*Solo*' selects just the channel in which the switch is located and feeds it to a main loudspeaker; '*cut*' takes out just that one channel, to hear what the mix is like without it. On some desks it is possible to solo one channel while retaining another channel or group for background or comparison: for this, press '*solo isolate*' on the second fader at the same time. Some desks also allow '*solo in place*' – that is, in its stereo position. If these operate before the fader, the individual balance can be checked; if after, that fader's contribution to the overall balance is assessed. If operated during rehearsal, these switches do affect studio output, so they are linked to the transmission or recording condition that is set up in the desk when the *red light* is switched on: solos, etc., will still be heard, but on the small loudspeakers, as for PFL. There will also be a headphone feed.

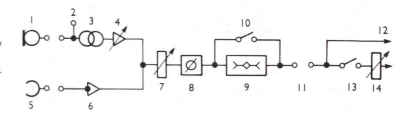

Mono channel

Some of the facilities that may be found before the channel fader. 1, Microphone (low level) input. 2, 48 V phantom power for electrostatic microphone. 3, Isolating transformer. 4, Microphone channel amplifier. 5, Line (high level) input: in this case, tape replay. 6, Buffer. 7, Fine preset gain (typically ± 15 dB). 8, Phase reverse control. 9, Equalizer. 10, Equalizer bypass. 11, Optional 'insert' routing: allows external module to modify channel signal. 12, Prefade-listen. 13, Remote cut, or 'cough key'. 14, Channel fader.

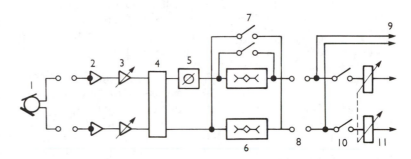

Stereo channel

Some facilities before channel fader. 1, Stereo microphone. 2, Buffers. 3, Coarse channel amplifiers. 4, Channel input switching (to mono). 5, Phase reverse in left leg only. 6, Equalizer. 7, Equalizer bypass. 8, Optional 'insert' for plug-in modules (e.g. graphic filter). 9, Prefade-listen. 10, Remote cut. 11, Ganged channel faders. Post-fader stereo controls are described later.

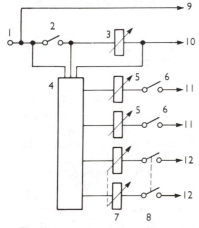

Flexible auxiliary send switching

This is used for public address, studio feedback, 'echo' send and feeds to remote locations which require a mix that is different from the main studio output. 1, Feed from channel (or group) input. 2, Remote cut. 3, Channel (or group) fader. 4, Auxiliary send switching. 5, Mono send output fader. 6, Send output switch. 7, Stereo ganged output fader. 8, Stereo ganged output switch. 9, Prefade-listen. 10, Channel out to group bus or main control. 11, Output to mono send buses. 12, Output to stereo send buses.

Control desk: channels

Sound control desks (otherwise called mixer desks, panels or consoles) range in complexity from the simplest that are used in radio for speech and record programmes to those used for pop music recording which may have a vast number of channels, each with the facility for a range of treatments and two separate main mixing systems so that one can be used to monitor (or broadcast), as the music is being played, while the other feeds a multitrack recorder, for subsequent remixing (or *mixdown*).

The size of desk reflects the maximum demands expected of the studio it is used in. In a large broadcasting centre, radio studios tend to be more specialized than those used for television. In any case, in television, the studio itself represents a large capital outlay and will often be used for a greater range of different kinds of programme; besides which, the demands of vision add to the complexity of sound coverage. As a result, television desks tend to be more complex than their radio counterparts. But, as noted earlier, whatever their size, they must all be designed for use by a single pair of hands. Much will be preset; other adjustments are made one or two at a time, and continuous control will be exercised only over a narrow range near the centre of the desk.

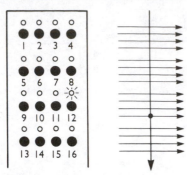

Routing matrix
At the top of each channel strip on a desk designed to be linked to a multichannel recorder, the mass of press-buttons looks more complex than it is. Press any button to send channel output to selected track.

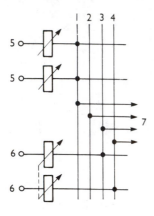

Groups
1-4, bus bars, actually wires that cross all channels. In this example two mono channels, 5, are both connected to bus 1; while the stereo channel, 6, is connected to 3 and 4, and nothing is fed to bus 2. The output of each bus (or stereo pair) is fed via group controls to a main control. Auxiliary sends also have their own buses and output controls. 7, Group outputs.

In some desks, more complex control can be exercised by computer: this can be used to set the desk to some preferred initial state, or to one of many that have been worked out stage by stage. In the most extreme case, much of the action can be reproduced, as though by many hands working with inhuman speed and precision – but still allowing manual over-ride (or minor 'trim' changes for evaluation), in order to create yet another generation of operational control.

The layout of most desks has many common features. Below are mentioned some of the things to look out for in the individual channels.

Even in a full stereo desk, many of the channels will be mono, for, as we have seen, a stereo mix often contains mono components. Mono channels may be linked (or 'ganged') as stereo microphone pairs: a clip is attached to the two faders, so they are never out of step by more than the fader manufacturer's 'tolerance': the difference should never be greater than 2 dB.

There are two distinct kinds of input. One is from the microphone: this will start at the point where the microphone feed is plugged into the desk, typically at a three-pin female XLR socket. There will often be provision in each microphone channel on the desk for a 48 V d.c. supply to be fed back to an electrostatic microphone, using the same three-conductor system that carries the signal (the phantom power supply). On some desks there is also remote polar response switching for several microphones, but this is not built into the channel strip. The incoming signal passes through an isolating transformer which has a coarse rotary gain control which may be preset to lift its level to close to the standard (*zero level*) used throughout the desk from this point on.

A second kind of input comes from *line*, and this is expected to arrive from outside the studio, or from reproducers and other equipment within it, at about zero level. For this, entry to the desk may be through a jack-plug, and the signal will not necessarily be balanced between the wires. So this signal is taken first through a small amplifier which electronically balances and feeds it forward into the channel.

A mono channel is usually switchable between line and microphone inputs. An additional small rotary fine control of about ±15 dB serves either, adjusting its level more precisely, so that the channel fader can be set to the zero marked on its scale when it is making a full contribution to the mix.

Stereo line inputs (including the return from artificial reverberation devices) may also be fed to their own special stereo channels, in which the faders operate equally on both legs; this saves space on the desk and saves cost; and treatments on the strip above it can be tailored more for the demands of stereo. On some desks, these 'line' faders have a 'remote-start' button, which can be set up to run tape replay, etc.

On both mono and stereo channels there should also be a phase control switch, often designated by its Greek symbol phi, ϕ. In mono, this simply switches the signal between the wires that carry it, so that a stereo signal can be corrected if the wiring on one leg is wrong (or to put the signal out of phase to achieve some specially distorted effect). On a stereo channel, the phase control reverses one leg only – conventionally the left, or 'A' leg.

From its position at the bottom of the strip it might have been assumed that the channel fader is one of the first controls the signal encounters, but this is far from the case. Before that is reached, the signal must still pass through (or bypass) an equalizer and filters (described later) and a point where other equipment may be plugged in, after which the channel is again electronically rebalanced.

Also before the fader, a side-chain feeds through the PFL switch to a small speaker. Back on the main chain after this, there may then be a switch which, if activated, permits the whole channel to be cut remotely: this is to allow for a '*cough-key*' to be operated by a studio presenter – or, in a more extreme case, a '*prompt-cut*', when a performer has to be given lines or instructions.

Close to the fader there should be provision to split the signal and send to *auxiliary* circuits – which may include feeds to artificial reverberation devices, public address (to audience loudspeakers), foldback (to the studio) or feeds to anywhere else. These may all require different combinations and mixes of the sources, which must therefore be controlled separately, in another part of the desk.

After the fader there will be controls (described later) which place the signal within the stereo image. Finally, a switch assigns the channel output to a group fader (usually stereo), or via a subgroup fader (which may be mono or stereo).

Desks for use with multitrack recorders are likely to have a row of small modules, each with a large number of tiny buttons (for selection of send or return) and also dedicated 'dynamics' (compressor, limiter, expander – see next chapter) associated with each channel return.

The controls for a channel are often mounted in line in a single strip consisting of one or several plug-in modules, and for a 'multitrack' desk (which also has the greatest number of channels) this covers a vast area. Even so, this strip layout offers operational clarity; and where (in all but the most sophisticated computer-controlled desks) the channel carries the audio signal, this keeps the wiring relatively short. Note, however, that the order of controls along the strip does not reflect their actual sequence; it is arranged for operational convenience, as determined by the desk manufacturer. In several types of desks, some of the modules, such as equalization, dynamics and multitrack routings, may be located on and selected from another part of the panel.

When a source has been selected and plugged to a channel, it is convenient to write its name, or microphone number, with a wax pencil or label in a strip along the desk just above the faders.

Group and main controls

For effective manual control, each of the mono and stereo channels is assigned to a group, which is usually given a number. Colours have also been used to identify group, and letters, too, though in this case at the risk of confusion with 'A' and 'B' stereo legs.

The switches for group selection may be at the top of the channel strips, or down near the faders; in some cases a display by the fader shows which group it belongs to. The selector may actually route the audio feed to the group fader strip, or alternatively the group fader may be part of a VCA (voltage-control amplifier), which acts as an additional control on all the individual channels. Whether or not the audio signal actually passes through the group faders (or for that matter any of the faders on the desk), it is convenient to imagine that it does. Even though on some 'mimic' or 'virtual' console systems most of the operations are remotely controlled, operationally the result is much the same.

Group faders require far fewer elements associated with them, mainly stereo balance controls, routing and feed switches, and some means of checking the signal at this stage: PFL and possibly a meter.

Signals are added together at each stage by feeding them to a common, low-impedance 'bus', from which they cannot return back down through the other channels feeding the bus, because these present a high impedance to the signal. The mixture is taken from the bus by an amplifier which feeds it to the next stage.

The groups are in turn mixed (on a stereo pair of buses) to the *master fader*, also known as the main gain control, then through a final compressor/limiter to the studio output, where a monitoring side-chain feeds the main loudspeakers and meters. Meters, compressors and limiters, and their role in volume control are described in the next chapter.

The desk will also usually have elaborate provision for treating and routing the various auxiliary feeds or 'independent main outputs' (taken from before or after the faders at each stage and passed through their own submixers). Some of the uses for these are described later. Alongside the main output there may be a second 'independent' or 'mix-minus' output, which was derived before one last channel (usually a narrator or announcer) was added.

Plainly, the capacity of its desk determines what can be achieved in the studio where it has been installed. Individual balancers may become proficient in operating and understanding the quirks of one desk to the almost total exclusion of all others; and conversion to another may be accompanied by a temporary loss of speed. This is an extra reason why innovation is often disguised within a relatively conventional layout: 'intuitive' layouts make for greater interchangeability. This is at a premium when working outside the studio using equipment that is hired by the day. The balancer would usually prefer to have a known desk even if it is a bit too big, but careful costing may require that something smaller is selected. Brief

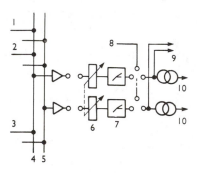

Main or master control
1-3, Stereo inputs from group faders. 4 and 5, Left and right main buses. 6, Ganged main control. 7, Limiter and make-up amplifiers, also linked for identical action on left and right channels. 8, Tone input. 9, To monitoring meters and loudspeakers. 10, Stereo output.

catalogue descriptions will include the number of mono/line, stereo, group and independent channels that are available; how much equalization, what auxiliaries, communications, and so on.

Faders will normally be linear, in the usual row along the front of the panel; but for some purposes, such as modest film or video recordings on location, a smaller mixer with just a few rotary faders and no in-line facilities may be all that is needed.

Faders

A fader is essentially just a variable resistance or potentiometer – sometimes called a 'pot' for short. As a contact moves from one end of its travel to the other, the resistance in the circuit increases or decreases. At one end of the scale the resistance is zero; at the other it is infinite (in practice it moves right off the resistive surface or otherwise interrupts the signal). The 'law' of the fader is near-logarithmic over much of its range, which means that a scale of decibels can be made linear (or close to it) over a working range of perhaps 60 dB. If the resistance were to increase according to the same law beyond this, it would be twice as long before reaching a point where the signal is negligible. But the range below −50 dB is of little practical use, so here the rate of fade increases rapidly to the final cut-off.

Faders have been produced in a variety of shapes and sizes. Those found in most professional equipment today are flat and linear (as distinct from rotary or quadrant). The biggest 'long throw' faders have an evenly spaced main working range (in decibels); where less space is available, 'short throw' faders have the same distance between 0 and −10 dB as the larger ones, but progressively less beyond that, with a more rapid fade-out below −40 dB. Music desks designed to work with two complete, parallel fader banks may have long throw faders on the nearest to the operator, with short throw faders above. (On computer-controlled desks, they may also have hidden motorized drive, either with linear motors or wires around high-torque rotary motors.)

As already stated, a fader is now usually associated with several amplifiers. One of them compensates for the action of the fader itself, so that the scale can be marked with a zero (for 0 dB gain or loss) at a comfortable distance from the top of the stroke – in fact, with some 10 dB of 'headroom' available above the zero. The layout can be the same whether it is a true audio fader or a VCA fader controlling the audio level from a distance. A minor refinement is that the knobs may be supplied in a range of colours, so that white might be used for channel faders, blue for groups and red for the master.

Over the years, fader design has advanced steadily, and this has led to changes in what is considered the most reliable for use in professional equipment. That now most commonly used is based on conductive

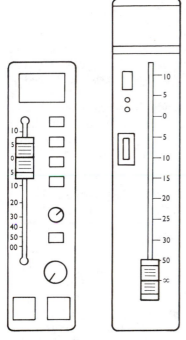

Faders
Short and long throw conductive plastic faders. These may be associated with channel monitoring and group selection controls. Some faders used for line input may also have remote start buttons.

plastics. Carbon granules (in a complex layout, to produce the 'law' which is scaled in decibels) are screen-printed on to a plastics base and heat-cured to render the surface smooth and hard-wearing. For ease and smoothness of operation and to reduce wear further, the slide makes only light, multi-fingered contact with the resistive surface. Such wear as there is may initially help to reduce any microscopic high spots left in manufacture. High-quality individual specimens may vary by up to a decibel from the average (a range that is within the typical limits to human awareness). One measure of performance is in the evenness of movement, the lack of any 'stick/slip' effect that the user might detect when making light, finger-tip movements of the fader.

Historically, simple faders based on carbon-track and wire-wound resistors both rapidly became worn and subject to unpredictable crackles as the slide contact was moved. For many years, the only satisfactory alternative was a fader with separate studs that the sliders engaged in turn. Between each successive pair of studs there was a fixed network of resistors. Unfortunately, however, even with the most expensive faders it proved difficult to create studs and sliders with surfaces that always gave perfect contact, unaffected by dust, moisture and other contaminants from the atmosphere, so these are now used only for coarse presets.

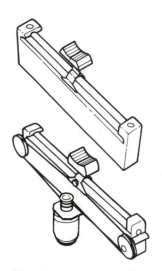

Manual and motorized faders

In the days when they were used throughout a desk, a rare crackle or two seemed more than offset by the capacity for the precise setting of levels, stop by stop in stages of $1\frac{1}{2}$ or 2 dB. Further, with the advent of stereo these also provided exact matching between the two signals, though with some danger of getting momentarily out of step when moving from one pair of fader studs to the next. This could result in stereo image *wiggle* or *flicker*. Some of the simplest designs have been used in low-level mono mixers (i.e. those in which there is no preamplifier, or internal amplifiers compensating for the loss in the faders) but these are unsuited to mixing where there is much movement of the faders, because as each additional channel is faded up it provides an alternative path for signals from the other sources. In consequence, the faders are not independent in operation: as fader 2 is brought up, there is a slight loss in signal 1, so that its fader must be adjusted to compensate. Such problems should be encountered only in simple, archaic or non-professional equipment. A simple mix for which it may be used satisfactorily is that of a subgroup of audience microphones. This can then be fed to a group or stereo 'line', so freeing several channels with full facilities for other uses.

Stops
The studs on older, stepped faders (many more than shown here) were arranged at intervals of approximately 2 dB, close to the limit of human perception. 'Stops' were the discreet jumps in level associated with these, and the term is still used. Rotary faders are now used mainly for coarse pre-sets, or for trimming over a narrower range. Studs have largely been replaced by conductive plastics.

One concept that has its origin in stepped faders is that of 'stops'. The studs in the faders that used them were separated by about $1\frac{1}{2}$ dB over the main working range, perhaps increasing to 2 dB over the lower half of the range, with broader spacing still for the two or three steps at the bottom end. A difference in level of $1\frac{1}{2}$–2 dB is only just appreciable on pure tones in the middle and upper middle frequency range, so nothing more precise than this was necessary for ordinary working. Nor is it still, even though faders are now (again) continuous. A bonus is that these newer faders are, in principle, less expensive.

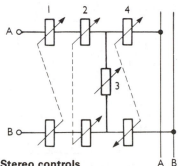

Stereo controls
1, Preset control to balance output between microphone elements. 2, Ganged channel fader. 3, Image width control. 4, Channel offset control (this displaces the complete spread image).

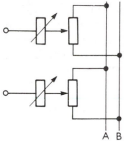

Mono channels in stereo
A panpot is used to place each mono component within the stereo image.

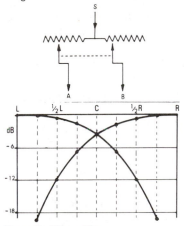

Panpot attenuation
A signal from a monophonic source is split and fed through a pair of ganged potentiometers. Step-by-step, the attenuation in the two channels is matched (dotted lines). The apparent volume will then be the same in all available positions across the stereo stage. L, Left. ½L, Half left. C, Centre, etc.

Stereo image controls

Rotary potentiometers are still used at many points on the desk where space must be conserved. In particular they are widely used (after the channel faders) for constructing or modifying the stereo image. These controls include:

- *Panpot,* short for panoramic potentiometer. This steers or pans a monophonic element to its proper position in the stereo image. The input, fed through the moving contact, is divided in two and fed to the A and B legs.
- *Image width control.* This cross-links the A and B signals, feeding controlled and equal amounts of the A signal to the B channel, and vice versa. As this cross-feed is increased, the image width narrows. However, if at the same time the phase is reversed, the image width is increased.
- *Displacement control.* This is a ganged pair of faders, one on each path, and working in opposition to each other, so that turning the knob has the effect of moving the image sideways. This control can also be used to give apparent movement to a static but spread source.

In a stereo channel the main pair of faders and the width control could also be put into the circuit in a different way: by first converting A and B into M and S signals and operating on these. When M and S signals are faded up or down there is no 'wiggle'; instead, there might be minor variations in image width – but this is less objectionable. In addition, image width can now be controlled simply by changing the proportion of M to S signal. In particular, image width can be increased without change of phase. However, after passing through the channel fader and the width control, the M and S signals must be converted back to the A and B form before passing through the displacement faders.

One stage in this process can be eliminated by using microphones which produce the M and S signals directly. However, this implies a radical change in working practices and the use of desks which are at present laid out primarily for AB operation. So it is usual to convert MS microphone signals to AB and work with them in the usual way.

Where two monophonic channels with panpots are linked together to form a stereo channel, image displacement and width are controlled by the relative settings of the two panpots. If they have the same setting as each other, the source has zero width and is displaced to whatever position the panpots give them. If the panpots are then moved in opposite directions from this point, the image increases in width. Full width is given by steering one of the signals to the A side and the other to the B side in the mixed signal; but this can, of course, be done without a panpot at all.

Quadraphony is no longer used for making records, but is still employed as a stage in treating film sound. A quadraphonic panpot is made up by combining two stereo panpots, one for side-to-side

Stereo signals
An electrical method for converting A and B signals to M and S, and vice versa. 'M' is equivalent to the signal that would be obtained by directing a single suitable monophonic microphone towards the centre of the audio stage. It therefore gives a mono output which can be used by those without stereo loudspeakers. The balance between centre and side subjects, and between direct and reverberant sound, will not necessarily be as good as for a normal mono balance. 'S' contains some of the information from the sides of the audio stage.

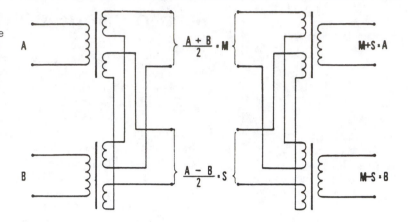

position and the other for forward-to-rear. By use of the two together a monophonic source may be steered to positions within the four corners. For special effects, such as a circling aircraft or aggressive insect, a joystick control gives continuous smooth movement, but for most purposes it is sufficient to have separate controls.

Desk checks

Operating a desk is like driving a car: easy, given training and experience – and provided that the car starts and is in working order. With hundreds of faders, switches and routing controls to choose from, if you cannot hear what you think you should be hearing it is as well to have a systematic strategy for recovery, which should include the apparently obvious. Desks vary, but a basic trouble-shooting checklist might start like this:

1 *Power.* Is the desk switched on? Is power reaching it? Has the supply failed?
2 *Microphone or line-source.* Is it working? Is it fully plugged in? If in doubt, try an alternative (but not an electrostatic microphone, if this requires power from the desk).
3 *Jackfield.* Is the routing correct? Is the path intercepted (and so cut)? Are the jacks fully in?
4 *Channel input selector.* Is it correctly set to microphone or line?
5 *Routing within desk.* Is it going to an audio group or VCA or as an independent direct to the main fader?
6 *Faders.* Are all those on the chosen routing open? Set them all to '0'.
7 *Solo, cut and AFL switches.* Has the signal been interrupted by a switch on another channel?

Make up your own checklist, or use one that others have written for that studio – which might also contain comments on performance limitations and how to make the best of them. Keep a fault, repair and maintenance log and, in a studio used by several operators, a suggestions and queries book.

With the growing complexity of studio equipment, maintenance is increasingly a matter of module replacement, of dealing with 'black boxes' rather than understanding the detail of their internal circuits and how to investigate and repair them on the spot. But it is essential to be aware of operational problems, such as any switches that make a click on the line, so that these can be avoided when on-air or recording. And the operator should know how and be prepared to route round any module that fails – especially during a live broadcast. It is very useful, therefore, to understand the signal pathways through the desk, and to be familiar with common deficiencies of audio quality and know how to localize and describe them.

Monitoring for quality

We have now been introduced to the desk and its monitoring loudspeaker system. Before we can begin to use this creatively, we must know to judge sound quality, if only to distinguish between artistic and technical faults. But it is also necessary to recognize that when a fault does occur, there is liable to be a frustrating and potentially expensive delay until it is tracked down to its source and cured or isolated. This brief account includes some of the 'traditional' problem areas associated with analogue audio signals and equipment. The use of digital audio bypasses some of these, but introduces others of its own.

One problem in some early digital equipment is found in analogue-to-digital converters. These must sample the audio signal at least twice as often as the highest audio frequency. In practice, a further margin must be allowed, so that the frequencies above the nominal maximum can be effectively filtered out – because, if they are not, any that are at more than the 'Nyquist frequency' (half the sampling frequency) will, in effect, be folded back into the top of the audio band. This spurious signal, which is called 'aliasing', is a form of distortion which is unrelated to the harmonic structure of the sound that is already there.

Where this fault appears, it can be eliminated by replacing the filter by one that has a better slope (the rate of cut-off above the maximum required audio-frequency). Aliasing is common enough for a number of manufacturers to offer replacement filters that can be retrofitted to their equipment.

A more common problem associated with computers is psychological: the danger of getting locked into a search for bugs (or simply for the correct instructions) that to any observer can seem incredibly time-consuming. It is, of course, essential to be sufficiently familiar with programming that no time is wasted, and that a crashed system can be reinstated quickly. Some desks with computers store information against this possibility, but even so, the recovery procedure requires enough rehearsal to ensure that it does not significantly reduce production time. Handled (and operating) well, computer-driven digital audio is a dream come true; handled badly or failing for perversely obscure reasons, it is a nightmare that makes the apparently simple, understandable faults of analogue audio seem almost forgivable.

Noise

Noise, for our present purpose, may be regarded as random sound of indefinite pitch. There are plenty of other forms of 'unwanted sound', but they can each be treated separately.

All irregularities in the structure of matter cause noise: the separate particles of iron oxide in magnetic tape, tiny roughnesses in the wall of an old record, the granular structure of carbon in a resistor, the random dance of individual electrons in a metal or semiconductor; all of these are bound to produce some level of noise in a recording or broadcast. These all create crackling, buzzing or hissing noises, which must be kept at low level compared with the audio signal; that is, we need a good *signal-to-noise ratio*.

Noise becomes serious when components before the first stage of amplification begin to deteriorate; at a later stage, when there are more drastic faults (e.g. a 'dirty' fader or a 'dry' joint); where vibration reaches a transducer (e.g. record-player motor rumble at 25–2000 Hz); or where there are long lines or radio transmissions broadcast in unfavourable circumstances.

In recording or transmission systems where noise levels increase with frequency in a predictable manner, the high-frequency signal is increased (*pre-emphasized*) before the process and reduced (*de-emphasized*) after it. For radio or television, such systems require national or international standards, so that professional and domestic equipment produce complementary effects.

A major advance in noise reduction came with the switch from amplitude to frequency modulation of broadcast audio signals; another, when signals could be transmitted along the audio chain in digital rather than analogue form. Even so, a great amount of analogue equipment remains in use.

For the recording industry a range of commercial noise reduction systems is widely used, with different standards applied to domestic recordings, professional audio, optical recordings on film and combined magnetic recordings on film. Most divide the sound into a number of frequency ranges and apply treatments to each, together with a control signal which allows the reproducer to decode it.

In one noise reduction system that is used in professional studios for analogue recordings, the signal is split by filters into four frequency bands: 80 Hz low-pass; 80 Hz to 3 kHz; 3 kHz high pass; and 9 kHz high-pass. Each band is separately compressed, so that very low level signals (40 dB down) are raised to 10 dB, while signals from 0 to 20 dB down are virtually unaffected. The recovery time is 1–100 ms, depending on the signal dynamics. There are also limiters in each channel: these are simple peak choppers. An exactly matching network is used to reconstitute the original signal on playback. The principle of the system depends on the fact that noise similar in frequency to the louder components of the signal is masked by it, while that which is widely separated in frequency from the louder components is reduced in level: the improvement is therefore partly real and partly apparent. As the system produces a coded signal which must then be decoded, it cannot be used to improve an already noisy signal. It is widely used to improve the quality of tape recordings, where an increase in the signal-to-noise ratio is used partly to reduce hiss, hum and rumble and also, by lowering recording levels a little, to produce less distortion and to lower the danger of print-through. It is particularly valuable for multitrack recordings, but can in principle be used around any noisy component or link. The signal is, of course, unusable in its coded form: characteristically it sounds bright, breathy, and larger than life.

Cassette recordings are better when made using an improved version of the domestic noise reduction system (with an extra frequency band).

Digital recording offers a 90 dB signal-to-noise ratio without further treatment. It is very convenient for multitrack work, as 32 tracks will fit on a 1 in (25 mm) tape, though at a high playing speed of 45 i/s (114 cm/s). Harmonic and intermodulation distortion and interchannel cross-talk are all very low; while wow, flutter, print-through and erasure problems are avoided completely.

Hum

One form of noise that does have definite pitch is *hum*. This nuisance is caused by the mains frequency (and its harmonics) getting into the audio signal. Mains hum may be caused by inadequate smoothing of the electrical power supply when it has been rectified from alternating to direct current, or by lack of adequate screening on the wires carrying audio signals (particularly those at low level, or in high-impedance circuits such as those from electrostatic microphones before the first amplifier). Magnetic microphones can be directly affected by nearby electrical equipment, such as mains transformers or electric motors. Hum may also be caused by earthing (grounding) faults; either the simple lack of an earth or, at the other extreme, the connection of an item by separate paths to a common earth, thereby causing an earth loop, which like any other loop can pick up signals from wiring or equipment carrying mains.

The second harmonic of mains hum is generally the most serious component except in certain types of hi-fi loudspeaker cabinet that provide a high output at low frequencies, so most hum filters concentrate on getting rid of 100 or 120 Hz. More elaborate 'notch' filters take out narrow slices of a wide range of harmonics without substantially affecting music.

It is, however, better to avoid hum in the first place. It can be a particular problem when recording pop music from public performances, where feeds are split from the system set up for the stage foldback loudspeakers. To identify the source of hum quickly, fit the earth (ground) leads for each circuit with a switch so that each can be individually checked. Where sound is combined with vision, lighting is the main culprit, and particularly certain types of dimmer. A full rehearsal will show up the danger points; but failing that, ask for a quick run through the lighting plot and see what it does to open microphone circuits.

Portable hum-detection devices are useful when setting up a studio – or any temporary rig that involves a lot of audio and lighting circuits, such as a pop concert.

Distortion

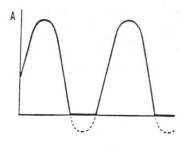

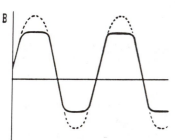

Harmonic distortion
The dotted portion of the signals is flattened. A, Bottoming causes each alternate peak to be clipped. B, Both peaks are flattened, as excessive swings in each direction cannot be followed by some item of equipment. In both cases the distortion produces tones which are exact multiples of the frequency of clipping.

If at any point in a broadcast or recording chain the volume of an analogue signal is set too high for any stage that follows, the waveform is changed – distorted. Typically, the peaks are flattened, introducing new frequencies that are harmonics of the original frequencies present. In musical terminology, new overtones are produced. However, most are added at precisely those frequencies that are already present in much sound, and most music. The introduction of harmonic distortion is therefore to some extent masked by sound already present; and where it is not is made somewhat more acceptable by its pseudo-musical relationship to the wanted sound. Speech sibilants, not already having a harmonic structure, bring it out strongly. Note, however, that human hearing is itself subject to distortion at high acoustic intensities.

Small amounts of distortion may be more acceptable than a high noise level: 2% harmonic distortion is generally noticeable and 1% represents a reasonable limit for high quality.

Unduly high distortion may be due to actual equipment faults. For example, it occurs if the moving parts of a transducer have worked loose or are in the wrong place, e.g. a ribbon or moving coil hanging out of the magnetic field, or if too much or too little bias is used when recording on magnetic tape.

Intermodulation distortion is far less tolerable than harmonic distortion: it occurs when two frequencies interfere with each other to produce sum or difference products. High-quality sound equipment is designed to avoid the production of intermodulation effects that occur in the audio range.

Distortion may arise right at the start of the chain, at points that are – in theory – under the control of the balancer. Some air pressures (perhaps caused by wind, or due to explosive consonants in close speech) may be such as to move a microphone diaphragm outside its normal working range. Or the output of the microphone may be at too high a level for the first pre-amplifier. Amplifier stages have a linear response to changes in volume (i.e. their output is proportional to input) only over a limited range. The sensitivity of professional microphones is related to the volume of expected sound sources in such a way that they mostly feed signals to the mixer at about the same range of levels. But there are notable exceptions in such sound sources as the bass drum, brass and the electric guitar, the loudspeaker for which may be set very high. If the pre-amplifiers are of fixed gain, a 'pad' (a fixed-loss attenuator of, say, 12 dB), may be required between microphone and amplifier. Alternatively, a microphone of lower sensitivity may be selected; or one with a head amplifier that can be switched to accept different input levels.

Another example of this form of distortion may occur in very close piano balance. Used for pop music in this way, pianos may be regarded as percussion instruments, deriving a great deal of their effect from the attack transient. The brief, very high volume of this transient may not be apparent from a meter that takes time to respond – but it is there, and is liable to distort if the signal is not reduced again by perhaps 6 or 12 dB before pre-amplification. In such a case the amplifier works on the lower part of its range for the main signal, and the overall signal must be lifted again later in the chain – where the same damage may still occur. There has therefore been controversy as to the value of this technique.

Loss of high frequencies

Most of the links in an analogue audio chain are capable of distorting the frequency response; and this is usually worse at the higher end of the scale. Whether anything can be done about it depends on whether or not there is a frequency at which the signal is totally extinguished.

When a signal passes along a line, the capacitance between the two wires carrying it provides a path whereby the high frequencies are selectively lost, but the damage can be repaired by boosting the high-frequency response from time to time – and the signal-to-noise ratio is substantially maintained because high-frequency noise is first reduced and then boosted in the same proportions. But the amount of high-frequency loss depends on distance and, with long lines, what compromises have been made. It may, for example, have been decided that 8 kHz is all that can be economically maintained and to install equalizers at the distances where the signal-to-noise ratio has fallen, say, from 55 to 40 dB at this frequency. Nor is this a complete cure, because at every stage where the frequency response is corrected a little of the noise on the line is boosted, and at the same time the many stages of distortion and subsequent correction inevitably begin to make the top end of the frequency spectrum erratic, so that eventually the signal begins to get lost even below the nominal 8 kHz.

In a building the size of a broadcasting studio centre there is a danger not only of high-frequency losses due to capacitance, but also induction of programme signals, hum, etc., from other lines. One way in which this is kept at least partially in check is by the use of standard impedances, usually of several hundred ohms. If a low impedance system is adopted, the resistance and capacitance of the wiring itself becomes significant in comparison, simply because there is so much of it.

The exact matching of impedances is important when *power* is being transmitted, e.g. to a loudspeaker, as mismatching produces power losses. Where the significant part of a signal is its *voltage*, mismatching matters less. Low into high may go; high into low may not: this is the rule. If a high-impedance input such as a crystal pick-up is fed to a low-impedance input there is loss of bass: if the output from a magnetic pick-up or microphone is fed into lower than its rated impedance there is loss of high frequencies. Even more important, do not feed a line signal (e.g. from a tape deck or a record player) to a low-level input (such as that for a microphone) without appropriate attenuation or there will be severe overloading, with consequent distortion.

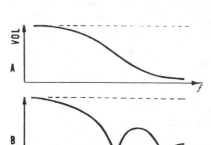

Restoring high-frequency losses
A, Top can be boosted and effectively restored but at the cost of increased noise. B, H.f. extinction at two frequencies: top cannot be restored effectively.

Cross-talk

A defect of quality of particular importance to stereo is *cross-talk*. This is measured as the level at which a tone at a particular frequency in one stereo leg can be heard on the other. What is regarded as unacceptable varies with both frequency and application. For example, for tape recordings used in broadcasting the limit might be set at −38 dB at 1 kHz, rising to −30 dB at 50 Hz and 10 kHz. These figures compare favourably with the −25 dB (at middle frequencies) available in the groove of a stereo disc, but are very poor compared with digital recordings. In a good desk, the specification will be much higher – perhaps with a separation of more than 60 dB (one desk offers better than 75 dB at 15 kHz).

A subjective scale for quality check

Judgements of sound quality are subjective, but it may nevertheless be helpful to try to apply objective standards that are common to any group of people who may be working together. In a system that has been used within the BBC, observations are made on general sound quality: background hum, noise, or hiss; change of sound level at programme junctions; and interference. In this last category is talkback induction which can sometimes be heard on radio as it is transmitted, but much more often on television because of the continuous stream of instructions and information that is being fed from the gallery to the studio floor.

Sampling is done for periods of 5 minutes on a wide range of material, including the programmes as they are transmitted (both before the transmitter and off the air), studio output (but taking into account conditions that may exist early in rehearsal before all adjustments have been made), material incoming from other areas (by landline, radio link or satellite) and all recordings.

Below are two subjective scales that are used:

Scale 1: Assessment of degradation

    1.1. Imperceptible
    1.2. Just perceptible
    1.3. Perceptible but not disturbing
    1.4. Somewhat objectionable
    1.5. Definitely objectionable
    1.6. Unusable

Scale 2: Assessment of quality
    2.1. Excellent
    2.2. Good
    2.3. Fair
    2.4. Fairly poor
    2.5. Poor
    2.6. Very poor

These divisions may at first glance seem coarse, but are readily assessed and easily understood.

Chapter 10

Volume and dynamics

By programme *volume* we usually mean the level of electrical signal that is indicated by some standard meter – of which there are two main kinds, with completely different principles. One responds logarithmically (as our hearing does); the other does not. *Dynamics* are the changes in volume from moment to moment, controlled primarily by the manual adjustment of faders, but sometimes also by automatic devices such as compressors or limiters. If these descriptions seem somewhat vague, this is intentional. Like so many in the study and control of sound, this is a realm in which subjective assessment and objective measurement can be matched roughly, but not absolutely.

For monitoring volume, meters can prove misleading because although they do indicate either near-instantaneous or peak values of the electrical signal, this is not what we actually hear, which is loudness. Some instruments, such as bagpipes, harpsichords, clavichords and virginals, sound very loud compared with the meter reading. So does much pop music. Balance of these against less noisy music and speech can and must be judged by ear – and this is only a particular case of a general rule: meters are a guide, but ears must be the judge.

Below the maximum indicated by meter, the volume and stereo spread are controlled according to the judgement of the balancer. This is a dynamic process, for not only must levels be set correctly for the various technical processes that the signal undergoes; individual passages may also have to be compressed (restricted in dynamic range) so that they are neither too loud nor too quiet in comparison with the rest of the programme.

Among the most important of all sound operations are volume control and compression. There are no dependable short cuts: automatic volume control, for example, would ruin Ravel's *Bolero*. Nevertheless automatic devices have become sophisticated enough to control specific types of programme material with reasonable adequacy. A disadvantage is that different control characteristics are required where several types of output are heard in succession, and another is that for some types of music in which subtle dynamic control by the performers is important to the artistic effect no compressor can do an adequate job unless it can be specially set for the individual item. Automatic control lacks intelligent anticipation.

The uses for compressors and limiters include short-wave radio where a continued high-signal level is necessary to overcome static and other noise; certain other types of high-intelligibility radio where a limited range of output is broadcast to a maximum service area for a given transmitter strength; and certain purposes within programmes themselves, e.g. as an element of pop music balance. But here we are concerned with the manual control of volume – using ears (both ears, for stereo), eyes (for meter, script or score and to observe actions which may affect sound) and brain, to anticipate events.

Programme meters

Although a meter cannot be used to judge such things as the balance of one voice against another or speech against music, it is nevertheless an essential item of studio equipment. Its uses are:

- to check that there is no loss or gain in level between studio, recorder and transmitter
- to provide indications of levels that would result in under- or over-modulation at the recorder or transmitter
- to compare relative levels between one performance and another
- to check that levels meet the prescribed values for good listening
- to provide rough visual checks on the relative levels (and correct phase) of stereo channels and mono compatibility.

A *peak programme meter* (PPM), well established in European practice, is a special type of voltmeter that is arranged to read logarithmically over its main working range. In order to give prominence to the all-important 'peaks' which may be of short duration, but vulnerable to overload and therefore distortion, it has a rapid rise characteristic and a slow die-away which makes it easy to read. The BBC version has a rise time-constant of 2.5 ms, giving 80% of full deflection in 4 ms (the ear cannot detect distortion of less than this duration), and a slow die-away (time-constant, 1 s; giving a fall of 8.7 dB/s. Traditionally the original electromechanical meter has a black face with white markings, for visual clarity.

Other European meters may have different rise and fall times and face-scales from those favoured in Britain.

The original mono PPM had a single needle on a meter that was placed prominently on the middle of the desk. For stereo this has been replaced by two meters, each with two concentric needles. One

Peak programme meter (PPM)
The markings are white on black. Each division is 4 dB; '4' on the meter is a standard 'zero level' and '6' is full modulation, above which distortion is to be expected.

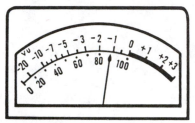

VU meter
The lower scale indicates percentage modulation and the upper scale decibels relative to full modulation.

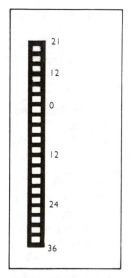

Simple LED bargraph
To give a rough guide to level (or overload) in an individual channel or group, a simpler scale with (here 3 dB) LED segments may be adequate.

is for the A and B signals. The colours follow the nautical 'red-port-left' and 'green-starboard-right' convention. The second PPM shows M and S signals, conventionally white and yellow, respectively (and not always easy to distinguish). Because M = A + B it normally reads higher than either; because S = A − B that is normally less. So S is to the left of M unless there is phase cancellation – the presence of which would suggest that one leg has been wired incorrectly. This is an immediate visual check for such a fault.

The *VU meter*, widely used in American equipment, has two scales, percentage modulation and decibels, and is different in operation. It is fed through a dry rectifier and ballast resistance, and (unlike the more complex PPM) may draw from the programme circuit all the power needed to operate it. For cheaper models the ballistic operation may allow the needle to peak substantially higher on programme than on steady tone. However, the more expensive, studio-quality VU meter overshoots by only a small amount when a pulse signal is applied (in which it compares favourably with a PPM). But the time-constant of 300 ms prevents transient sounds from being registered at all, and the meter under-reads on both sharp percussive sounds and speech (at its '100% modulation' level, these are distorted).

Though it may have a linear calibration in percentage modulation (the rectifier characteristic should ensure this), more than half the scale is taken up with a range of 3 dB on either side of the nominal 100% modulation. Little programme material remains consistently within such a narrow range, so the needle is likely to register only small deflections or flicks bewilderingly over the full range of the scale. For maximum volumes, continuous sounds (including most music) may read '100% modulation' on the VU meter; however, for staccato sounds such as speech a true maximum volume is given by an indicated '50%', because the meter does not have time to register a full swing to the 100 mark on the louder individual constituents of speech.

Note that only those VU meters which conform to the relevant American standard can be used to compare programme material.

On a radio control desk, the meter (of whatever type) is likely to be either centrally placed or roughly in line with the main control fader; a television sound meter is mounted in line with the transmission monitor and, to avoid eye fatigue, at much the same distance from the operator.

Line-up tone

Meters are used to establish common levels at all stages in the chain by using a pure tone at 1000 Hz and of known volume as a reference. Organizations with a central control centre for signal routing will have a reference tone source at that point, and most desks and some other items of equipment such as recording machines also have their own. A studio will therefore often have the choice of using the central source

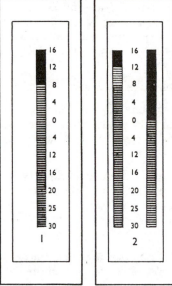

Bargraphs
1, Mono. 2, Stereo (or any pair of channels). The simplest design has LED segments. The more complex, bright plasma display can be programmed to offer a variety of displays including both PPM and VU. Above a given level (here +8 on the PPM), the plasma glow is brighter still, to indicate overload. A third type of display simulates the bar on a VDU.

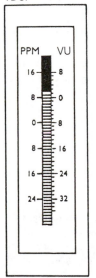

Switchable bargraph
Here, zero level on the PPM scale is indicated by 0 not 4, as on the older, mechanical PPM. 100% modulation is therefore +8, which corresponds to zero on the VU.

or its own local tone – which, in principle, should be the same. In the procedure adopted by the BBC and many other organizations it is arranged that this always registers '4' on a monophonic PPM (or as the M reading in stereo, with nothing at all on S). Stereo A and B signals should each read 3 dB below '4'. Tone from other sources will be faded up until they reach this same reading, to establish the incoming zero level; any losses on the line should be made good at the input to the desk.

The standard used is 1 mW in 600 Ω, and is equivalent to 40% modulation at the transmitter (so that 100% is about 8 dB above this). At all stages in the chain the tone should give the same steady reading on all meters, and when a recording is made, some of it is recorded at the start at the appropriate level for the equipment and recording medium, so that the process can be repeated when the tape is replayed.

Multitrack meters

In music recording studios, multitrack techniques ideally require separate metering of each channel that is being recorded. Ideally, there will be in-line displays on all channels; they may also be supplied for auxiliary 'sends'. Mechanical PPMs are unnecessarily complex, expensive and also difficult to follow in such numbers. If needle-pointer displays are used, the simpler VU meters are adequate, indicating which channels are not only faded up but also passing a signal. In America, these have sometimes been arranged in pairs, one above the other for each stereo channel, all across a large desk.

A better option is *bargraphs.* These are vertical strips ranged side by side, presenting a graphic display which is much easier to follow. With processor control, they can be switched to show a variety of VU or PPM recordings, with stereo in AB or MS pairs, and may also have overload or 'hold' options. In one example, they can even be switched over to show a frequency spectrum, analysed as one-third octave bands.

A bargraph display may take one of several forms. A strip of LED segments is perhaps the simplest. More flexible, in that its brightness can be controlled, is the plasma bargraph – especially bright in the 'overload' range. This takes more power and is more expensive. A third type merely simulates the bar, and is displayed on a colour VDU.

Controlling stereo

Where the stereo broadcast is on FM, control is easy. It can be shown that the deviation (size) of the signal does not exceed the greater of the A and B channels, so all that is necessary to avoid overmodulation is that these do not exceed their normal limits. The same is true for analogue recordings on magnetic tape, where the A and B channels are kept separate. For systems like these, it is necessary only to keep the separate A and B signals below the standard 'full modulation' mark.

In broadcasting, however, several problems may arise:

● in a large network some of the transmitters may be in mono
● some of the listeners to the stereo broadcast have mono receivers
● some of the items on the stereo service may be broadcast in mono only.

For practical purposes it is assumed that the average level for A + B is 3 dB higher than A or B on its own; accordingly, the stereo and mono parts of the system are lined up with a 3 dB difference in level. Unfortunately, this is not a complete answer to the problem of relative levels.

There would be no difficulty if the signals added together simply, so that the maximum volume of A + B really were always no more than 3 dB higher than the separate maximum volumes for A + B, but in fact this happens only for identical signals, such as tone or a source in the centre of the sound stage. If A and B are both at the same volume but are different sounds, the sum varies from + 0 dB to + 6 dB. So where A and B are at maximum volume the M signal may be overmodulating by 3 dB. This should therefore be controlled for the benefit of mono systems.

On the other hand, if the signal is being controlled primarily on the M signal, the stereo signal can overmodulate. This happens when A + B is at its maximum and all of the signal is in the A or B channel, which is then 3 dB over the top.

A comparison of the M and S signal levels is less crucial, but can give useful information about the width of the stereo image. If S is close to M, the stereo is wide and a significant amount of information may be lost to any mono listener. If S is 3 dB lower, the stereo content is modest, and the width of the image is relatively narrow.

PPM peak levels
(dB relative to maximum fed to transmitter)

Talk, discussion programmes	0
News and weather	−6
Drama: narration	−8
Drama: action	0 to −16
Light music	0 to −16
Classical music	0 to −22 or lower
Harpsichords and bagpipes	−8
Clavichords and virginals	−16
Announcer between musical items	−4 to −8*

*Depending on type of music

Programme volume: relative levels

A peak programme meter can be used to suggest or check levels. For example, if ordinary talk or discussion is allowed a normal dynamic range with occasional peaks up to the maximum, then newsreaders (who generally speak very clearly and evenly) usually sound loud enough with peaks averaging about 6 dB less.

A whole system of such reference levels can be built up; and the following are some specimen PPM peak levels for speech and music that experience indicates are about right for listeners hearing radio programmes in reasonably good conditions.

Even if such a list is adopted as a general guide, it does not solve all the problems posed by junctions between speech and music or, on the larger scale, between programme and programme. For radio stations with a restricted range of programming, there are a limited number of types of junction, and it is fairly easy to link item to item. For a service that includes news, comedy, various kinds of music, religious services, magazines and discussion programmes, problems of matching between successive items can become acute. Unfortunately, there can be no rule-of-thumb answer. It is a question of judgement; of trying to imagine what the audience will hear when listening to your programme for the first and (almost certainly) only time. Remember that your own ideas are considerably coloured by long acquaintance with the material: familiarity breeds acceptance.

A further slight complication is suggested by the results of a study of listeners' preferences for relative levels of speech and music. Taking a large number of pieces in different styles (but excluding modern pop music) and with different linking voices, the results were:

- speech following music to be (on average) 4 dB down
- music following speech to be (on average) 2 dB up.

But this apparent contradiction is easy enough to resolve: it simply means that announcements linking music items should be edged in a little.

Preferred maximum sound level
(dB, referred to 2×10^{-2}Pa)

	Public M F	Musi-cians	Studio mngrs M F*
Symph. mus.	78 78	88	90 87
Light mus.	75 74	79	89 84
Dance mus.	75 73	79	89 83
Speech	71 71	74	84 77

*M, male; F, female.

Maximum volumes

When a broadcast or recording is being made, the balancer monitors at a fairly loud listening level with full attention. How does this compare with most of the audience? To find out, tests were made on the maximum sound levels preferred by BBC studio managers, musicians and members of the public. It was found that studio managers preferred louder levels than the musicians, and very much louder than the general public (see table).

The figures for the public correspond reasonably well with what they would hear in real life in a seat fairly close to the players. Musicians, who would be much closer to the sound sources, might reasonably be expected to choose higher levels. But the studio managers (sound balancers) chose unrealistically high levels. Why?

Part of the reason for the diversity must be that people professionally concerned with sound are extracting a great deal more information from what they hear, and the greater volume helps them to hear the finer points of fades and mixes, etc., and check technical quality. Musicians, on the other hand, draw on a vast fund of experience of musical form and instrumental quality, and are listening for performance, often disregarding technical quality almost completely.

Those preferred maximum levels refer specifically to the case where the listeners, like those concerned in creating the programme, want to listen attentively and have reasonable conditions for doing so. But there are very many occasions when this is not the case. For example, part of many people's early evening listening may be done while driving or working in the kitchen. Where the listeners' attention is limited and background noise may be high, speech may be peaked 8 dB higher than music. If, when listening to a programme controlled in this way, you decide that a particular item of music appeals to you and you turn it up, you will find that the announcement following appears unduly loud. As a general rule, the listener can do little to combat the levels as transmitted.

Television sound in the home is often set at levels comparable to that which would be appropriate to background radio listening. This may be satisfactory for some programmes, but if there follows a play with a wide dynamic range of sound, the overall level drops still further, and may remain uncorrected by the viewer. The television sound supervisor (monitoring at high level) may well have set relative levels of speech to effects, or singer to accompaniment, that are inappropriate to such quiet listening. Alternatively, the viewer may raise the volume, and then the subsequent announcement and following programmes are too loud for him: indeed, they seem all the louder because of his irritation that he has to adjust the level at all. For the broadcaster, all this requires a fine judgement of compromises. A start may be made by listening to part of the rehearsal at realistic domestic levels on small loudspeakers.

Linking together speech items of varying intelligibility presents a special problem – particularly in the case where an item containing noise and distortion has to be matched to studio quality. Here, the least unsatisfactory solution is found if the noisy item is matched for loudness at the beginning and end, and lifted for intelligibility in between: the item must be slightly faded in and out.

There is a further way in which listening conditions may vary. On the one hand, there are small portable radios that demand high intelligibility and, on the other, wide-range hi-fi, demanding not only high-quality transmission but also a wide dynamic range. The two sets of requirements are largely incompatible. In the USA there are different radio stations tailored to fit the two audiences and many other countries have gone much of the way in following this lead.

Dynamic range – and the need for compression

The ear can accept an enormous range of volumes of sound. At 1000 Hz the threshold of pain is 110 dB or more above the threshold of hearing – that is, a sound that is just bearable may be over a hundred thousand million times as powerful as one that is just audible. Studios or microphones, with noise limits that are 10–20 dB above the ear's threshold level, are good for a range of perhaps 100 dB or more. Digital recordings have a maximum signal-to-noise ratio of 80–100 dB; but analogue recordings are limited to 50 or

70 dB (the upper figure demands a noise reduction system). FM broadcasting at 50 dB was a great advance on AM at 30 dB. Even that may be greater than some audience preferences or listening conditions warrant.

Plainly, some compression of the signal is usually necessary. For this, the old concept of 'stops' remains useful. 'Taking it down a stop', a reduction of 1½–2 dB on the fader, amounts to a drop in level that will be just perceptible to the operator, but which the audience will not associate with a hand on the control. A script or score can be used during rehearsal to note settings and other information that will be useful on the take.

The standard procedure is roughly as follows:

Set the gain at a suitable average level for the type of programme, and adjust the loudspeakers to a comfortable volume. Judge divergences from the average level on the grounds of relative loudness and intelligibility – sometimes these two factors do not suggest the same results, as we have seen. Glance at the meter occasionally – mainly to watch for very quiet or loud passages. When recording, the effects of any high peaks can be checked on replay, to see whether there is appreciable distortion. On a broadcast, too, an occasional peak may be risked in preference to reducing the effective service area of the transmitter by undermodulating a whole programme; compressors or limiters deal with any peaks big enough to cause technical problems. On speech programmes also check the meter from time to time in order to maintain the same average volume. The ear is not reliable for this – it tires.

At the end of the rehearsal of a record programme, say, it is worth checking the levels of the first few records again, to see if one's judgment of the appropriate level for them has changed. It often has.

Compression of music

Manual control of programme volume
1, Highest permissible level. 2, Lowest acceptable level. Method of compressing a heavy peak to retain the original dramatic effect and at the same time ensure that the signal falls between maximum and minimum levels. Steps of about 1½–2 dB are made at intervals of, say, 10 s.

Overmodulation on music that has loud and quiet passages must be avoided not by 'riding the gain' but by careful preparation – gradually pulling back on the gain control for half a minute or so before the peak is due. During the rehearsal, the appropriate setting for handling the peak will have been noted on the score, and after the loud passage there is a slow return to the average level. Each operation should consist of one or more steps of about 2 dB. A similar procedure may be used to lift long periods of low level. It may be necessary to vary the level continuously throughout a programme, so that all the elements of light and shade fall within the acceptable dynamic range.

Such techniques as these are most likely to be necessary when recording orchestral works, but – depending on the nature of the music and the closeness of the balance – control may also be needed with smaller combinations or groups, or even a solo instrument such as a piano.

There is one other possible way of dealing with the control of orchestral music, and this is to leave it in the hands of the conductor. It has even been known for a conductor to have a meter on his desk as he rehearses and records; but this technique has not been generally adopted. On the whole, it seems preferable that the conductor should concentrate on the performance, bearing the limitations of the medium in mind if he can, but leaving control to the balancer.

However, music programmes specifically designed for background listening may have some measure of compression built in by the arranger; and also, of course, smaller musical forces will be employed.

The further the microphone is away, the easier it is to control music: studio or concert hall acoustics smooth out the peaks. The balancer who adopts a bright, 'tight' balance has more work to do without necessarily earning the thanks of listeners – many of whom prefer a well-blended sound for classical music, as BBC listening tests have shown.

Control of music affects the reverberation just as much as the direct sound. The effect may be disconcerting: the microphone moves closer, but the perspective remains the same. The two may be controlled more effectively by using two microphones (pairs, or arrays), one close and one distant, and working between them on changes of music volume. Alternatively, there may be reverberation microphones facing away from the orchestra: they are controlled independently to achieve the best effect as the acoustic volume of the music varies.

Compression: speech and effects

In controlling speech the problems are somewhat different, as are the methods employed for dealing with them. Studio interviews and discussions for radio and television also benefit by careful control, particularly if one voice is much louder than the others, or habitually starts off with a bang and then trails away. Here, control is a matter of anticipation: whenever the needle-bender's mouth opens, drop the fader back – sharply if in a pause, but rather more gently if someone else is speaking: a slight dip in level is more acceptable than the risk of a sudden loud voice. Laughter may be treated similarly if the general level is held well up. But edge the fader up again as quickly as possible after the laugh.

Careful control should be unnoticeable: do not miss the first high peak and *then* haul back on the fader. No control at all is better than bad control.

Do not discourage speakers from raising their voices, unless they actually ask for advice; even then the best advice is, 'Be natural'. A natural but controlled programme is better than one in which voices are never raised and the meter never has any bad moments.

A good programme depends on variety of pace and attack. It should not be necessary to hold the overall level back to accommodate possible overmodulation. But if much editing is envisaged it would be better to deal with the more violent fluctuations of level afterwards, otherwise there may be difficulty even when using a studio in which extraneous noise is kept at a low level.

A completely different situation arises when recordings are made on location, and *subsequent editing is envisaged*. In this case it is better that, if possible, a level should be set so that background remains constant throughout a sequence. If necessary, some control can be exercised by slight movement of a directional microphone held by hand.

For scripted drama the BBC regards something like 16 dB as being an adequate range for peak values of speech – the average being about 8 dB below the maximum permissible. Even so, scenes involving shouting have to be held back, possibly making them sound rather distant – which may not match a television picture. In cases where close shouting must be mixed in with more normal levels of speech, it is up to the actors to hold back on volume and project their voices.

This upper limit makes even more difference when it comes to loud, percussive sound effects – a gunshot, for example, or a car crash. Even a sharply closed door is too loud.

It is therefore fortunate that radio and television drama have established a convention in which such effects are suggested by character rather than by volume. Volumes have to be evened out and held back – which explains why unlikely-looking methods for making sound effects often produce the best results.

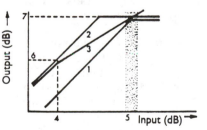

Limiter and compressor
1, Limiter used for overload protection. 2, Limiter used for programme compression. 3, Compressor. 4, Lowest level of programme interest. 5, *shaded area:* Range of expected maximum level. 6, Lowest acceptable level in output (defining lower limit of required dynamic range). 7, Maximum permitted level.

Compressors and limiters

The proper use of any recording or transmission medium requires that some passages are close to the maximum permissible level, and that they may momentarily or unpredictably exceed it. In such cases, the danger of frequency distortion is replaced by controlled (and usually less noticeable) distortion of dynamics. After most corrections have been made manually, the few that remain can be 'cleaned up' automatically by a *compressor* or by its more extreme form, *limiter*. In popular music, these can also be used to raise the average level of the individual components of the mix. Essentially, a compressor works as follows:

Below a predetermined level (the threshold or onset point) the volume of a signal is unchanged. Above this point the additional

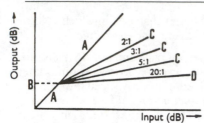

Compression ratios

A, Linear operation, in which input and output levels correspond. Above a given threshold level (B), various compression ratios (C) reduce the output level. In the extreme case (D) the compressor acts as a limiter holding the maximum output volume close to the threshold level, but still permitting a small proportion of any high peak to pass.

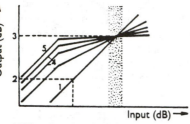

Lifting low levels with a compressor

1, With the compressor out of circuit, only a small proportion of the input range goes into the desired narrow dynamic range (2–3). 4, With increasing degrees of compression more can be accommodated. 5, The effect of a limiter. In these examples the choice of threshold setting has been governed by arranging for the expected maximum level to produce full modulation. The greater the compression, the less the effect of overmodulation beyond this expected level.

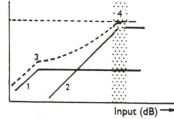

Compression by two limiters

1, Effect of low-level limiter. 2, High-level limiter. When the two signals are combined, the resulting compression characteristic is different from that of a normal compressor which would give a straight line from 3 to 4.

volume is reduced in a given proportion, e.g. 2:1, 3:1 or 5:1. For example, if the threshold were set at 8 dB below the level of 100% modulation and 2:1 compression selected, it would mean that signals which previously overmodulated by 8 dB were now only just reaching full modulation. Similarly if 5:1 had been chosen, signals which previously would have overpeaked by 32 dB will now only just reach 100%. What this means in practice is that the overall level may be raised by 8 dB or 32 dB, so that relatively quiet signals are now making a much bigger contribution than would otherwise be possible.

If the compression ratio is made large, say 20:1, and the threshold level raised close to 100% modulation, the compressor now acts as a limiter. In this condition it can be used *either* to hold individual unexpectedly high peaks *or* to lift the signals from virtually any lower level into the usable working range of the equipment. Note, however, that high level signals may be excessively compressed and that background noise will be lifted as well – of which, more later.

The effect of a 2:1 compressor with a threshold at 8 dB below 100% modulation is to compress only the top 16 dB of signals while leaving those at a lower level to drop away at a 1:1 ratio. An interesting variation on this would be to place two limiters together, working on a signal in parallel but with the threshold set at different levels. If one is set for 2 dB below 100% and the other for (say) 24 dB below, but raised in level, the effect is to introduce a variable rate of compression that is greater at the lower end of the range than toward the top, where something approaching the normal dynamic range is used until the upper limiter operates. A further variation is that the lower limiter could have a weighting network to arrange that quiet high- or low-frequency signals were lifted more than those whose power was primarily in the central (500–4000 Hz) part of the audio range.

Expanders operate as compressors in reverse, allowing further fine control of dynamics. They can also be used for purely technical reasons, for example as part of the *compander* of a radio microphone, where the signal may be compressed before the local transmission, then expanded in the receiver to restore the original dynamics.

In large desks (especially those used for popular music) a range of controls is included in a dynamics module in line with each channel fader (or in 'assignable' desks, available for programming it). In simpler desks they may be offered only on output, and if required for individual (usually mono) channels, they must be inserted into the signal routing.

Compression using digital circuitry is very versatile. One such device, called an *Omnipressor*, has simple controls that can be set for expansion, compression or dynamic inversion. The expansion ratio starts at 1:10 then proceeds by discrete steps to 1:1 after which it becomes normal compression. The scale continues to infinite compression (i.e. a ratio of 1:0, a 'perfect' limiter) beyond which the ratio becomes negative, so that the quiet sounds are made louder and vice versa. This carries the compressor principle into new realms, as an instrument of creative sound treatment and synthesis.

Compressor design and use

Compressors and limiters work by sampling the signal at a particular point in the chain, and if it is above a designated level, deriving a control signal which is fed through a side chain to reduce the overall volume in the main programme chain. It is usually arranged that this control signal operates with a fast attack and slow decay (release or recovery) – though both 'fast' and 'slow' may have a very wide range of actual durations, with some controlled by the balancer. A typical (compressor) attack may vary from 0.1 to 30 ms, and the decay may take from 0.1 to 3 s. Other controls include threshold and compression ratio, and there may also be a meter to show the degree of attenuation.

In one compressor/limiter the signal is momentarily earthed (grounded) at intervals of 250 Hz, reducing its power, and so volume, by the necessary amount. At 6 dB above the onset point the selection of a 2:1 compression ratio means that half the power must be removed from the signal – to reduce it by 3 dB. It is arranged, therefore, that the control switch spends half of each successive 4 µs open and half closed. Such a high rate of switching has no audible effect on the signal. A signal 12 dB over the threshold has to be reduced by 6 dB, so that the switch must be closed for 3 µs of every 4; for signals at or below the threshold it does not operate at all.

Other systems have also been used for compressing and limiting signals: but a necessary criterion for any is that it should not operate selectively on that part of the wave which exceeds the threshold level (which would introduce harmonic distortion), but should reduce the whole wave in proportion.

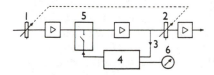

Simple compressor/limiter
1 and 2, Threshold level control (ganged). 3, Sample signal fed from the main programme chain. 4, Side chain with compression ratio and recovery time controls. 5, Control element operating on the main programme chain. 6, Meter showing the degree of attenuation. If a small amount of delay is introduced into the main chain, the side chain can operate into a *feedforward* mode, intercepting peaks before the first transient arrives.

Signal chopping to reduce power
1, Compressor threshold. 2, When the signal does not exceed this, the compressor remains inoperative. 3 and 4, For higher signals progressively more of the power is removed. Note that the chopping frequency is high compared with the highest audio signal and that it operates over the whole waveform, not just on the peak.

One of the problems in design (or operation) is choosing a suitable attack time – which can be reduced to a few tens of microseconds. However, such a very sharp change of level is liable to introduce an audible click. The answer is to delay the whole signal for about half a millisecond and then make a gradual change of level over the same period – in advance of any sudden heavy peak. This eliminates the click as the background changes level, but cannot affect the fact that if the background is there, it is suddenly going to change level – and that this is bound to be in some degree objectionable. The particular quality of unpleasantness is also affected by the rate of decay of the

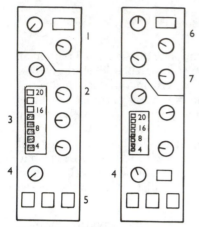

Compressor/limiter and dynamics modules
In-line channel controls. 1, Limiter threshold and release (decay time) settings and 'gate' key.
2, Compressor threshold, compression ratio, attack and release settings. 3, Gain reduction due to compressor or limiter.
4, Make-up gain amplifier setting.
5, Switches for compressor, limiter, or external control. Compressor/expander modules may also be available for fine control of popular music dynamics. 6, Expander controls. 7, Compressor controls.
Left, mono; *right*, stereo.

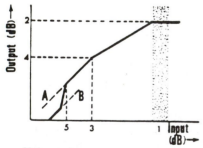

Noise gate
1, Range of maximum expected input level. 2, Corresponding maximum output level. 3, Lowest level of interest. 4, Lower end of designed dynamic range. 5, Gating level: if level of signal drops below this, the input characteristic falls from level A to level B.

control signal. Heavy limiting, or strong compression with a moderate recovery time, produces a 'pumping' effect that is mechanical-sounding and ugly.

Early limiters had gain-recovery, or control signal decay rates of the order of seconds: this allowed a reasonable time for the overload condition to pass so that it would not be pumped by too rapid a train of high level signals. The more sophisticated limiters (described above) have a selection of decay times available, typically 100, 200, 400 and 800 ms and 1.6 and 3.2 s. The fastest of these can hold isolated brief excessive peaks without having a severe effect on the background; but they are not so satisfactory for a rapid series of short peaks: any background is given an unpleasant vibrato-like flutter.

However, if there are only occasional peaks, and they are not too big – say, of not more than about 3 dB or so – the recovery time is probably not critical, and something like half a second is generally adopted: it is only at high levels of compression that the exact choice of recovery time becomes important. An alternative is to allow automatic variation between about 30 ms and 10 s according to the level and duration of the overload signal.

Discrimination against background noise

A major problem with limiters and compressors working on high gain is that background noise is lifted as well, unless something is done to prevent it. In particular, when there is no foreground signal to regulate the overall level, that is, during a pause, the noise is liable to rise by the full permitted gain. Such noise will include ventilation hum and studio clocks, movement and other personal noises, traffic or other more distant sound, electronic or line noise, and so on. Even where such sounds are themselves not raised to an objectionable level, certain consonants at the start of subsequent speech are not powerful enough to trigger the compressor, the most unpleasant result being the excessive sibilance at the start of certain words, though this can be avoided by high-frequency pre-emphasis in the input to the limiter: this is sometimes called 'de-essing'.

One device that can be used to discriminate against noise is a *noise gate*, a specialized use of the expander in a dynamics module. In a gate, when the sound level falls below a second, lower threshold setting the gain is allowed to fall to a parallel but lower input/output characteristic: this might, for example, be 8 dB below that which would apply without a gate. At this setting, a device that could not previously have been used for more than 8 dB compression on a particular signal without the background noise becoming excessively loud in the pauses could now be used for 16 dB compression.

Noise gates are often made fast operating, so that in even the slightest pauses between the syllables of a word the noise is sharply reduced.

But for such an operation to be successful, the device has to be set up with care: the gate must be *below* all significant elements of speech, but *above* all of the basic noise. Indeed, if both of these conditions cannot be met the result could be unpleasant, with either the speech or the noise wobbling up and down on the edge of the gate. But in any case extraneous staccato sound of a higher volume or a rise in the background to an unexpectedly high level will fool the gate; as also (in the opposite sense) would a very quiet passage at the start of a piece of music.

In the 'automatic level control' available on some high-quality portable recorders used for professional sound radio and film location work, yet another type of limiter is used: one that avoids most of the rise and fall of background noise that is heard with other limiters. A short delay is introduced before the control signal begins to decay. The duration of this delay can be adjusted to allow for different lengths of pause between syllables or words as spoken by different individuals or in different languages, but is normally 3 s – after which the 'memory' is erased and a new level of adjustment selected. This same recorder has a separate circuit for the control of very brief transients, which do not therefore affect the basic level. Obviously, such a device still has limitations: there would undoubtedly be many cases where a human operator would adjust the level, for example, on a change of voice; so it will probably only be used where it is inconvenient to employ a human operator. Like the fast-operating gate, the device can all too easily be fooled by unexpected events.

Chapter 11

Filters and equalization

Changes in audio-frequency response, which might also be called frequency attenuation or distortion, have been one of the most important obstacles to be overcome as audio engineering has developed. Once it is in the form of an electrical signal, audio information can be handled with little significant change, or with changes that can readily be compensated. On the other hand, transducers – devices that convert the signal from one medium to another – often introduce marked changes, differentially responding to the frequencies present. The worst occur on conversion of the electrical signal back to sound. All loudspeaker designs differ, and domestic loudspeakers are totally outside the control of the professional originator of the sounds they reproduce. Further selective emphases are imposed by the rooms in which the material is finally heard.

In fact, the expected variability of acoustics, together with the wide range within which sounds remain recognizable for what they originally were, is a psychological asset that makes or excuses changes in frequency response at all stages in the chain: the untrained ear is remarkably tolerant of this form of distortion. The tone controls of early audio equipment were blunt instruments, coarse in their operation, accommodating the need to eliminate hiss or rumble, or to excise frequency bands containing other forms of distortion that were far more annoying than simple truncation of the reproduced audio range.

The ear is more objective about direct comparisons between sounds with different distributions of frequency distortion if they are presented one immediately after the other or both at the same time. In the early development of audio equipment the most fallible items were microphones, and their matching became an important component of radio and television sound technique. For example, for a long time BBC radio used little but ribbon microphones for speech: monophonic drama was balanced using several, differing in their acoustic surroundings but not in their basic response. Stereo drama has used either one main matched pair of microphones throughout a production, but changing the acoustics from scene to scene, or similar pairs in a range of acoustics. These techniques remain valid, but an alternative is to use a range of microphones chosen according to different criteria of, say, sensitivity or handling characteristics, and either bear with any differences in frequency response or change them.

For television drama, the main boom microphone is the standard against which other sound is judged by the listener. Other microphones picking up the same voices have to be matched to that standard or the transition is obtrusive. This demands a capacity for changing each microphone's frequency response in detail. So, for film documentary, does the jump between a personal microphone on location and the studio microphone in a commentary booth. In both radio and television, an even more drastic effect is the simulation of telephone quality, for which the normal frequency range is curtailed by means of filters.

For music, it became part of the balancer's skill to choose a microphone with frequency characteristics complementing those of the instrument. Today, some music balancers use an expensive condenser microphone with a response that is substantially flat throughout the audio range for nearly all instruments. Comparable results can, in fact, often be achieved at lower cost by the earlier technique, with the added advantage that the microphone's response is reduced outside the wanted frequency range of the instrument, thereby discriminating against noise and spill. Indeed, there has been a resurgence of interest in 'tube' microphones − early electrostatic microphones in which the head amplifier contains a valve (vacuum tube) − which are treasured for their vintage quality. In general, it is pop music balance that takes the greatest liberties with frequency response, with bass and treble emphasized or rolled away, and selected bands of mid-range frequencies lifted to lend presence to particular components of the mix. If frequency bands containing the characteristic elements of musical quality of each instrument are raised (which is the same as taking down the remainder of each sound), the result is a musical caricature or cartoon. In fact, that is precisely the object intended − and when it succeeds, the selected essentials do, indeed, seem to press themselves upon the listener. Historically, the capacity to achieve such effects was provided by equipment originally designed to correct the deficiencies of microphones, and to allow for their matching more smoothly, one to another. It is curious, indeed ironic, that deliberate distortion of the sound of an expensively engineered smooth response has usurped the name of *equalization* (EQ).

In this chapter we are concerned with equalization and related techniques, and their creative use in radio, television and music recording.

Filters

Older audio equipment often had bass and treble tone controls, as they were called, to add or subtract 10 dB (or more) at 100 Hz and 10 kHz, doing so in continuously variable or switched increments. The name has gone, but the function remains. There may also be a range of sharper top- and bass-cut filters (with or without variable slope). Narrow-band 'notch' filters are used to eliminate mains hum at 50 and 100 Hz, or 60 and 120 Hz, and also the television line frequency at about 15 KHz. It is also convenient to be able to switch any of these in or out, so that direct and corrected sound can be quickly compared. In most audio control desks today there is still provision for varying top and bass as part of the equalization equipment in each channel, while other filters with more limited uses are provided as patch-in facilities as, for example, in the case of the sharp filters that are used in drama to distort speech.

An advance on the original bass and top-cut filters was those used for *midlift*, now considered a vital part of many music balances. For example, a singer's voice may be raised by 6 dB at 2.5–3 kHz. A snare drum or woodwind might also benefit from lift within the same frequency range. Note, however, that recommendations here and elsewhere in this book are necessarily imprecise, since the final effect must be selected by ear, in the light of the characteristic qualities of the instruments, the relationship of one to another, and the overall melodic or harmonic effect desired.

Midlift is a name that is appropriate only in the sense that above and below the frequency at which it is applied, the response returns to normal. The middle of the audio range is generally taken to be around 256–440 Hz; but midlift is often applied a few octaves above that, acting on the harmonics of the notes produced, perhaps enriching and extending them. As different notes are played by a particular instrument, the midlift frequency remains constant, so it behaves rather like a physical part of that instrument, creating a formant – or, more likely, enhancing a formant already present that helps to define the character of the instrument.

Vocal formants, in the middle range, are changed continuously in the process of speech, but the higher ones resonate in cavities of less flexible or fixed size; and for midlift on vocals it is usually this upper range that is enhanced.

Early midlift devices had a simple range of possibilities, offering perhaps four or five nominal frequencies separated by intervals of rather less than an octave, together with several degrees of lift, switched at increments of 2 or 3 dB. Later designs have made the range more flexible in both frequency selection and volume, and may also allow for a dip or *shelf* as well as a peak: by analogy with *presence* the effect would be *absence*, and this too may be used in popular music mixes. A desk in which equalization is programmed digitally can have a wide range of settings that it can reproduce reliably. In early midlift systems there was no choice of the width (known as the

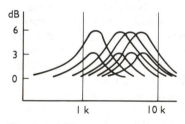

Simple bass and treble control
The effect of early 'tone' controls.

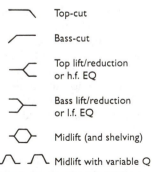

Simple midlift
The effect of an early two-stage, four-band system.

⌐_	Top-cut
_/⌐	Bass-cut
⤳<	Top lift/reduction or h.f. EQ
⤳>	Bass lift/reduction or l.f. EQ
⟨◇⟩	Midlift (and shelving)
⋀ ⋀	Midlift with variable Q

Some commonly used EQ symbols

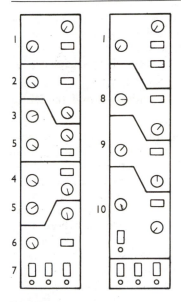

Equalization modules
In-line channel controls on a desk designed for mixing popular music.
Left: Mono channel. 1, High and low pass fitters. 2, HF equalization.
3 and 4, High and low mid-frequency equalization. 5, Variable Q (band width). 6, LF equalization. 7, Routing switches.
Right: Stereo channel. 8-10, High, mid and low frequency equalization. All equalizers have a wide choice of centre frequencies and boost or shelf levels.

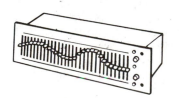

Graphic filter
In the most usual design the slide faders are at third-octave intervals, and control the gain or attenuation in a band of about that width. In principle, the overall effect should match the line shown on the face of the module, but this may not be achieved in practice. The result should be judged by ear.

Q) of the peak; in a more flexible system this, too, may be varied and the peak may be replaced by a shelf, also of variable width.

In most desks today, a group of controls for bass and treble cut and several stages of midlift is provided on all of the individual channels, mono and stereo, and also in group fader channels.

At the top of the analogue-controlled range, each mono channel may have as many as a dozen rotary controls and a variety of other switches. The high- and low-pass filters remain relatively simple, with (in one example) an 18 dB/octave cut at a choice of ten turnover points from 3 kHz up and 12 dB/octave cut at another ten from 350 Hz down. (The nominal cut frequency is generally taken to be that at which output is reduced by 3 dB.) The equalizers operate in bands of 1.5–10 kHz, 0.6–7 kHz, 0.2–2.5 kHz and 30–450 Hz, each with a range of peak and shelf settings. In addition, the two in the middle have variable Q. There is an overload indicator light, monitoring the signal at three points within the module.

The stereo modules have fewer controls, but each is available over a wider range; and overload is indicated in the two channels separately.

If the signal passes through a filter network with everything set to zero, an analogue-controlled setting can never be quite as smooth as the original, so it is usual to include a bypass switch which, in effect, switches the whole module on and off.

In computer-controlled systems, the settings may be selected and assigned from a single central module which is separated from all of the individual channels, but stored in computer memory to operate upon them.

Graphic equalizers

Even in combination with bass and treble filters there is a limit to the flexibility of equalization based on midlift controls. In a television studio or when filming on location, the degree of mismatch between microphones may be more complex. For this we must take the principle of midlift a stage further and split the whole spectrum into bands, typically about 30, each of one-third of an octave, and apply correction separately to each. Equalizers of this kind are cross-plugged (patched) into the channels where they are needed.

If all the controls are set at zero, forming a horizontal line across the face of the panel, the response, too, is approximately level throughout the audio range; in practice, the actual line may be distinctly bumpy – and there is a surprising amount of difference in performance between units from different manufacturers. In any case, the unit should be switched out of circuit when not in use.

If individual sliders are raised or lowered, their associated circuits form peaks or troughs that overlap, each blending into the next. The selection of the desired effect is made by setting the slides for the

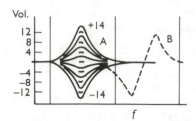

Audio response control
Ideal effect of (A) a single fader and (B) two adjacent faders set at opposite extremes. In practice, sharp jumps in the response can only be achieved at the expense of a more erratic response when the sliders are set for a smooth curve. Apparently similar units may differ in their capacity to compensate for this.

various component bands to form what appears to be a graph of the new response. In fact, the actual response achieved may be different: it depends on the design of the filter and the size of the fluctuations set. The true arbiter – once again – must be subjective judgement.

Graphic filters are required in film dubbing theatres, on television control desks and also in creative sound studios such as the BBC's Radiophonic Workshop.

When speech is put through a graphic filter it is possible to make severe changes, for example, several dips of 30 dB, without significantly reducing intelligibility. If the filtering is intended not simply to restore the deficiencies of a balance made under adverse conditions, but instead to achieve some positive and marked dramatic effect, the imposed distortion has to be drastic. However, the selective emphasis of particular frequency bands is noticeable where a dip would not be. This is because the brain reconstructs defective harmonic structures, perceiving the sound as experience says it should be, but finds it somewhat more difficult to disregard that which should not be there, but manifestly is.

Given that peaks are more noticeable than dips, a useful strategy for locating and reducing distortion or noise which affects a limited but unknown frequency range is to lift one or several faders successively, until the problem sounds significantly worse – then reduce their settings by as much below the median.

Telephone effects

Telephone circuits usually have a bandwidth of some 3000 Hz with a lower cut-off at about 300 Hz, but this should not necessarily be copied when simulating telephone quality. The selection of suitable settings for bass and treble filters should, as always in such cases, be made by ear. And the degree of cut that seems most appropriate will vary from one play to another – or even between different simulated telephone conversations in the same play – as, in real life, the amount of the lower frequencies may vary with how closely a telephone is held to the ear.

Simple effects unit
The unit has four degrees of bass cut (B1–4) and five of treble cut (T1–5).

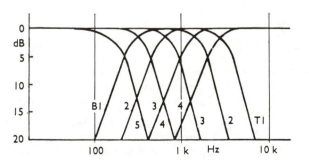

One factor that has a bearing on this is the difference in voice quality between male and female speech; and between different examples of each. If the fundamental of a male voice is at 110 Hz, then a cut at 220 Hz will give a small but appreciable change of quality. For a female voice with a fundamental an octave higher, a 440 Hz cut is more suitable. In fact, 440 and 880 Hz are more often used: the loss of bass is stressed for dramatic effect. The top-cut will probably be closer to that of the telephone circuit. A microphone used in conjunction with such circuitry is called a *filter microphone* or, in BBC jargon, a *distort microphone.*

In a full-scale studio telephone conversation set-up it is usual to have two microphones, one normal and the other with a filter in circuit. In radio, these two can be placed in different parts of the same studio, the idea being that they should not be separated so much that the actors cannot hear each other direct. This arrangement avoids the necessity of wearing headphones.

In such a set-up it is important to avoid acoustic spill, which in this case means pick-up of the supposedly distorted voice on the normal microphone: otherwise the effect of telephone quality will be lost. If the 'near' microphone is directional, and it is placed with its dead side toward the 'far' voice, the pick-up of direct sound is effectively prevented. But discrimination against reflected sound presents more of a problem, so it sometimes helps to use the deadest part of the studio for the filter microphone. As spill from the full range to the filter microphone does not noticeably change the quality of the 'near' voice, it does not really matter very much whether the second microphone is directional or not.

The two microphones are mixed together at the control desk and the relative levels are judged by ear. When this sort of distortion is being introduced into a programme, meters are even less use than usual in matching the two voices, as the narrow-band input contains less bass, and therefore less power, for similar degrees of loudness. At normal listening levels there is very little difference in intelligibility between full-range speech and the narrow band used to simulate telephone quality, so a fairly considerable reduction in the loudness of the distorted speech may be tolerable in the interests of creating a feeling of actuality.

In a television studio the two microphones should be well separated in space. Then, as the picture cuts from one end to the other, the normal distort arrangement can be reversed, without spill in either direction. It is useful to have the selection linked to the vision cutting buttons so that sound and picture changes are synchronized. But it is best to avoid cutting during a word as this sounds unpleasant and draws attention to the technique being used.

An alternative arrangement uses auxiliary circuits. Each channel in the desk has provision for a feed to be tapped before the fader and fed to an auxiliary bus, in which the output of all channels selected is combined in a rough mix which can be sent to any local or remote point. Many desks have two or more such auxiliary 'sends', to allow for *multiway working.*

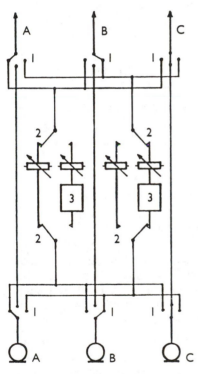

Telephone effects switching unit for television

1, Pre-selector switches. 2, Switch linked to video cut buttons, any sound routed through this is automatically switched in and out of the 'distort' condition as the pictures are cut. 3, Bass and treble filters. In this example, microphone A is undistorted only while B is filtered and vice versa, and microphone C remains undistorted whatever the picture.

are not, coloration can affect both artists (distant, reverberant but otherwise undistorted sound is superimposed on the telephone quality speech). Also, if one performer speaks more loudly than the other his voice is likely to be heard over the quiet speaker's microphone. It may help for the speakers (in their separate sets) to be facing each other in the studio layout, so that cardioid boom microphones are directed away from each other. On the other hand the use of wide shots makes matters worse, because boom microphones have to retreat from the speaker, thereby making it more likely that the distant voice will be heard as the channel gain is raised.

Background effects added to a telephone conversation may make a considerable difference to the necessary balance between voices. The degree of distortion and its volume relative to the undistorted speech must depend on both the loudness of the effects and their quality. In particular, background sound that is toppy in quality tends to blot out quiet speech, whereas if the backing is fairly woolly or boomy the shriller telephone effect has little difficulty in cutting through it. In either case, the mix must be very carefully monitored to check that intelligibility is reasonably high – unless, of course, the script requires otherwise.

A fairly complicated scene may require varying degrees of intelligibility – as for example in a radio script like this:

EFFECTS	(*Loud party chatter background; phone dialling close, ringing heard on line: it stops.*)
GUEST	Hey, is that you, Charlie?
CHARLIE	(*On distort, barely audible against the party chatter.*) Yes. This is Charlie.
GUEST	Say, this is a great party; why don't you come on over?
CHARLIE	What are you at, a football game or something?
GUEST	What am I what?
CHARLIE	I'm in bed.
GUEST	Hang on, I'll shut this door. (*Door shuts; chatter down.*) S'better! Now what was all that about a football game? It's three in the morning.
CHARLIE	(*Now more distinct.*) Yeah. It's three in the morning. I'm in bed.

Two different mixes are required for this; and the balance and pitch of the near voice must change on the cue 'door shuts'. In addition, the backing should change in quality, too: heard through a real door, such effects would become muffled and muddy, as well as being reduced in volume. Some degree of top cut might be introduced at this point.

There are many other dramatic uses for filters besides telephone quality: for example, to simulate station or airport announcements (adding a touch of echo), or intercom (at the same time simulating a poor radio link by playing in a crackling, hissing background). A radio play (or recording) may include the effect of a radio announcement (or a tape recorder) for which a convenient convention is that the reproducer has little better than telephone quality. This again is a caricature in which one type of distortion, reproduction of sound through a single loudspeaker, is represented by another, limitation of frequency range.

Chapter 12

Artificial reverberation

'Artificial reverberation' is often shortened to the more convenient 'echo', but that is something of a misnomer for a studio technique that serves to extend reverberation without (it is hoped) introducing any noticeable echoes. It is used on occasions when more reverberation is wanted than the built-in acoustics of a studio can supply, or (for music) to rebalance the contribution of spotting microphones, or more widely with a multimicrophone close balance. In the past, various forms of artificial reverberation have been used, including echo chambers, plates and springs. These all differed in quality from one another, and from the real thing, as did early attempts to use digital devices.

This is one of those areas where increased processing power at reduced cost has completely defeated the opposition, to the extent that 'plate' may be just one of a hundred descriptions offered in a display window – retained for the benefit of those who recall its characteristics and can still find a use for its distinctive response. But now that artificial reverberation can be constructed to mimic the characteristics of natural acoustics so much more accurately, most users, for most purposes, will now prefer to head for another window – if only as their starting point.

Components of reverberation

First, a distinction. A true *echo* may be heard when reflected sound – a discrete repetition – is delayed by some twentieth of a second or more. As the separation is reduced there comes a point when even a staccato original and its repetition are no longer perceived as separate entities, but as a single sound, a composite that is different in character. There is, in fact, partial cancellation at certain frequencies.

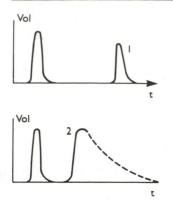

Echo and reverberation
1, An echo is a discrete repetition of a sound after a delay of some 20 ms or more. 2, Delays that are shorter than 20 ms, associated with the dimensions of all but the largest halls, are perceived as components of the reverberation.

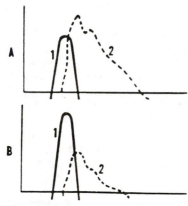

Echo mixtures
1, Original signal. 2, Reverberation. With separate control of direct sound and artificial reverberation, a wide range of effects is possible. A, Reverberation dominates the sound. B, Reverberation tails gently behind it.

Delay effects may fall into either category: they may be short, to make use of the phase cancellations; or long, producing perceptibly separate echoes, perhaps with further repetitions.

Natural reverberation begins with *first reflections* that arrive from the roof and walls – in the centre of a concert hall, after a substantial delay – followed at progressively shorter intervals by many more, smaller products of multiple reflections until, vast in number, they trail away below the threshold of hearing. This is the process that much Western instruments and ensembles were designed to exploit: it is an unwritten part of the music itself. When considering alternatives to natural reverberation we must therefore ask how their decay processes compare with the natural form. In the music room or concert hall, some frequencies are absorbed more readily than others at each reflection; if after multiple reflection some frequency bands are sustained while others diminish, the result is *coloration*.

Coloration occurs in both natural and artificial reverberation, but note that the build-up and blending of natural reverberant sound depends on a three-dimensional system of reflections and decay which greatly enriches the quality at just that stage when coloration is likely to be most severe. When the artificial reverberation differs from natural acoustics, its use is justified if the new pattern of coloration and density of decay products is still perceived as musical in quality (and if its origins are not obtrusively apparent). In many cases the natural acoustics are themselves unsatisfactory, and at worst, the artificial reverberation available to augment them may be justified on the grounds that it is better than nothing; at best, it may be indistinguishable from the real thing. In some cases, it may provide a distinctively different, but desired effect.

It is best to feed an audio signal to echo at fairly high volume so that noise from the echo circuit itself can be kept low, though this has become less critical now that digital echo is used; ambient noise was a marked limitation of plates and chambers.

Most control desks have provision to split the feed from each source channel fader, diverting part to a range of 'auxiliaries', one or more of which can be fed to artificial reverberation devices which are located outside the desk. In many desks auxiliaries can be fed from a choice of points, either before or after the fader. For 'echo', the most convenient arrangement is for the signal to pass through the channel fader first. If this is not possible because the only auxiliary feed is before the fader, the echo return level may have to be adjusted every time the individual channel is faded up or down.

If the whole of the sound from a particular (single) source is to be passed through the echo device, this may be cross-plugged (patched) into the channel: this is likely to be before the fader.

For stereo echo, A and B signals are often combined before being fed in, but A and B reverberation is extracted separately. This also means that mono sources taken as part of a stereo mix automatically have the stereo reverberation that is essential to a realistic sound. The reverberation from each device is returned to the desk as a line source (usually stereo) with its own fader (or faders) and equalization.

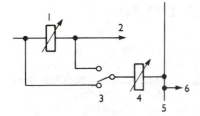

Flexible 'echo send'
1, Channel fader. 2, Channel output.
3, Pre- or post-fader switch. 4, AR
send fader. 5, Bus-bar allocated to
AR. 6, To AR device, outside control
console. On many consoles, the feed
to AR passes through auxiliary 'send'
circuits that may also be used for
many other purposes, which in the
simplest case may be fed from a
point after the channel fader, so that
when a channel is faded out, the
reverberation goes, too. In a complex
balance, several AR devices may be
used, each fed from a separate bus-
bar.

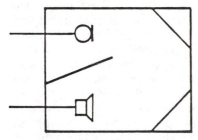

Echo chamber: size
A U-shaped room may be used to
increase the distance the sounds
must travel from loudspeaker to
microphone.

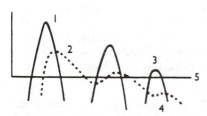

Artificial reverberation: level
1, Loud direct sound. 2, Audible
reverberation. 3, Quiet direct sound.
4, Reverberation now below
threshold of hearing (5).

The echo chamber

Historically, the first way of enhancing reverberation was to use a separate room, formalized as an *echo chamber*. This had the advantage of real acoustics, although usually in far too small a volume.

It has reflecting walls, perhaps with junk littered about at random to break up the reflections and mop up excess mid-range reverberation. It might have a wall down the middle, so that sound has to follow a U-shaped path in order to delay the first reflection. The loudspeaker itself can be made directional so that no direct sound is radiated towards the microphone, which in turn can be directional, discriminating against both sound from the region of the loudspeaker and coloration due to structural resonances (between parallel roof and floor, for example).

A humid atmosphere gives a strong high-frequency response; a dry one absorbs top. An echo chamber linked to the outside atmosphere therefore varies with the weather.

In fact, when a large hall is being simulated, there should actually be a reduced high-frequency response, corresponding to the long air-path. In some chambers, the bass response itself may be too heavy; or alternatively the original sound may already have sufficient bass reverberation, so that none need be added. EQ and bass-cut controls in the return channel allow the necessary adjustments. Note, incidentally, that if less bass is fed to the chamber, its loudspeaker can be used more efficiently, but if bass is reduced for this reason it needs subsequent equalization.

Because it takes up valuable space within a building, an echo chamber is often rather small, perhaps 4000 ft^3 (113 m^3) or less. As a result there is broadly spaced coloration, due to the natural room resonances in the lower middle frequency range. Like radio studios, it needs to be isolated from structure-borne noise, and even then is likely to be put out of commission by building or engineering works unless all noisy work is stopped.

Yet another disadvantage is that, once laid out, its reverberation time is fixed. If this is about two seconds it may be suitable for music but little use for speech. If, however, only a touch of echo is used it is noticeable only on the louder sounds; a situation that corresponds well with what we hear in real life. In this case an overlong reverberation time is apparently reduced.

For music the reverberation time itself may be right, but with too early a simulated first reflection (because the shortest pathlength between loudspeaker and microphone is far less than in the concert hall). To restore both clarity of attack and a sense of spaciousness, a delay circuit should be inserted into the echo return feed.

Far better results have been achieved by using a spare empty studio or full-sized hall as the echo chamber, but this is rarely practicable.

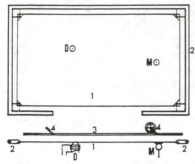

Reverberation plate (mono)
1, Metal sheet. 2, Tubular steel frame. 3, Damping plate, pivoted at 4. (Here the spacing is shown as being set by the handwheel, but many damping plates have been motor driven from a point on the sound control desk.) D, Drive unit. M, Contact microphone. For stereo, two are required, asymmetrically placed.

The reverberation plate

The first substantial advance on the echo chamber – more versatile because its reverberation time may be varied – is the reverberation plate. In principle, this is rather like the sheet of metal that has been used in the theatre to create a thunder effect, except that, instead of being shaken by hand and delivering its energy directly to the air, it has transducers. One vibrates the plate, rather as the coil vibrates the paper cone in a moving-coil loudspeaker; and (for stereo) two others, acting as contact microphones, pick up the vibrations.

Reverberation plates have a tinned steel sheet suspended in tension from a tubular steel frame at the four corners. To reduce the metallic quality of the resonance to proportions that are acceptable for most purposes, a minimum size of 2 m^2 is combined with a maximum thickness of 0.5 mm, which gives it good transverse vibrational properties. Unlike the echo chamber effect, its natural resonances at low frequencies do not thin out to sharp peaks in the bass, but are spread fairly evenly throughout the audio range.

A moving-coil drive unit is anchored to a bridge across the frame and piezoelectric contact microphones pick up the sound; these are asymmetrically placed on the plate. For stereo, the A + B input is fed to the same drive unit, but two pick-ups are used, so placed that they do not respond to a common set of vibration antinodes of the plate. The frequency characteristics of the plate itself, the two transducers and their amplifiers are arranged to give a response that has its maximum duration at mid-frequencies, with some roll-off in the bass and rather more at higher frequencies (15 dB at 10 kHz), thereby simulating the high-frequency absorption of the air of a room of moderate size. To achieve this, some damping has to be applied: without it, the response of the plate would rise to very high values in the extreme bass.

To produce this damping, and to control the reverberation time, a thin, stiff, porous foil is rigidly held at a controlled but variable distance from the plate: typically, 0.8 mm of a compressed glassfibre material is used. With the broadest separation of the damping plate (120 mm) there is a reverberation of 5.3 s at 500 Hz, dropping to 1.5 s at 10 kHz, as in a cathedral. The narrowest separation between the plates is as small as 3 mm: this gives a nominal reverberation of 0.3 s – a figure that may prove optimistic; if so, the plate is better used with reverberation times suited to music rather than speech. The damping plates may be operated mechanically or motorized, with push-button control and a reverberation time indicator on or near the control desk: this allows experimental changes to be made at any time during rehearsal.

The two-dimensional structure of the plate allows the wave to radiate outward and break up, producing reflections that arrive in increasing numbers as reverberation continues and dies away; a 'natural' effect, though not so marked as with three-dimensional reverberation, from which it is distinctively different – most notably in its effect on string tone.

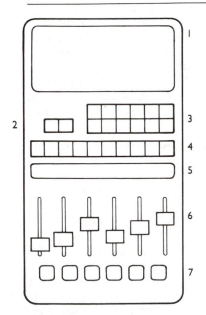

Digital AR control unit
The deceptively compact controls for one 'digital signal processor' which offers a wide variety of AR and other effects. 1, LED display showing a descriptive name of the selected program, its main characteristics, and the volume (including overload warning). 2, Program or register keys. 3, Number of program or register selected. 4, Function keys. 5–7, LED display, sliders, and keys associated with the currently selected control 'page'. Operation is not intuitively obvious: new users should follow the step-by-step instructions, guided initially by the supposedly 'user-friendly' program names.

Plates vary both in the quality of the sheet and in the tension that may have been applied to them when they were set up. This is a specialized skill, and maintenance of old plates is a problem. Individual plates have their own characteristic response.

With a plate, having a short direct path between the transducers is unavoidable, so, again, to simulate a large hall tape delay is required.

There may be difficulties in finding a suitable place to stand the plate: it needs to be kept fairly quiet as it picks up noise from the air. In particular it cannot be kept in the control cubicle, as with very high loudspeaker levels howlround may occur at 200 Hz. The studio lobby is a reasonable place if there is room; or some separate quiet room – perhaps with other plates for other studios (they do not affect each other).

A third medium for artificial reverberation has employed what are usually described as 'springs'. Although that is what they look like, they work quite differently: the acoustic input is applied torsionally to one end of the helix, and sensed in the same manner at the other. The metal is differentially etched and notched along its length, so that its transmission characteristics change, with the effect that at each discontinuity there is a mismatch, and part of the sound is reflected. Each such reflection fathers its own family of subsequent echoes from all the other irregularly distributed discontinuities, and so the characteristics of reverberation are produced.

Separate springs are required for the stereo A and B channels, and these can either be totally independent of each other or their inputs may be linked so that A + B is fed to both. One example has a built-in effective first-reflection delay of 20 to 50 ms. When set to have a nominal reverberation time of 2 s, its response is level to 5 kHz (the reverberation time itself gets shorter at 5 kHz and above).

Early, inexpensive systems of springs were developed for use in electric organs. With improved engineering, they produced a quality of reverberation comparable with chamber and plate, but with reverberation times of 2.0–4.5 s were not suited to speech, although their treatment of string tone could be more pleasant than that of their early rivals.

Digital echo

Echo chambers, plates and springs may still be encountered and used for particular special effects, but are no longer worth the effort or expense of maintaining them. Digital techniques now cost less and are far more versatile.

A typical good digital 'echo' device may have a wide range of different reverberation characteristics, many of which are preset, but with the additional capacity for customized programs which can be saved and re-used. As with other advances in computer technology, this has depended in part on vast increases in random access memory (RAM), together with high-speed processing and improved software. As with all such equipment, its operation – or, at least, choice of setting – is a matter of becoming familiar with the opportunities offered by the original programmer, together with any deficiencies or 'bugs' that may emerge during use. Few operators use more than a limited range of the facilities made available by any computer system – life is too short – but it is worth experimenting.

The manuals and instructions issued with computer equipment are often full of jargon that is likely to put off all but enthusiasts, in addition to which they may be appallingly badly written. Unfortunately, it is necessary to struggle to understand them, as somewhere among the wild swings between the incomprehensible and the oversimplified there may be genuine nuggets of indispensable information which apply to this device and no other.

To learn to operate a relatively complex (and therefore versatile) device, first run through the existing programs in memory. As well as displaying strings of numbers (the parameters selected within them), these may have been assigned 'user friendly' names; but it is best to take these purely as a general indication of content and as a starting point in your search, rather than expect the programmer's idea of a particular kind of acoustic to match your own. Check how to select from what is on offer, and also how to store your own modified (or originated) acoustic programs. The simplest strategy is to use only those already available, of which you may feel there is an adequate range available for most purposes. But you will probably soon want to start 'tuning' them. Plainly, it is essential that this should be done without significant incursions into production time, so it is essential to familiarize yourself with the effect of what may seem to be an all-too-wide variety of options. A simple way of learning to use the controls (press buttons and sliders) is to convert one existing program until it matches another already in memory: the match is confirmed by the series of numbers – 'parameters' – which define the acoustic effect. Add the original and the intervening stages to a short, rather dry recording in order to monitor the changes.

The next stage is to alter the parameters to something completely new. This can be more complicated, as there may be an inhibitingly large range to choose from, and you may be justified in your suspicion that some offer different routes to the same objective. One device offers 22 parameters which can be altered to modify reverberation, and another 13 to develop 'effects'. The aim is to control all stages in the simulated acoustic reinforcement of a sound, beginning with the time and level of first reflections, the onset and character of more continuous reverberation, and the subsequent dynamics of various frequency components and the change in density of echoes that go to make it up – all within a stereo image. The 'reverberation' parameters may include a 'chorus' program to increase the apparent number of sources within the originating section, and also offer flanging and pitch-changes. The controls actually under the heading of 'effects' include differential delays for variously defined bands or tones within some of the more complex programs offered, and move on from those to go even further beyond the realm of natural sound modification.

Variable reverberation

The most important single measure of an enhanced, natural-sounding acoustic is its reverberation time. Using electronics, this can be set for different values in several frequency ranges. Depending on its period and style, classical music generally sounds best in a hall with a mid-range decay of 1.8–2.5 s. However, the low frequencies may require longer, say 2.5–3.0 s, and it will be necessary to set a crossover frequency, at perhaps 600–800 Hz. Treble decay may need to be shorter, with a crossover at 6 kHz or higher.

In addition, the length of decay in one or more bands might be made to depend on the presence or absence of gaps, so that it is relatively short when the sound is continuous (for vocal or instrumental clarity), but longer (and so richer) in pauses. This is a modification that will be used cautiously: on lead vocals it helps to define phrasing, but if it also applied to supporting lines that have gaps in different places, it may muddy and confuse the overall effect. In a reversed, or 'stooped' version of this, a component can be given rich reverberation as it runs continuously, but rapid decay to clear any pauses in order to allow intervening figures to stand out in them.

Delay, sometimes confusingly called 'pre-delay' (it means the delay before reverberation begins), is another essential, natural-sounding quality, at best creating the effects of distance and spatial volume, and allowing the transients of the original sound to be heard clearly before the arrival of any reflected sound (in which the initial volume – or so-called 'attack' – may be another separately controlled parameter). Delay, too, must be used with care. In a large concert hall the delay between the original sound and the reverberation may be considerable: 100 ms corresponds to an apparent difference of some 110 ft ($33\frac{1}{2}$ m). When the original sound is not clean (not completely dead) this allows ample time for the original acoustic to bridge the gap. One remedy is to trim the delay, so that the original and artificial acoustics overlap more; another, probably better, is to modify the distribution of early reflections – in effect, bringing the listener forward, but within the same apparent size of hall – using a 'depth' control, which provides an appropriate arrangement of the pattern in the early stages of reverberation. It may also be useful to have control of the earliest reflections in order to simulate nearby reflecting surfaces, while the density of later reflections can be altered to suggest various patterns of diffusion.

This catalogue of controls could continue, and the ultimate possibilities seem virtually endless – but already there is some overlap between the functions offered even for extending reverberation that sounds natural. For treatments that do not occur naturally, plainly the range may be much wider; there is more than sufficient to feed a creative imagination.

Advances in computer technology have been such that extremes of complexity are now available in the professional studio for less than the price of a plate – versions of which are mimicked by the electronics among the set programmes, so that balancers who had learned to make constructive use of its severe limitations need not feel deprived and will also have a known (if simulated) reference point to move on from.

Less complex devices can now, in principle, be provided at much lower cost, and will still do far more and with better quality simulations than in the earlier systems. But while the choice of 'echo' effects has grown so much, the microphone techniques for joining them with live sound have remained much the same.

Using 'echo' on speech

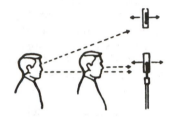

Echo for perspective
The use of a second directional microphone, suspended above the main and having a stronger echo feed, helps to give depth to crowd scenes (for stereo, this may employ two crossed pairs).

Because the plate and other early devices were so poor at providing short reverberation times, one of the greatest improvements offered by the use of digital reverberation is found in its application to speech – usually for dramatic effect. The main limitation is now due to the studio itself, and the performer's working distance. If an actor works close to the microphone and whispers, he will, for the same setting of the echo fader, produce just as much reverberation as when standing 10 ft away and shouting. So watch perspective effects: as an actor moves in toward the microphone the echo has to be faded down.

However, if there are other voices or sound effects in a different perspective to be taken at the same time, a more complex set-up is necessary. One useful arrangement has a second directional microphone (or stereo pair), with a strong echo feed, suspended above the main microphone, which itself has little or none. As the actor moves in close to the lower microphone, he is no longer in direct line with the upper one. Thus, perspectives can be emphasized without constant resort to the echo fader. This refinement is more likely to be of use in radio than in the visual media where there are restrictions on the ideal placing of microphones.

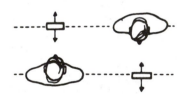

Echo on two voices
If differing amounts of echo are needed on two voices taking part in a conversation, it helps if opposing microphones and voices are placed dead-side-on to each other. For stereo, the two can be steered to suitable places within the image. This layout might be adopted in a radio studio; in a television studio there is normally more space available, making separation easier.

Feedback and delay effects

Some of the less natural effects of feedback and delay had already been explored and developed by more laborious techniques long before the present digital systems became available. For example, a simple form of delay could be achieved by employing the distance between the recording and reproduction heads of a tape recorder, though for any flexibility, variable-speed drive was needed.

If (in whatever system of delay is employed) the output is mixed back into the input, the sound circulates, dying away if the feedback signal is attenuated, or becoming a howlround if the signal is amplified. This effect is called recirculation, *spin* or sometimes *slap echo*, and has a mechanically repetitive quality. The effect depends on the delay interval: at a fifth of a second, for example, there is a marked flutter effect.

It is possible to start with a staccato effect such as a tap on a water-glass and let it spin indefinitely by controlling the feedback level manually. Before the advent of digital technology, any slight unevenness in the frequency response of the circuit was quickly disclosed: using tape, even a high-quality professional recorder soon reveals its deficiencies and within half a minute the characteristic of the system completely swallows the original sound. This effect can now be controlled.

It was also found that delay could be used to double the apparent numbers of instruments – now more flexibly achieved by digitally randomized 'chorus' effects. Another practical application may be to resynchronize sounds that have been produced a little out of step owing to the time a sound cue takes to travel through air.

The short-delay effects in which phase is important were originally explored in pop music by tape techniques before pure electronic devices were introduced. Two recordings were played, initially in synchronization, then one was slowed by hand against the flange of the tape spool – from which the technique took the name of *flanging*. The sounds were separated by a progressively increasing interval, causing phase cancellation to sweep through the audio range. This could be made more manageable, first by the use of variable speed control and then by switching to fully electronic techniques.

One device produced a shifting phase-cancellation effect without the use of delay. The signal was fed through a series of simple circuits (rather like filters) which passed all of the frequencies present but changed their phase relationship. When this output was recombined with the original signal there was partial cancellation at some frequencies, which could be altered continuously by varying the components of the phase-change circuit. An interesting side-product was that the frequency of the signal as a whole also shifted a little, while any change in the phase-shifting network was being made, so permitting deep vibrato effects. But the phasing effect was different from the 'flange' effect in other ways, too, of which the most important was that cancellation was distributed through the audio range in an exponential manner (in the ratios 1:2:4:8:16, etc.) compared with flanging's arithmetical array (1:2:3:4:5, etc.). The flanging effect was therefore directly related to the harmonic tone structure used in many

Flanging and phasing
The change in amplitude (vertical scale) with frequency. A, The effect of flanging – here with a delay of 0.5 ms – is evenly distributed throughout the frequency range, but is not related to the musical scale, which is logarithmic. B, Cancellation and enhancement produced by the use of a phase-changing network: the dips are distributed evenly over the musical scale.

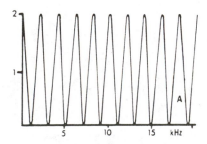

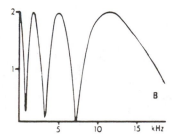

strong, dense, high-frequency overtone structures whether it is harmonically related or not.

With techniques based on the use of delay circuits, phasing reverted to the pattern of cancellation originally found in flanging, which many users still preferred. One digital example has two outputs, one of which has constant delay, and the other, which is continuously variable, can be swept through it, with the signals either added or subtracted. Note that when delay is changed, some data is either sampled twice or omitted and that this produces a characteristic form of distortion (see later).

Flanging is most effective when the interval between the signals lies in the range from 50 μs to 5 ms; as it increases beyond this to about 15 ms, the two signals separate completely to give a doubling effect. The technique works well with sounds that are sustained and have a broad frequency spectrum: guitar, cymbals or vocal-backing chords are typical applications. Other interesting effects are found when it is applied with a 1–5 ms delay to transients.

Like all gimmicks in pop music, phasing can easily be overused; it is perhaps most effective when applied to a single phrase in a piece, repeated only when that phrase recurs.

Chapter 13

Line sources

Record players and tape reproducers (plus, in television, videotape and film sound), together with lines from other studios and remote locations, are all fed to the control desks of radio or television studios. They are programme sources comparable to the studio microphone, but differ from it in that they usually arrive at zero level: some minor volume adjustments may be required, but no great pre-amplification. Most will go straight to stereo channels.

This chapter deals with the operation and handling of these sources, apart from echo which has been described already. The techniques for tape and disc replay described here are valid equally for monophonic or stereophonic sound; they are valuable to television and in the film dubbing theatre, and vital to radio, which often consists of a sequence of records or recordings.

Each small segment of this type of programme is collected, selected or pre-recorded individually. The creative process is spread out in time, and each element may have lavished upon it all the care that it demands for perfection – within the limits of human fallibility and commercial pressures. So far, the technique is very similar to that of film-making and, indeed, the final assembly of the programme has much in common with the film dubbing session. With a sound production there often comes a stage where all the pre-recording is complete, and all the 'insert' material is taken back to the studio to be re-recorded into a complete programme, using live links.

In order to match levels, mix in effects, introduce fades and avoid abrupt changes of background atmosphere, tapes where possible should be played in, not physically cut in to the programme (as they may be if the segments are recorded with this intention in the studio itself). Each insert must be controlled afresh, so that it is fully integrated into its new setting – and particularly in terms of the relative levels at the beginning and end of each insert, as discussed in the next chapter. In the operation of the reproducing equipment,

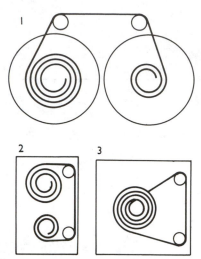

Tape transport systems
1, Reel-to-reel. 2, Cassette.
3, Cartridge, a continuous loop.
Pressing the cartridge record button
automatically puts a tone (below the
audible frequency range) on the
tape. On replay, when this point on
the tape returns to the head, the
tape is stopped automatically, then
recued ready for immediate replay.
The recycling time depends on the
length of tape in the loop. In a four-
track configuration the first two
tracks are available for programme
material, the third has auxiliary tone
or data as required, and the fourth
has the stop tone. On a cassette,
the auxiliary tone can be used to
stop the tape and switch to rewind
back to the leader, at which point
the tape stops and runs forward to
recue at the first stop tone. Tape cue
tracks may also have time code
recorded on them for automatic
synchronization with videotape, etc.

perhaps the most important single element is good timing, for where successive sounds are linked by mixing it is unlikely that faults of timing can be put right by subsequent editing; in any case, on a radio transmission there is no calling back miscues.

In older broadcast automation systems analogue tape replay is controlled by cueing systems that sense tone (or silence). In one example each recording has, superimposed at the end, a 25 Hz signal that provides two cues. As the sensing amplifier detects the beginning of the 25 Hz tone it starts the next tape in the playback sequence; and at the end of the 25 Hz signal it stops the first play-back. The most extreme form of digital automated broadcasting system employs a single 1 Gb hard disk with 4:1 digital compression at the recording stage for the whole of its time-code controlled output.

These devices are capable of cueing the routine transmission of a large number of items of a random nature very effectively; but manual techniques are still used for much creative production work – in fact, to build the programmes or items that can subsequently be broadcast automatically – and they are also the techniques that, in their most developed form, the automatic equipment must simulate. An alternative to all this is to assemble material by using a hard disc computer-controlled editing system, which can achieve much the same results, but by more analytical techniques. If a studio is still used at the end of this process, it will be for fine-tuning and review.

But whatever the technique, the objective is the same, and this is described here in terms of the original studio techniques.

Live-and-insert programmes

The linked actuality programme is a typical example of what might be termed 'live-and-insert' work: location recordings of sound effects and interviews arranged in a sequence that helps to tell a story. Unless it is arranged that each collected item leads into the next, linking narration is almost certainly necessary. Very similar in construction is the illustrated report, in which summary, comment and discussion in the studio are mixed with extracts from recordings of the event. Disc-jockey shows, whether of popular or classical music, may also be regarded as live-and-insert programmes. Here, we can be sure that exquisite care has been exercised in recording the 'inserts': their arrangement within a programme deserves care, too.

These are just a few examples that combine tape and disc with live narration. The script of such a feature might look like this:

1.	TAPE	(BELLS. *Actuality: peal from beginning, 15″ and fade under.*)
2.	NARRATOR	Bells. Bells to proclaim to the world that a new cathedral stands ready for the praise of God.
3.	TAPE	(HYMN, *Actuality. Lose bells behind. Fade singing on end of first verse, and out before start of second.*)

4.	NARRATOR	But this is a story which began one quiet night (*Begin to fade in grams*) in 1940.
5.	GRAMS	(NIGHTINGALE. *Fade in drone of* BOMBERS *until nightingale drowned out. Add some distant* ANTI-AIRCRAFT FIRE. *Hold bombers behind speech.*)
6.	NARRATOR	The bombers swept across the quiet countryside to the city. In the centre of the city, the old cathedral . . .
7.	GRAMS	(BOMB *screams down and explodes.*)
8.	NARRATOR	. . . a glory of Gothic architecture. Windows of richly stained glass, set in delicate traceries . . .
9.	GRAMS	(BOMB, *tinkle of* BROKEN GLASS.)
10.	NARRATOR	. . . carvings and statuary that were the life-work of countless medieval craftsmen . . .
11.	TAPE	(*Pre-recorded effects sequence:* 2 BOMBS, *begin to fade in* FIRE: *hold this, but gradually fade out bombers.*)
12.	NARRATOR	. . . the hammer beam roof a triumph of human ingenuity . . .
13.	TAPE	(*Effects continue: several* FALLING TIMBERS *then* CRASH *as roof falls.* FIRE *up to full, and fade to silence.*)
14.	NARRATOR	With the cathedral had gone many other buildings. The centre of the city was largely a wasteland. Rebuilding could not be started until after the war. But nothing could stop people dreaming of the future . . . and planning for it.
15.	TAPE	(*Interview, as edited. Quick fade in on atmosphere.*) At first there was quite a lot of opposition to the idea of a new cathedral: well, of course, people said houses first and the new city centre, shops; the cathedral could come later . . .

This script requires two tape insert machines, a bank of turntables and a microphone. Among the techniques employed, the most important is one that costs nothing: good timing.

Preparing tape inserts

Each insert must first be considered as though it were a programme in its own right: it must go through such processes of rough and fine editing as may be needed to give it point and clarity. This can often be more drastic than would be possible in editing a complete programme: anything that does not quite come off may be cut out without necessarily being completely lost; the essence of confused or muddled actuality can generally be tied into the script in a way that makes it both lucid and concise.

Heavy background effects should be recorded for long enough before and after speech to allow for fades (or, alternatively, a loop or an effects record can be used to cover the ends and provide the fade). Allow a few seconds of atmosphere for fades, even in cases where this is not heavy. The intake of breath before the speaker's first word may be left on the tape. At normal listening levels neither atmosphere nor breath are particularly noticeable, but their presence is natural. Take particular care with any background atmosphere when speech is interpolated between music recordings on which the background of studio atmosphere may be low.

The various pre-recorded segments of a programme can be made up into an insert tape, with a second for any items that have to be mixed or crossfaded from a separate replay machine. Short spacers, one or two seconds long, give a visual indication of where one insert ends and the next begins. If atmosphere has been left on the front of the inserts, some sort of mark is necessary to indicate the point at which the tape should be faded full in. Use a felt marker pen or a wax pencil. The end of the spacer indicates the point at which the fade-in starts.

At this stage of a production a convenient form of script gives narration in full but the inserts in cue form (and shows their duration) – something like this:

NARRATOR	. . . spoke on his return to London Airport.
TAPE	*Insert* 10. *Dur: 27"* *In:* What we all want to know . . . *Out:* . . . or so they tell me (laughter)
NARRATOR	The minister stressed the need for unity between the two countries and then said:
TAPE	*Insert:* 11. *Dur: 30"* *In:* It's all very well to be critical . . . *Out:* . . . there's no immediate danger of that.

Note that on this example there are two different types of cue-in from narration. The first needs about the length of a normal sentence pause, but the link ends with a *suspended cue*, which demands to be tied closely to the sentence that follows. In the first case the fade-up of atmosphere may follow the cue; in the other it needs to go under the last words of the cue. In the second case the timing also depends on the pace of the recording that follows. For instance, 'It's (pause) all very well . . .' would sound ludicrous if tied too close to the cue '. . . and then said:'.

Other more complex joins between narration and insert might be needed for artistic effect: how such fades and mixes can best be arranged is considered in more detail in the next chapter. At this stage, note the importance of marking the script in some bold private shorthand to indicate the peculiarities of each individual cue. The purpose of a rehearsal is to examine these in detail, and unless the programme is being assembled by trial-and-error, or in short sections, cues may come up so fast on the take that some quick visual reminder is needed for each in all but the simplest sequences, and used as a musician uses a sheet of music. This marked-up script is in a real sense the score – and there is no virtue in trying to play it from memory.

The main things to mark on the script are as follows:

- location of material – which replay machine
- any peculiarities of timing

- volume control settings and type of fade in
- duration (or other notes on when the out-cue is coming up)
- type of fade-out

This will offer a standard to work from and perhaps improve on.

Tapes may carry either digital or analogue signals. If quality is required above all else, only digital signals will be used, on digital audiotape (DAT) with time code, or by digital storage on disc – floppy or hard (magnetic) disc, or optical disc: see later. But for flexibility of tape operation, analogue still has great advantages, though with the limitation that it takes time to run down to a required internal cue, and is fussy to locate the first time.

Replay of analogue tape on cue is simple enough. Rehearse to get the feel of the timing. There will always be a pause – however short – between starting the tape and beginning the fade in, otherwise you may get a wow, if only on atmosphere. Do not rely entirely on remembered timings: watch as the tape runs up to speed and fade up as the marker passes its reference point – different machines may have different run-up times. With many reproducers the whole operation can be completed with one hand, leaving the other free for operations which have to be carried out at the same time – as when mixing from one tape to another.

A typical case for a mix is presented by speech tapes with different background effects. If the changeover point is internal to both tapes, there are several ways of doing the mix, depending on the length of pauses between words. In the simplest cases there is a full sentence pause on both tapes, offering the option of retaining the whole of this at its natural length.

In preparing to make the mix, check the lengths of the pauses, perhaps marking their beginnings and ends. Set back the normal run-up from the start of the pause on tape two. Watch or listen to tape one as the cue comes up, and at a predetermined point in the last words from it, run tape two, so that the start of the pause arrives on both reproducers at the same time. Then fade in tape two to full (or nearly full) before starting to fade the other one out. (If the two fades were made simultaneously there would be an unwanted dip in atmosphere.) By varying the time of entry of the second tape the pause can be lengthened or shortened.

If one of the originals has no pause, it should still be possible to arrange a tidy mix: for example, if there is no pause at the outpoint of the first tape, then the second can be cued and faded up early. Where there is no pause on either side, fill in with matching atmosphere from somewhere else on the tape.

In all cases match the levels with care (a reason why it may be necessary to mix, rather than edit): edge one side of the join in or out a few decibels to avoid a sudden doorstep effect on the voice. The same technique can be used where heavy atmosphere ends too close to the last or first word.

A *tape cartridge* offers near-instant replay: like a video-cassette it is made ready by simply pushing it into the slot. Cartridges contain loops of lubricated tape in standard lengths of say 10, 20 or 30 seconds which allow various durations to be recorded, switching off if there is a pause of more than say 2 seconds, after which the tape carries until it has returned to the starting point to await re-cue. Cartridges have found their greatest use as a vehicle for spot announcements such as station identification, jingles and commercials where their automatic reset is ideal for run-and-forget operation.

Cueing discs

Compact discs are always to be preferred for technical quality, but are designed to be run only from the points programmed on to them, as listed in the notes in the box. When setting up a CD, time must always be allowed for these cues to be read automatically into memory; then the track can be selected and held on 'pause'. The delay after pressing 'play' is checked, and the track reset on 'pause' until required. On cue, it is started early enough to allow for the delay.

For an extract from a CD that does not start from a programmed cue, it must be transferred to another more suitable medium, which process may (in principle) violate copyright unless permission to do so has been negotiated.

Far more flexible in operation is a recording stored as a digital signal on hard disc or recordable optical disc – the latter being a slightly more substantial operation at the recording stage: a laser heats the surface for recording, then when cool again it is read like a CD. Both can be set (in appropriate equipment) for instant replay. Also, since the reset time is low, they can be set to run a series of recordings in rapid succession, perhaps using a MIDI (keyboard) controller to cue them in an arbitrary order. In principle, these devices can readily be combined with other digital operations such as sampling (re-recording between preset cues) and pitch changing. They could be used (for example) to create a tune out of a single toot on a car-horn, or to comply with a director's request to 'make the footsteps a bit deeper'. They are therefore particularly useful for sound effects (see later).

For standard announcement, jingles and commercials a digital disc, perhaps with a MIDI controller, may now be preferred to the banks of 'cart machines' once found in many American studios.

Cueing grooved discs

On a grooved disc of music or effects, it should be possible to set up a cue within it accurately and quickly, provided that it has been located in advance. This means being able to put the disc on the turntable, locate the right groove, and the right part of that groove, and set the record up ready for playing in, all in a matter of seconds.

In a complex programme there may be many cues, creating extra problems of storage and location of material, and some may have to be used several times at different places in the programme. Some simple system of quick identification is necessary. Start by numbering the spaces in a record rack, and as the sequence of discs is established number these too, and also the script at the corresponding points. Except when actually needed for playing, keep the discs on the rack. They must, of course, be treated with care: to avoid touching the grooves hold a record with two hands by the outer edge, or one-handed by supporting it at the label with the fingers and at the rim with the thumb.

Number discs with a wax pencil by writing on the smooth part just outside the label, so it can be rubbed off later with a paper tissue. If there is room, number only alternate spaces in the rack, to allow room for extra discs to go in. On discs, successive excerpts can be found quickly without having to run through intervening material: the head is lifted, relocated and the stylus lowered. With a steady hand it can be placed manually within a few grooves of a predetermined cue. It is convenient to have a pick-up head below which the stylus is easily visible.

Another advantage of a grooved disc is that the modulation can be seen. A point source of light reflected in an unmodulated part of the record shows up as a smooth-edged radial line; the heavier the modulation the more it is diffused sideways along the grooves. The progress of a symphonic work can be 'read' as easily as the volume markings in the score.

In the days of 78s it was possible to mark the actual groove with a wax pencil sharpened to a knife edge. For microgroove, however, some additional location device is necessary, such as an optical groove-locating unit with a mirror beneath the pivot which throws the image of a scale on to a ground-glass screen. This also shows any 'swing' in the record due to the grooves not being accurately centred on the hole, a condition that can cause wow on sustained notes of music.

The type of locator which sits on the spindle and extends in an arc round in front of the pick-up head (the other end being held in the hand) is less satisfactory for quick operational work, but useful for scripting disc programmes where selections are made on a record player away from the studio. A ruler, indicating distance from the centre, is better than nothing.

For a clean start in the middle of a work it may be necessary to fade up within a fraction of an inch of the start of a note. For real accuracy of timing the disc is stopped, and then, on cue, set in motion and faded up, as with tape.

Having found the place roughly, lift the stylus back a few grooves and run up to the point again, switching off the motor before reaching it, and bringing the turntable to halt as the in-cue passes under the stylus. Indicate the start of the required sound by putting a wax-pencil mark

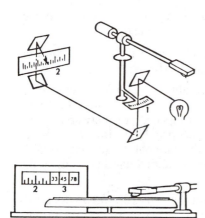

Optical groove locator
1, Scale fixed to the pick-up pivot. 2, Ground glass screen displaying projected image of scale. This indicates the exact position of the needle on the radius of the disc. It is accurate to within the error produced by groove-swing (which can also be seen very clearly). 3, Illuminated panel indicating the speed chosen.

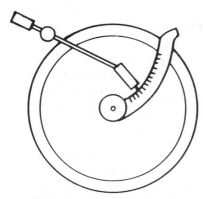

Simple arc scale locator
This can be used to find a position rapidly within an unmarked disc. The scale is rested over the turntable spindle at one end and in the hand or on a block at the other. Such scales have been manufactured in plastic, but can also be made very simply from card.

opposite it just outside the label. Check the mark by running the disc again: the sound should start just as the mark is opposite the stylus. The head is then set back once again and the disc stopped at an angle appropriate to the runup time. Listen on **headphones** before the disc is faded up (either locally or at the main desk).

Quick-start techniques depend on the type of drive employed by the turntable. With a rim drive, the turntable gets up to its full speed quickly after the motor is switched on. Set up with the drive disengaged and then re-engage with the motor switched off. On cue, switch on and as the mark passes the stylus, fade up.

Do not leave the turntable standing with the drive engaged for long periods: this can cause a 'flat' on the idler wheel.

If it is driven from the centre, the turntable acts as a flywheel, to even out speed variations. If the motor is small compared with the power needed to get the turntable quickly up to speed, try using a spin start – a sort of assisted take-off. The disc is set up and left with the motor off. On cue the motor is switched on with one hand and the turntable flicked into motion with the other; then the first hand fades up at the marked point.

During replay, there is always a danger that a pickup will run forwards (a groove run) or, worse, back. If a repeating groove occurs during a radio transmission the cure may have to be as violent as the effect: quickly fade out, jump the stylus on a groove, and quickly fade up. Alternatively, as the stylus generally requires very little persuasion to return to its original path, or at worst to jump forward, simply rest a pencil very gently across the pick-up arm or head and 'squeeze' the stylus toward the centre of the record.

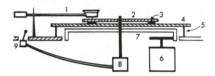

Quick-start record player
A form of drop-start mechanism in which turntable and motor assembly can be lowered away from the record. When a record is set up, the pick-up (1) is at rest in groove of the record (2) which in turn is supported by rubber studs (3) on aluminium plate (4). This rests on tripod (5) when motor/turntable assembly (6 and 7) is dropped away by raise/lower cam (8) operated by lever (9). In other designs there is no aluminium plate and the record itself is raised and lowered on a moving tripod.

Archive recordings

Archive recordings in outdated formats have to be transferred to a more suitable medium before use. This will mean finding vintage replay equipment or a modern replica, or an adaptation of it that is designed to convert the recording into an electrical signal without unduly wearing the original. This is followed by such treatment of the signal as is practicable, in order to clean up the sound.

Among archive material, an operator may encounter discs with *overlap changeovers*. Before recording tape came into wide use recordings were made on direct-cut discs that lasted between four and five minutes. About half a minute before the space on one disc ran out another was started, and the overlap was marked on both by a *scroll* – advancing the groove a little to mark the end section on one and the matching opening section of the other. To replay, the second disc is set up near to the start on a recognizable cue, and started as this comes up on disc one. The two are now running almost in synchronization. To complete the job, the speed of disc two is varied slightly by hand, or by the speed regulator: listening to this disc on pre-fade, the operator brings it into step with disc one. When the two are synchronized, number two is faded in and number one out.

An alternative method of replaying an overlap is to convert it into a *butt changeover* by setting up for a drop start just before a pause, and changing over during that pause. This method is not satisfactory for continuous music or absolutely continuous speech, unless the timing of the drop can be made perfect.

Butt changeovers on disc are necessary for playing records of works that continue over several sides. The timing between movements of a symphony must remain consistent and (if possible) the fades linked, again bringing in the second almost to full volume before fading out the first. A well-produced recording should in principle have sufficient atmosphere recorded to fill the gap. Where orchestral movements or operatic scenes are split internally, perfection counsels that a score should be consulted to check the timing: does the next side follow after a beat, or a bar; or is the last note on one side repeated at the start of the next? A score was essential with 78s, and blanks, discs of unmodulated surface noise, were sometimes used to ensure continuity of imperfection in the gaps between movements. For butt-joins that are internal to one or both discs, the cueing is as for tape.

These techniques should still be used today when transferring old 78s to tape.

Cueing videotape and film

Videotape machines and videocassette players are generally allowed a few seconds to run up. Though the speed pick-up is almost instantaneous, time has to be allowed for the picture to stabilize, and then to be locked to the same picture scan as the studio, so that a dissolve is possible or a cut is clean without a picture bump. Telecine machines are traditionally run from the figure 10 on the leader. This means 10 '35 mm-feet', i.e. 10 measures of 16 frames, and is the same for both 16 mm and 35 mm film. It is equivalent to about 7 seconds, to which an extra second is added for reaction and run-up time. Some telecine machines can be run from a lower number; even so, the delay is much longer than for audiotape or disc.

This long run-up makes exact cueing difficult: it is generally done on word counts in the first instance (two words per foot or three words per second for a fairly fast speaker with no pauses), and corrected after each rehearsal. This method of successive approximations goes for nothing if the performer suddenly produces a performance on the take that has radically different (and usually slower) timing. The director has either to make sure this will not happen or to outguess the performer.

Even where the rest of the material is ad-libbed, the film or videotape cue is generally scripted and learned, or perhaps written out on cue cards and held near the camera: or run on one of the systems for displaying script close to or optically superimposed in line with the camera lens. Alternatively, with practised performers, a cue to run is agreed, and then the performer matches the introduction to the time available.

All such systems fail occasionally, so there is a tendency to pre-record with a 'soft' start to the incoming material rather than one that crashes in with important picture and sound from the first second, as a disc or sound tape can. In consequence, the sound mixer (who has nothing to do with the cueing) has to be prepared to do either a straightforward match of sound to sound or a first-aid rescue job.

The easiest line is to follow the picture: fade in if the picture fades in or mixes; or make a quick mix if the picture cuts (on a miscue, a planned mix may be changed to a cut). But this assumes that no essential sound information is going to be lost by waiting for picture: sometimes it is best to anticipate by fading up the new sound in advance of a delayed cut. On an overlap of different voices it may be best either to clip a few words of the incoming speech or, more likely, to hold the first word or two at low level behind the last words of studio. Neither is elegant, but the latter makes it easier to make sense of what is said subsequently and is therefore preferable.

Where 'hard' cues are needed within a recorded programme because the insert has been designed to start with a bang and exact cueing cannot be guaranteed, the whole of this clumsy procedure is thrown overboard, the cue made loose, and the two parts subsequently edited together (which means that the sound change will probably have to be resynchronized with the cut – see later).

In other circumstances the cueing is predetermined by cue dots. When switching from one reel of film to the next cue dots are shown in the top right-hand corner. The first is the cue to run film and the second is for the film projectionist to switch over sound and picture together.

Cue dots in the top edge of frame are used in television for continuity purposes, giving exact synchronization to everybody on 'stand-by', 'run' and 'fade-up' cues.

Outside sources – remotes

Another type of source that may appear on a desk at a radio or television station is a line from some other studio or remote location (in Britain these are called *outside sources*). A local radio station might have lines to the regional weather bureau and recreational areas, local government offices, traffic monitoring points and vehicles, sports arenas and concert halls, together with other information centres for news, business or entertainment reports. Long distance and international links, network lines, other local studios and the telephone system may also be used. Contact between the central studio and the remote source may take a number of forms. Often there are, in fact, two lines: the programme circuit, which may be one- or two-way, plus a control line which is effectively a telephone link between the control points at the two ends.

Consider first the case of a one-way *music line* (i.e. broadcast quality line) plus control circuit. It is usually possible to prove the line fully before the programme goes on the air. Contact is first established on the control circuit, and the distant contributor is asked to send line-up tone and to identify the line by speaking over it. Both of these can be checked with the channel fader open, to that the level of the incoming material can be accurately set and its quality is known.

In the event that the line is not available until the programme of which it is to form part is on the air, the line can be fully tested to the last switching point prior to the active part of control desk. On its assigned channel within the desk it is checked on *prehear* (also called *prefade* or *PFL*) using a small loudspeaker or headphones.

The control line may be made double-purpose: as well as being used for general communication between the two points before, during and after the contribution, the return circuit may have cue programme fed to it at all other times. Other switching arrangements can also be made – for example, for the producer's talkback circuit to be superimposed on cue programme. In each of these cases the cue feed and the results of any other switching are checked by the contributor as part of the line-up procedure.

What happens at the contributor's end is again a matter for local arrangement: it depends, for example, on whether the source is a self-drive studio or not. If there is a separate, manned control point the cue programme may terminate there and the contributor be given a cue to start by the operator. Or it may be fed to the studio for the contributor to hear on headphones or a deaf-aid type of earpiece (better than headphones in television, but still inelegant). Either of these should be preferred to a loudspeaker, which might cause coloration. The feed might be for just the beginning and end of the contribution, or, for an interview, will continue all the way through. If there is no alternative to a loudspeaker being on while the microphone is open, ask the contributor to use the directional properties of both to minimize spill, and then to reduce his loudspeaker level until it is only just clearly audible to him. Then, as he speaks, test the circuit by raising the incoming level until coloration occurs: this establishes a level to stay below. During the programme it may also be possible to reduce the volume of the cue feed while the distant contributor is speaking, but be careful not to give him reason to fear that he has been cut off. Anti-feedback pitch-spreading would also help.

An apparently simpler arrangement is for there to be not three channels of communication but two: a music line in each direction, instead of a single music line and a two-way telephone circuit. However, this is a more expensive arrangement and is therefore used mainly when there is a need for high-quality sound to be transmitted both ways (as when the programme is being simultaneously broadcast or recorded at both ends), or when the facility is in any case available on stand-by. The arrangements prior to and during the broadcasts are broadly similar to those already described, except for the obvious limitation, that it is not possible for a line to be used for control when it is on the air.

A further possibility is to have two-way programme lines with a completely independent control circuit. This, the most versatile arrangement, is also the most expensive.

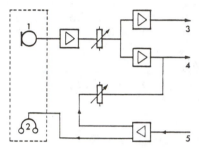

Clean feed circuit
For multi-studio discussion a speaker in the studio can hear a distant source on headphones without hearing his own voice. 'Clean' cue is also provided. 1, Microphone and 2, Headphones or 'deaf aid' for speaker in studio. 3, Clean feed to distant studio. 4, Combined studio output. 5, Incoming line. This arrangement is essential for both radio and TV when studios are linked by satellite, which introduces a relatively long delay to any sound that is returned to its point of origination. A loudspeaker used in place of headphones could easily wreck the arrangement unless it is carefully placed for minimal pick-up on the studio microphone and kept at low level.

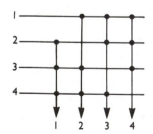

Multiway working
In this example four remote sources each receive a broadcast-quality mix without their own contribution. This matrix can be achieved by the use by 'auxiliary sends' or 'independent main output' facilities in the central mixer console.

Clean feed and multiway working

The simplest form of *cue programme* consists of the output of the mixer studio, which is satisfactory if all that is required is a cue to start. The remote source can then switch to monitor local output until the end of the contribution.

In the more general case the participants must be able to hear the originating studio and contributors from other sources throughout. Some speakers like to hear their own voices coming back, and feel that it helps them judge their delivery and effectiveness – but most do not, and at worst it may inhibit their performance completely. It is therefore desirable that if possible all speakers who hear cue on headphones should have the choice of hearing themselves or not.

The choice is not always possible: if, say, a speaker at the New York end of a transatlantic discussion were fed the full programme from a mixer in London he would hear his own voice coming back, delayed by the time it takes for a telephone signal to travel there and back. By cable this was a thirtieth of a second, bad enough to make speech decidedly uncomfortable, and if the return signal was loud enough it becomes impossible to carry on without stuttering. Satellite links introduce delays of a quarter-second in each direction.

The alternative to this is *clean feed* or *mix-minus*, an arrangement whereby participants hear all of the programme except their own contribution. Most desks provide the appropriate circuits. For two-way working where the outside source line is added to the studio output after the main gain control, a clean feed can be taken off just prior to the main control and fed via a parallel fader: in effect, there are two main gain controls, one with and the other without the remote contribution. A typical use for this is to provide an actuality feed of an event to which various broadcasters add their own commentary. The actuality sound is fully mixed at the originating control desk, at which the commentary for the home audience is added through an independent channel which is fed in to the parallel output. This arrangement is similar to the provision of mixed music and effects tracks in film dubbing: there, too, commentary may be added separately by the user, perhaps in a language different from that of the original production.

An alternative arrangement uses auxiliary circuits. Each channel in the desk has provision for a feed to be tapped before the fader and fed to an auxiliary bus, in which the output of all channels selected is combined in a rough mix which can be sent to any local or remote point. Many desks have two or more such auxiliary 'sends', to allow for *multiway working.*

An application is the 'independent evidence' type of programme where a compère questions two people in separate studios who cannot hear each other (e.g. to examine the reactions of various partners who knew each other well: brother and sister, secretary and boss, or husband and wife). Three studios are required: one for the compère, who, listening on headphones, hears the full programme output, and two remote studios in which the 'victims' sit, and are fed nothing of the other person's answers, and only that part of the questioning that is directly addressed to them.

A complex form of clean feed is used when recording sporting and other events of international interest. Picture and sound are recorded on videotape and a synchronized multitrack sound tape. Many commentaries in different languages can be recorded at the same time and then recombined individually with the actuality sound on replay of the tapes.

Telephone circuits used in radio and television programmes

Telephone circuits are not routinely used in all broadcast programmes for several reasons:

- The speech quality of the microphone is satisfactory for communication but below that normally required in broadcasting. Distortion and noise levels are relatively high, and the frequency response is not flat. Nor are telephone lines designed to the same sound-quality standards as are those intended specifically for broadcasting.
- The bandwidth is relatively narrow: 300–3000 Hz in the USA and 300–3300 Hz in the UK. This contains more than enough of human speech to permit full intelligibility, but, again, less than is desirable in broadcasting.
- The telephone system cannot reasonably be expected to guarantee the same standards of freedom from interruption that are required on lines used for broadcasting.
- Connection by automatic equipment at any given time cannot be guaranteed: junctions, overseas circuits – or the number itself – may be engaged.

Clearly, since broadcasting is geared to different technical standards, telephone circuits are used only for good reasons – which should not normally include the economy of operation necessary to a telephone system. Valid reasons are:

● *Extreme topicality.* With no time to arrange high-quality lines or to get a recording or the reporter to the studio, this is the only way of obtaining a contribution. The justification here is that the same programme value could not have been obtained in any other way.
● *Extreme remoteness.* Here, physical inaccessibility is the problem: a distant country from which high-quality lines are not available or would be so prohibitively expensive or inconvenient as to limit the availability of information. Or the speaker may be aboard a boat, train or aeroplane; or in an area that has been struck by natural disaster.
● *Audience participation.* Here, the justification is that the ordinary citizen at home may without elaborate preplanning be invited to take a direct part in proceedings, to which the audience is normally only a passive witness.
● *The phone-in.* An extreme example of audience participation, in which telephone quality is an intrinsic part of the show.

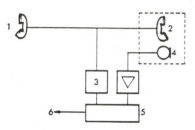

Broadcasting live telephone conversations
1, Remote telephone. 2, Studio telephone. 3, Constant volume amplifier. 4, Studio microphone. 5, Mixer. 6, Studio sound output.

None of these, however, excuses a change in quality of the studio voice, which is what would happen if a normal two-wire circuit, carrying both the 'transmit' and 'receive' elements of the connection, were used on its own for both sides of the conversation. In practice, a feed is taken from the telephone circuit through an amplifier to the studio sound-control desk, where it is mixed with normal studio output to add the full frequency range of the voice at the near end of the conversation.

Obviously, for the studio sound to mask the telephone quality version of the same voice the latter must not be mixed in at too high a volume. Special measures may be necessary to avoid this, for it is all too likely that, tapping directly from the telephone circuit so close to the studio, the volume of the distant voice will be reduced by losses on the line, and may therefore be low in comparision with the near voice. For a long-distance call this depends on the line quality and the distance between the last signal amplifiers and the studio. In certain studio situations it may be better not to use the local exchange, but instead to have a low-loss line installed to a point nearer the centre of an area system.

The use of a constant-volume amplifier or (essentially the same thing) a combination of amplifier and limiter will ensure that incoming speech is as loud as that originating locally. Note that the gain of a simple limiter should not be set too high, say, a maximum of 8 dB, or line noise may be boosted to an objectionable level during pauses in the conversation. However, if a noise gate is also incorporated in the circuit (and assuming that there is adequate separation between speech and noise level) the gain of the limiter can be substantially increased.

The line may be specially provided for programme use, and have its own number independent of that listed for the studio and office complex, or it may be routed through the local switchboard. In either case it is advisable that rather than go directly into the studio proper the line should terminate at a control point under the supervision of a switcher who may feed it either to a control telephone, which may be used for engineering tests or production purposes, or to the performer in the studio and, in parallel, the control desk.

In this system the control phone is used to establish the call, which is then checked for quality before transmission: in particular, it is essential to establish that no cross-talk is intelligible. When the circuit is finally switched through to the studio it must be arranged that the control telephone microphone is out of circuit, or noise from the control point will also be heard. As the call is switched to the studio it is also fed in parallel via the 'receive' amplifier to the control desk and to a separate loudspeaker or headphones to pre-hear the incoming circuit, so that while the circuit is faded down it may still be monitored.

It should be arranged that the call will remain connected as long as it is routed through to the studio, even if the studio telephone is not picked up, and will not be disengaged at the end of the conversation by putting the studio phone down. It is finally disconnected by switching back to the control phone when this is down (and therefore open-circuit) or by the distant caller replacing that receiver.

Special services may sometimes be arranged; in particular, fixed time calls may be booked in advance. Or timing pips that would normally be superimposed on a call may be suppressed by special request.

From the point of view of the engineers of the telephone service there are two rules that must be observed:

● no equipment from which there is any possibility of a dangerous high-voltage signal arising should be connected to the line
● no unusually loud signals should be transmitted as these would inconvenience other users of the system by inducing abnormally high cross-talk levels.

The way in which calls are used in programmes is normally subject to some code of practice that is agreed with the telephone administration office. Whether the following are regarded as rules or merely as suggestions for such a code depends on factors such as the arrangements between the broadcasters and the telephone service concerned, and regulations imposed by the licensing authority. This may vary from country to country and also with the method of use (recording or live broadcast) and to a lesser extent the type of programme. Some are formulations of common courtesy or common sense; others are designed not to bring into disrepute or draw complaints to cooperating organizations whose service is inevitably seen at a technical disadvantage; yet others are designed to limit intrusion:

● Obtain the consent of the other party before a broadcast is made.
● Avoid the use of party lines.
● Do not, in the context of a call, broadcast any criticism, or any comment that may be interpreted as criticism of the telephone service.
● If calls become unintelligible or subject to interference do not persist with their broadcast, but bring them to an end as quickly as possible.

- Immediately abandon calls that are interrupted by the operator or by a crossed line. Someone should ring as soon as possible to explain to the other party what has happened.
- Do not broadcast telephone numbers, or permit them to be broadcast, except by prior arrangement. Special measures may have to be taken to avoid exchange congestion and the blocking of emergency calls: this may prove expensive.
- Do not involve operators or other telephone service staff in the broadcasting of a call.

Where a conversation is recorded the same result must be achieved as if the item were live – but in this case editing may be used to help achieve it.

In order to ensure that quality is as high as possible, speakers at both ends may be asked to speak clearly but without shouting, not too fast, and with the mouthpiece by the mouth and not under the chin. Broadcast sound should not be loud near the telephone. And if there are parallel extensions at the caller's end these should not be used by others listening in, or this too will reduce the quality.

Chapter 14

Fades and mixes

The fade is one of the simplest operations but, as much as anything else, it is the way that fades and mixes are carried out and timed that distinguishes the polished, professional product. Not every fade consists of a slow, steady movement in or out. Everything from disc-jockey shows to high drama on film or television depends on a sensitive hand: a good fade has the qualities of a well-played piece of music. A good operator does not wait for a cue and then act on it, but rather, anticipates it so that cause and effect appear to coincide – which is, after all, the purpose of rehearsal.

What constitutes a smooth fade? Since the ear detects changes of volume according to a logarithmic scale (the decibel scale), all faders are logarithmic over much of their working range of 30–40 dB, below which the rate increases – again, roughly in line with our expectations. At the lower end of a fade, as the listener's interest is withdrawn from the sound, the ear accepts a more rapid rounding off. The fade is, in fact, a device for establishing a process of disengagement between listener and scene: once the process is accepted, it only remains to get rid of the tail-end as quickly as possible. An even movement on a standard fader does most of this, but the rate of fade in or out can often be increased at the lower end, and reduced to allow fine control in the final few decibels at the upper end.

A more complex use of faders overlaps their operation as a mix.

The fade in radio

Consider first the use of fades in radio drama, which may employ a convention that each scene starts with a fade-in, and ends with a fade-out. Narration, on the other hand, is cut off almost square – so that there need never be any difficulty in distinguishing between story and storyteller. The type and rate of fade, and the length of pause, each have their own information to give to the listener; but in creating a sound picture the resources available are much more slender than those for film, which can employ cuts, fades, dissolves and wipes, plus all the elements of camera technique: pans, tracking shots and so on. A lot has to be conveyed in radio by slight variations in the technique of fading.

For a scene change, the gentlest form is a fade to silence over about seven seconds, a pause of two or three seconds, and an equally slow fade-in. This implies a complete discontinuity of time and action. Faster versions imply shorter lapses of time or smaller changes of location. For instance, the 'Maybe it's down in the cellar, let's go and see if it's there . . . well, here you are, have a look round' type of fade can consist of a very quick out and in, just enough to establish that a slight time lapse has taken place.

Moving off and back on microphone, perhaps helped by small fades, is used for still smaller changes of location – say, from one room to the next. In general, it is convenient to assume that a microphone is static within the field of action, although this itself may be moving, such as a scene in a car. Any departure from this convention must be clearly signposted with dialogue or effects, or the listener will be confused. A scene that starts near a waiting car which then moves off may create a moment of confusion during which it is not clear whether the microphone is with the car or not. Dialogue pointers are often used to clarify this. The deliberate misuse of pointers can be funny, as in the ancient gag:

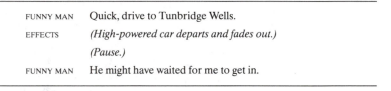

FUNNY MAN	Quick, drive to Tunbridge Wells.
EFFECTS	*(High-powered car departs and fades out.)*
	(Pause.)
FUNNY MAN	He might have waited for me to get in.

To a limited extent, moves on and off microphone are interchangeable with fades. According to tests carried out in the USA, listeners found that a fade gave just as good an impression of moving off as did actual movement. However, it only works this way round: if a move is used to represent a fade, the change in acoustic perspective usually shows, and can make the effect sound odd. But this may be deliberately exaggerated, for example by combining a fade with increasing artificial echo to suggest a move off in reverberant surroundings.

How deep should a fade be? Plainly, if a scene has a good curtain line it must not be faded very much; or if the first line of a new scene contains important information a quick fade-in is necessary. In both cases the adjacent fade is conventionally slow, otherwise the impression of a scene change is lost – the result could sound like a pause in the conversation.

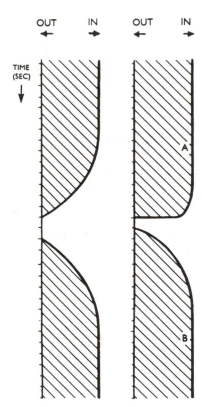

Fades in drama
Left: Simple fades between scenes – smooth fade out, smooth fade in. *Right:* Fades between narration and scene. Slight edge out at end of narration (A) and smooth fade in (B). (In these diagrams, the base line represents silence and the upper limit is normal volume.)

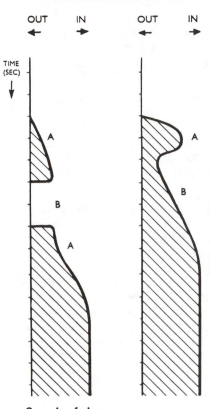

OUT   IN   OUT   IN

TIME
(SEC)

Complex fades
Left: A, Speech. B, Silence. Do not
fade in on silence. *Right:* Complex
fade to accommodate quiet effect at
start. At (A) fader is open for 'Door
Opens' then taken back for fade-in
(B) on speech.

Schools' radio broadcasts in Britain employ relatively shallow fades, partly because of the listening conditions in classrooms; the receiver may be poor, and the room boomy. The producers of these programmes make a point of listening with school audiences from time to time – and as a result favour high intelligibility with low dynamic range and background effects held well down so that they do not interfere with speech.

Typical listening conditions in many parts of the world may be expected to be poor, either because crowded short or medium wave transmissions are heard by many at the limit of their range, or on car or small, cheap radios. For these, too, a form of presentation that uses fairly shallow fades is necessary.

Difficulties arise when a scene starts off with what on the stage would be an actor walking in: in radio, this will be reinterpreted using dialogue, acoustics and effects to establish a scene, together with deliberate fades for exits and entrances within them. This often means breaking a fade up into several parts, for there is not much point in fading during a pause. So, if a new actor has two lines to speak during an approach, most of this should be while talking and not during an intervening line.

Similarly, it is better to hold up a fade-in momentarily if there is a pause in speech. It may even be necessary to reset, jumping up or down in level between lines from different speakers if the overall effect of the fade is to be smooth.

'Door opens' is another instruction that may head a scene. If such an effect is to be clear it has to be at close to normal level, and the fader then has to be taken back a little for the fade-in proper. In general, a short spot-effect of this sort is not recommended for the start of a scene.

Mixing in effects

If a convention is adopted in which each scene opens and closes with a fade it is apparent that some sound is required. The use of unimportant dialogue is dramatically weak. To fade out on laughter (where appropriate) is better. But more generally useful are background effects. As the scene comes to a close the speech is faded down and effects are lifted to swamp the line; then after a few seconds they too can be slowly faded out. Similarly, a continuous background of sound may be used for a fade-in – such effects are useful for establishing new localities – and here again the sound may be peaked for a short while, and then faded gently down as the dialogue is brought up.

Another way of handling a fade in is to bring up speech and effects almost together, but with effects slightly ahead of speech – just enough to be clear to the listener what they represent, but not enough to draw attention away from the first words. The more complicated

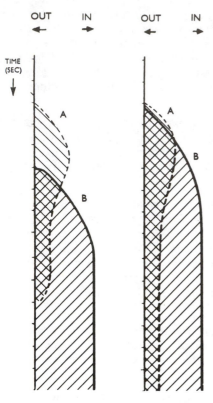

OUT IN OUT IN

TIME
(SEC)

A

B

B

A

B

Fades with effects
Left: Establishing effects (A) before speech (B), on separate faders.
Right: Fading up effects (A) with speech (B), and then holding them back to avoid an excess of distracting backing sound.

the background effects, the more they will need to be established first. In either case they nearly always have to be dropped to a lower level behind the main body of the scene than at beginning and end. The level may in fact have to be unrealistically low – more so in mono than stereo, where the spread background interferes less with intelligibility.

It is best to vary the level of effects behind a long scene enough to prevent the sound from becoming dull: it should be lowest behind speech that is important to plot, and louder when dialogue is inconsequential or there is reference to the backing sound, perhaps when characters are pitching their voices against it. In any case, actors should be encouraged to sharpen the tone of their voices if in similar natural circumstances they would do so.

Effects can often be peaked as a bridge between two parts of the same scene: different conversations at a party can merge into and emerge from a swell of background chatter. In film terms, this may be equivalent to either a slow move from one group to another or to one group moving off and another group on before a static camera: note how easy it can be, when constructing a scene for radio, to imagine that you are being more precise in depicting movement than in fact you are. Fortunately it is not usually important if the listener has a different picture from you. Nevertheless, it is sometimes possible to be more precise and give the impression of a tracking shot by peaking effects and then gradually changing their content.

Beware the over-use of background effects. Used for occasional scenes, they enliven a production; used for every scene, they become tedious – the more so if they are too loud.

Fades in television and film

For television and films the rate of fade and timing of any pause is normally set by picture: in the simplest case a fade of picture is matched exactly by fade of sound, but if there is a low-level but significant sound during the fade it can be emphasized disproportionately. Similarly, if there is an effect of little significance, but which would be too loud, it can be reduced.

Hardly less obvious are the uses that can be found for effects and music during fades of the picture. For example, to bridge a time lapse in a party scene, it may be appropriate to peak the chatter background up, rather than fade it out, implying firmly that action and location are unchanged. If the new scene then turns out to be a different party, a deliberate incongruity is introduced. A change of scene from one noisy location to another might be made by peaking effects against the fade and mixing while the picture is faded out. This works best if the two sounds are characteristically different.

This technique should be used only with calculated and deliberate intent. It is self-conscious in that it draws attention to technique as such – at least for as long as it takes to establish a convention. The dramatic reason must therefore be strong, and the sound important to the action.

The rather less obvious use of effects as already described for fades in radio also have their uses. When mixing from one picture to another it is normal to mix backgrounds (or fade the louder of the two in or out) to match picture. Here a slight peaking of effects to bridge the mix may be readily accepted.

If the scenes are staged in a television studio, the background will be on film with its own sound, or with effects from audiotape or disc, and would be whipped out on the cut.

More interesting – because more open to the creative use of sound – is the picture cut from one location to another. The simplest case, however, uses a straight cut with dialogue (or strong action) to carry it. For example:

Vision:	Sound:
Midshot oilman: turns on first words to see distant figure striding out towards burning rig.	*Effects: Burning oil rig* OILMAN: Sure it's easy. When you've stayed alive as long as this boy has the hardest job is working out what to write on the paper.
Paper in manager's hands: fast tilt up to C. S. face.	MANAGER: Two and a half million! . . .

Here the effects behind the first scene must disappear fast. The next example uses a variety of techniques on three shots:

Vision:	Sound:
Through archway to cathedral tower; tilt down to low angle, boys walk through arch and past camera.	*Effects: Singing of distant choir: chatter, footsteps* COMMENTARY: In the shadow of the medieval cathedral this group's interest in electronics complements traditional activities like choral singing . . . RUGGER PLAYER: Break!
Wide shot, football field, cathedral, and old school buildings in background: pan left as ball and players cross diagonally towards camera: when centred on science block, zoom past and in to windows.	COMMENTARY: . . . and Rugby football. Perhaps rather surprisingly, the work we're going to see started not with physics, but here in the chemistry lab.
Wide high shot, inside laboratory: two boys and glassware, dominating foreground.	BOY: Oh no! If we've got to do this, it's set all wrong. . . .

The picture has been shot with the commentary in mind. The music was recorded separately; but chatter and footsteps, football effects and laboratory sound were all recorded in sync with the picture. The chatter and footsteps track was, however, displaced a few frames because the approach matched better with the tilt down when this was done. The music was held low from the start and chatter and

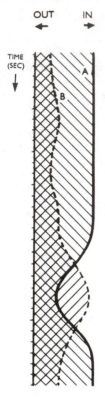

OUT IN

TIME
(SEC)

A

B

Mixing effects
Background effects (B) can often be
varied continuously and can
sometimes be peaked to bridge
scenes, as fader for speech (A) is
dipped.

effects faded in to dominate it with the tilt down. The shout 'break'
cued the cut to the football field and killed the music. The football
effects that continued on the same track had therefore been brought
up to their full volume before the picture cut. With the first zoom the
effects dipped sharply to a level matching that of the laboratory, so
that a quick mix on the cut was all that was required.

The key device here was that of sound anticipating picture. Here the
result was natural – rather like someone hearing a noise in a street
and turning to see where and what it was. It can also be used for
strongly dramatic cuts. More noticeably to the viewer, the
anticipation of sound can be used stylistically, as when the football
fan who is talking about the big match lapses into a daydream during
which we hear the sounds of the crowd distinctly as though in his
thoughts, followed by a cut to the crowd scene at the match and the
roar of the crowd at full volume.

The overlap of dialogue is now a widely practised convention, but still
needs to be used with care, as it can call attention to technique or,
worse, cause confusion.

A whip pan generally implies a spatial relationship, which may offer
interesting possibilities for the use of sound, as here:

Vision:	*Sound:*
Low angle through rigging to flag on mast.	*Music: military band starts playing*
Low angle, figurehead of Nelson, band crossing R–L.	COMMENTARY: And the martial music, the uniforms, the discipline, and the impressive reminders of a glorious past,
Closer shot figurehead . . . whip pan R. tilting down.	all tell us that this school trains its boys for the sea. . . .
Cut in to matching whip pan to boys turning capstan.	*Effects: capstan, foot-steps, creaking ropes*
Closer shot, pairs of boys on spokes of capstan, passing through shot.	Their project, too, is suitably nautical.
M.S. boy; in background, boat being hauled up slipway.	BOY: With two of our boats at school we had a problem. . . .

In this case the whip pan was not a single shot of the two objects: the
figure-head and the capstan were half a mile apart. But they were all
part of the same general location which was suggested by a series of
atmospheric detail shots, with sound overlaps. The sync music from
the second shot was laid back over the first (mute) shot and also
carried on through the whip pan, though dropping sharply during the
camera movement. On the capstan shot the associated effect
(recorded separately) was allowed to dominate, and the music
eventually faded out under the boy's speech. The noise, relevant to
what he was saying and to subsequent shots, was held low behind
him.

The use of sound in these various ways is suitable to both television
and film. So, too, is the use of music links between scenes. Indeed, the
music bridge is a device common to radio, television and film.

Theme, background and linking music

Few things about radio or television plays attract so much controversy as the use of music. This is because to be really effective – and indeed, to get a return on the cost and effort of using specially composed music – it is often loud. In the home, on the other hand, many people like speech loud and music soft (the cinema audience, however, is less likely to complain). Most people agree that where it succeeds, music helps to create works of the highest quality; but at worst it may merely be used to smudge a supposed elegance over a scene that would have been as effective with good natural sound.

In so far as television shows works that are the product of, or derived in style from the older traditions of the film industry, music is still definitely 'in'. But with financial economy as the keynote of modern television, simpler styles have been evolved, artistically justified and found to be effective. An increased use of well-recorded effects is linked with this: they increase realism as well as being cheaper, so serve many of the purposes that music was formerly used for. It follows – even more than ever it did – that music must be selected with the utmost care (and rejected if the perfect fit is not forthcoming), and must be integrated into programmes with considerable delicacy.

The ideal, for those with sufficient resources of money or talent, or both, is either specially composed music or well-chosen selections from commercially released records. Sometimes suitable background music may be found in the 'mood' record catalogues. However, it is as easy to lose tension and coarsen the texture by adding music as it is to enhance it. So, when in doubt, don't.

Signature tunes and theme music must be in harmony with or provide a counterpoint to the qualities of the programmes they adorn. At the start they must suggest the mood of what is to follow, and at the end, sum up what has gone before. But another criterion that might be applied to choosing linking music is 'fadeability'. Whatever other quality it may have, there must be points at which it can be got in and out tidily – and this in a piece of the right duration. At the end, to reflect the progress of the plot, the music chosen must be a fairly complex development of that heard at the start. For some works this rather conveniently means that the end of the record can be used for the end of the programme. But it is not so common for the start of a record to provide a good crisp opening signature; more often it is necessary to disregard any introduction and start clean on the main theme.

For the close of live radio programmes, the closing music is often *prefaded:* this is a device used to ensure that a programme slot is filled exactly to the last second. The term 'prefade' here means 'not yet faded up'. An American term for this is *dead-roll.*

In the case of a typical prefade (that used for the BBC's Radio Newsreel) it was known that the duration from a certain easily

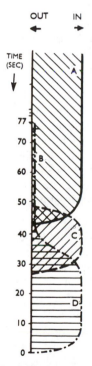

Prefading music
The prefade time is the number of seconds from the scheduled end of programme (in this case 77 seconds). A, Last item of programme. B, Closing music playing but not faded up. C, Announcer. D, Music faded up to close programme.

recognizable point to the end of the record was exactly 1 minute 17 seconds. So the record was started exactly 1 minute 17 seconds from the end of the programme. It was usually arranged that no more than about 15–30 seconds was actually needed; and this last remaining part of the record was faded up under the announcer's last words – or a little earlier if a suitable phrase presented itself to the disc player, who listened on headphones.

Mixing from speech to music

To examine the manner in which music and speech may be linked together, consider first a few of the ways in which it can be done in a disc-jockey show – taking the typical case of speech leading into a pop vocal which has a short instrumental introduction:

● *Straight – or nearly so.* The 'intro' on the record is treated as a music link between the preceding speech and the vocal that follows, and its level can often be altered with advantage to match the speech better. After this has been done, the record may be faded up or down to a suitable overall level. The first notes of the intro can sometimes be emphasized to give bite to an unusual start.
● *Cutting the introduction.* The first word of the vocal follows the cue fairly quickly, as though it were a second voice in a briskly conducted argument. It may be necessary to edge the record in a little.
● *Introduction under speech.* Here the start of the music is timed and placed at an appropriate point under the cue, which must be specially written. The intro has to be played at a low level, so that it does not distract from the intelligibility of the speech; but there may be a break in the cue in which the music is heard on its own for a few seconds – it is *established*. The fade-up generally starts just before the final words, to be at or near full volume for the start of the vocal, which may follow the cue after no more than a breath pause.

A coarser technique employs compressors arranged so that speech pushes down the volume of music: an American term for this is *ducking*. With good-quality reception, the effect is unattractive, but it is suited to transistor radio background listening or car radio reception in noisy conditions and is operationally simple.

OUT IN

TIME (SEC)

Mixing from speech to music
A, Speech. B, Pause in speech. C, Music 'intro'. D, Music at full volume. During the pause the music is started and then dipped behind continuing announcement. It is finally lifted on cue to its full volume.

Joining music to music

The simplest music link is the *segue* (pronounced 'segway' and meaning follow-on), in which the new number follows either after a very short pause or in time with the previous piece. Sometimes an announcement may be spoken over the intro at a convenient point, or over the join.

In some cases it may be better to avoid endings and introductions by *mixing* from one piece to another. Each two- or three-minute number on a pop record is, naturally, written and recorded as though it were a complete programme in itself, with a beginning, a middle and an end. In a fast-moving record programme it may be only the middle that is wanted, or the intro (or ending) may be too big or weak to go into the programme without throwing it out of shape. If the rhythm is strong in both pieces, precise timing is essential. It is easier if the music reaches a point where its form is not too definite; where, for example, one number can be faded on a long sustained note and the other brought in on a rising arpeggio or glissando. But whereas this sort of programme can be great fun to put together, it may not give the listener an increase in listening pleasure, so do not mess discs around just for the sake of doing so.

A segue or mix is only possible if keys are the same or suitably related. Indeed, it may be that keys that are nominally the same are not close enough in pitch without a slight adjustment of the playing speeds. Expert musical knowledge is not necessary for this, but good relative pitch is essential.

In any record programme, too, give occasional thought to the key sequence (this, too, can be done by ear): a lively variety of keys adds vitality. The question is, how to get from one key to another. Within a piece of music it is generally done through a progression of related keys. That may not be possible when linking together several pre-recorded items, so unless a deliberate dissonance is wanted, some other means of softening the transition must be found. There is a story of a music-hall conductor who, whenever he was faced with the problem of a quick modulation, simply wrote in a cymbal crash and immediately led off in the new key. This sudden loud noise of indefinite pitch was designed to make the listeners forget the key that had gone before. In radio work a marginally more subtle means to the same end is used – the announcement.

With even a short announcement of five seconds or so, difficulties over pitch tend to disappear. The ear does not register a dissonance if the two sounds are spaced and the attention distracted by speech. Happily this means that programmes can be compiled according to content, with suitable contrasts of style and pace between items, and without worrying too much about keys.

Film music and commentary
Left: Normal cueing. Music (A) is gradually adjusted in volume so that commentary or other speech (B) is clearly audible when it starts. After the speech ends, the level is adjusted gently. Alternatively the music may be written or chosen specially to give this effect. *Right:* Newsreel cueing. Music (A) is held at high volume until a moment before speech (B) starts, and is raised to high level again immediately speech ends. Adjustments are rapid.

Adding speech to music

Where speech is added *to* music, as in the case of theme music with announcements or introductory comments superimposed, the voice is normally chosen for high intelligibility. The music can therefore often be given its full value and dipped only just in time to accommodate the first words. Similarly, ensuring only that the last word is clear, the music is brought up close on the tail of speech. Also, in anything longer than a breath pause in speech, the music is peaked up. This is called a 'newsreel' commentary and music mix after the old style of filmed newsreel in which the inconvenience of synchronized speech – or even sound effects – was avoided.

At the end of a programme, an up-beat finish may demand that music, applause and speech all fight each other: here, similar techniques may again be used. But in neither case should it be forgotten that for many of the audience disengagement is accelerated only by what at home just sounds like a loud and not particularly pleasant noise.

An alternative, and much more relaxed way of mixing music and speech is to employ gentle music fades before and after the speech.

Chapter 15

Sound effects

Sound effects are of three main basic types – 'spot' or manual effects, 'library' recorded effects and actuality recorded effects.

Spot effects are those that are created live in the radio studio at the same time as the performers speak their lines, out of vision in a television studio, or live in a film dubbing theatre to match action on the screen. An American film term for synchronous spot effects is 'foley', named for a 1950s practitioner who bridged a gap during which the technique had gone out of fashion in Hollywood. Spot effects include such noises as doors, telephones, bells, tea-cups, letters being opened, crashes, bangs, squeaks, footsteps, and even that old standby, horses' hooves. They include sounds it just is not worth recording, and sounds that are so much part of the action that to have them anywhere but in the studio would leave the actor confused and uncertain of his timing.

Recorded effects consist principally of those that cannot conveniently be created in the studio: cars, aircraft, birdsong, weather, crowd chatter, and so on. Their field overlaps with that of spot effects in many instances: for instance, as an alternative to coconut shells, a recording of real horses' hooves may be used. But curiously enough, some very realistic horses' hoof sounds were actually recorded using coconut shells. It may seem odd that an effect may be created by an apparently – and in the case of coconut shells, even ludicrously – inappropriate sound source. But it is important to realize that the purpose of sound effects is not, in general, to re-create actuality, but to *suggest* it. Realism is not always necessary, and indeed at times may even be detrimental to the final result, in that some inessential element of the 'real' sound may distract the listener.

Library effects are those already available on disc or tape; *actuality effects* are those specially recorded (more often for film than for any other medium).

For any radio production in which effects appear, it is probable that both spot and recorded effects will be needed, and a brief examination of the script will indicate into which category most cues fall. The division of labour will probably work out about fifty-fifty, with most sounds obviously belonging in one group or the other. In a few cases the final decision depends on factors such as whether the studio is equipped with the heavier pieces of spot effects furniture or whether a particular recording is available.

In Britain the most common arrangement is for three people to divide the work into 'panel', 'grams' and 'spot' – that is, the team leader is responsible for balance and the operation of the control desk, the second looks after all recorded effects, including on-site editing, and the third creates effects in the studio. In many European countries the team is smaller, often a single individual who must add effects in post production; in America, staffing depends on the economics of the production as a whole.

The semi-realistic use of effects

Effects are rarely used in a strictly realistic way – and even when this is attempted, it is generally a heightened realism that owes a great deal to art. This is true enough in film or television; it is particularly the case in radio.

As an example consider an eye-witness account of a 'Viking Raid' – a highlight of a traditional festival. The tape received was not usable in the form in which it had been recorded. The reporter had, very rightly, concentrated on getting good speech quality, so that the background noises were too low. There were some good shouts, screams and crowd noise, but when the words 'You can probably hear the clash of their swords now' came up, the swords themselves were neither clear nor close enough. Also, there were breaks in recording and interruptions: public address loudspeakers were being used to explain the scene to onlookers, and whenever these were switched on the atmosphere of the event was lost.

To re-create the event in a form acceptable to the listener, the tape had first to be cut together into the form of a continuous narrative, deleting inessential expressions of enthusiasm – the report conveyed this without the necessity for words. As reconstructed, the first sound heard was a recorded sound effect, heavy breakers on the beach (the report referred to gusts of wind up to forty miles an hour), and these were held behind the studio narrator as he quickly sketched in the necessary background information. Toward the end of this the actuality screams were faded in, and the tape report took over. The effect of sea-wash was gradually lost as the description of the invasion brought the Vikings on to and up the beach, but an occasional squawk of a seagull was touched into the clamour to keep the sense of location strong. After the first reference to swordplay the sentence starting 'You can probably hear . . .' was cut and in its place the clash of swords was inserted in a much closer perspective. The final result was highly realistic: an actual event (with the actual shouts and screams) had been brought into sharp focus by a little touching up on the sounds which could not have been recorded at the same time without an array of microphones dotted about the beach, and a complicated piece of mixing.

So much for the neo-realist technique, one which is well suited to the sound medium when a programme consists largely of edited actuality. But where the programme is created entirely in the studio, we must live by a series of highly artificial – but effective – conventions; we rarely strive after the sound that would be found in a recording of an event as it happened.

The conventional use of effects

Sound effects are usually held quite deliberately to an unrealistically low level. There are many reasons for this, apart from the risk of overmodulation. One is that, for many listeners, sudden loud noises are unnerving or irritating, and not just dramatic.

A second reason for holding effects back – and a very important one for all continuous effects – is the masking that is caused by even moderately quiet background sound. This is worse in mono than in stereo: the *cocktail party effect* is often quoted as a demonstration of one of the basic qualities of binaural perception – the ability of the ear and brain to fasten on to a single sound source in a crowded room, rejecting and almost ceasing to be aware of the excessive surrounding noise level. But if a monophonic link is introduced into the chain (microphone, recorder or loudspeaker) a great deal of this ability to discriminate is lost, and the effort to concentrate on the main stream of dialogue becomes a very conscious one. To demonstrate this in a programme about hearing aids, one was plugged into a recorder to pick up a discussion between six people in a half-empty bar. As may be expected the din as recorded on the tape was appalling, with only a word here and there intelligible. It was clear why some deaf users preferred to switch it off and lip-read.

Record in stereo and the situation is vastly improved: one of the greatest benefits that stereo offers to drama is freedom from the absolute necessity to hold effects so far back as to separate them from the rest of the action. Unfortunately, the need for compatibility with mono often makes it impossible to take full advantage of this. When adding stereo effects, listen occasionally to the 'M' signal to check the maximum acceptable level.

Loudness is only one element in the make-up of a sound, and one that is nothing like so important as its *character*, the evocative quality that makes it visual in effect. This is true even of effects for which loudness appears to be a particularly important part of the sound – for example, a car crash. Analysis of a car crash shows that it consists of four separate elements: skid, impact, crunching metal, and broken glass falling to the ground. In a real crash the second of these is the loudest part: so loud that if it were peaked exactly to the loudest permissible volume for no appreciable distortion the rest of the sound would be much too quiet and lacking in dramatic effect. So the whole sound has to be reorganized to ensure that no part is markedly louder than any other. Even more important, during this recomposition certain of the characteristics have to be modified to create the right mood for the sound to fit into its context. The comedy actor driving into a brick wall makes a different noise from a payroll bandit doing exactly the same thing. One is ludicrous, the other vicious and retributive.

The various stages of the crash may be made up as follows:

1 The skid. This is essential for a dramatic effect. A crash without a skid is half through before we can decide what is going on. The scream of tyres is a distinctive noise that may or may not happen in real life: whether it happens in radio depends on mood. The skid is nemesis, doom closing in on the bandit and preparing to take him.
2 A pause. The listener, having been drawn to the edge of his seat, must be given half a second to hold his breath and . . . wait for it . . . .
3 Impact. A hard, solid thud: difficult to get just right, but not so important as the other sounds.
4 Crunching metal. In a real crash this usually sounds like a couple of dustbins being banged together, sufficiently ludicrous for the comedian, but not sinister enough for the bandit. A more purposeful and solid sound, and one with less of the empty-tin-can quality is needed.
5 Glass. Like the skid, this is an identifying sound, and also one which clearly puts an end to the sequence. For the listener it plays an essential part in the dying away of sensation.
6 A period of silence.

The car crash illustrates one of the main principles underlying the creation of sound effects. No drama sound effect should be simply the noise that would naturally accompany the action being played, unless it is merely included as part of the counterpoint of sound and silence, i.e. for variety of backing. Rather, it must be related to the dramatic content. This applies just as much to the more important actuality sound effects used in film. They must be recorded and treated with creative care if they are to work well.

Unrealistic and surrealistic effects

Effects that are intended to be funny are very different from normal dramatic effects. The latter may be exaggerated, but this is done in a subtle way, by slightly emphasizing the characteristic qualities and cutting away the inessentials; but in the comic effect the same process of emphasis of character is carried to its illogical extreme. A whole canon of slapstick, or what in Britain are called *cod effects* is created by this method, and in a fast-moving comedy show is deliberately overdone.

Consider an effect in the old Goon Show, for which the script simply said 'door opens'. An ordinary door opening may or may not produce a slight rattle as the handle turns, but for dramatic purposes this characteristic is seized upon: indeed, if it were not for this, it would be difficult to find any readily identifiable element in the otherwise rather quiet sound – a faint click would be meaningless. But for a cod door the volume and duration of this identifying characteristic is exaggerated beyond all normal experience – though not so much as to obstruct the flow of dialogue.

Such a method of codding effects works only if the normal spot effect makes use of some slightly unreal element: where the natural sound cannot be conveniently exaggerated, other methods have to be found. For example, little can be done with the sound 'door shuts', which in any case has to be toned down slightly to prevent overmodulation. So other methods must be examined: for example, the creation of completely different and totally unreal sounds. These might include such cod effects as a 'swoosh' exit noise (created, perhaps, by speeding up the change of pitch in a recording of a passing jet plane). Or an object (or person) being thrown high into the air might be indicated by the glide up and down on a swannee whistle (which has a slide piston to govern the pitch). This is just one example from the rich field of musical spot effects: the villain in a comedy show may be dealt a sharp tap on the Chinese block; or in a children's programme, Jack and Jill may come tumbling down a xylophone arpeggio.

As with dialogue, timing can be used to make noises comic. Typical of this is the way many a crash ends. Don Quixote being thrown from his horse by a windmill offered an example of this: after the main pile of 'armour' had been thrown to the ground and stirred around a little there followed a slight pause, and then one last separate piece of 'armour' was dropped and allowed to rock gently to rest. For this effect a pile of bent metal discs was used: they were threaded at intervals along a piece of string – except for the last single. Any similar pile of junk might be used. The timing of such a sequence is its most vital quality; the essence of its humour could almost be written out like music – and, indeed, an ear for music is very useful to an effects operator.

Normally a script should be concise, but just occasionally a writer may be allowed a flight of fancy: for one Goon Show the script demanded 'a 16-ton, 1½-horse-power, 6-litre, brassbound electric racing organ fitted with a cardboard warhead'. Here anything but sheer surrealism was doomed to remain stickily on the starting-line. The first stage in the creation of this sound was to dub together a record of an electric organ playing a phrase of music first at normal speed, then at double, and finally at quadruple speed. Three spot-effects enthusiasts dealt with gearchanges and backfires, engine noises and squeaks, and hooters, whistle, and tearing canvas – mostly produced by rather unlikely looking implements. The whole effect was rounded off by a deep, rolling explosion.

Spot effects

A radio studio needs a supply of 'sounds' readily available. Some items may be kept permanently in each drama studio; and others may be obtained from an Aladdin's cave of sound-making equipment – a spot-effects store.

The studio equipment may include a spot door, a variety of floor surfaces, a flight of stairs, a water tank and perhaps a car door. It is difficult to find an adequate substitute for this last, though a reasonable sound can be got with a rattly music stand and a stout cardboard box. Place the stand across the top of the box, and bang the two down on the floor together.

A well-filled store may take on the appearance of a classified arrangement of the contents of a well-appointed rubbish tip. The shelves are loaded with stacks of apparent junk, ranging from realistic effects such as telephones, clocks, swords, guns (locked in a safe), teacups and glasses, door chimes and hand-bells – none of which pretended to be other than they are – to musical sound sources such as gongs, drums, blocks, glocks, tubular bells and the rest of the percussion section – which may appear as themselves or in some representational capacity – and on again to decidedly unobvious effects, such as the cork-and-resin 'creak' or the hinged slapstick 'whip'. In this latter field, ingenuity – or a tortured imagination – pays dividends. One can, for example, spend days flapping about with a tennis racket to which 30 or 40 strands of cloth have been tied (to represent the beating wings of a bird) only to discover later that the same sound could have been obtained with less effort by shaking a parasol open and shut. A classic example is the French Revolution guillotine: slice through a cabbage.

But whether or not such ingenious improvizations have a certain period flavour, there are certain basic qualities of spot effects that are at least as valid today as they were 50 years ago. Certainly, the acoustic requirements of effects are now, in the days of high-quality audio transmission, a great deal more stringent than they have been in the past. Good timing, however, is the most important quality of a well-performed spot effect – the more so because it is now likely to be a staccato interjection, the continuous background sounds being more often in the form of recordings.

Where feasible, the operator works with the actor, at his side or behind him (relating to his position in the stereo image): this helps to match acoustic and perspective, as well as timing. Even large pieces of equipment can be set in the correct perspective. Where this proves impossible – for example for footsteps that are restricted to an area where a suitable surface is provided, or for water effects that depend on the use of a tank – the effects have to be panned to the correct position and if necessary given stereo spread by using a pair of microphones or artificial reverberation.

If a separate microphone is necessary, this will often be a hypercardioid for best separation. Where the same studio is used as that for action and dialogue, it is usually fairly easy to match the acoustics. Matching that of film has sometimes required more effort: the studio effect can sound too 'clean'. Here a reverberation device with a large number of preset acoustics may be the best solution; and atmosphere of some kind will be added.

The basic item of studio spot-effects equipment is a large square box on castors, generally having a full-size door set in it. The handle is a little loose, so that it can be rattled as it opens. Other door furniture may include a lifting latch, a bolt, a knocker, etc. There can be a sash window frame fitted in the back of the box.

The sound of opening has to be exaggerated somewhat, as in real life it may be too quiet to register. The sound of closing may have to be softened a little to avoid overmodulation. Close the door lightly but firmly. The characteristic sound is a quick double click.

Bad timing will impede the flow of a programme. The sound of a door opening should generally come slightly ahead of the point at which it is usually marked in a script; this allows dialogue to carry straight on without pause. If the scripted cue follows a completed line of dialogue, superimpose the sound on the last word, unless there is good reason for leaving that word cold (e.g. to follow a direction to enter). If the door is going to be closed again, allow sufficient time for the actor to 'get through' it. Also, the perspective of actor and door should match, and they should stay matched until the door is shut again, if that is part of the action.

Some types of door have to be pre-recorded: prison-cell doors, castle gates, and so on. Sometimes heavy doors of this sort can be made up from an ordinary sound by recording it with reverberation and playing back at slow speed. Halving the speed doubles the size and weight of the door, but does not change the characteristic resonance of a material, such as wood.

When knocking, take care to find a place where the door or frame sounds solid enough. Metal knockers, bolts, latches, lock and keys, etc., should be heavy and old-fashioned; they are best attached to the effects door or, if loose, held against it. The larger area of wood acts as a resonator for the sound and gives it true solid quality. Loose bolts, etc., should be screwed on to at least a small panel of wood.

One last note about doors: when in doubt, cut. Some scriptwriters overdo the mechanics of exits and entrances. An approach on dialogue is often as effective as door-plus-approach. It is better to keep doors, like all other effects, for the occasions when the sound has some actual significance in relation to the story. Otherwise, when the time comes that it does, it will not be so easy to establish its importance.

Sound effects door
This is used in radio drama studio.

Footsteps

What was a footnote on doors becomes a point of primary importance with footsteps. Are your footsteps really necessary? Go through a script with a blue pencil and make sure that every remaining footstep has direct relevance to the action. In films this effect is often used merely to fill a silence, to reassure the audience that the sound has not broken down at a time when attention must be concentrated on the picture. There is no direct equivalent to this in the pure sound medium, in which footsteps can all too easily draw the listener's attention in the wrong direction.

Such footsteps as remain may be created as spot effects or played in from a recording – the former is preferable, for the sound effect should tie in exactly with the mood of the moment, as well as reflecting character. In the studio the exact pace can readily be set and changes of pace and a natural stop produced exactly on cue.

A well-equipped drama studio has a variety of surfaces available for footsteps: the most important are wood, stone and gravel. Boards or stone slabs (laid out on top of the studio carpet) should be as hefty as is convenient, in order to give weight to their resonance – the carpet prevents them from ringing too much. The gravel effect can be obtained by spreading sand and gravel on the stone slabs.

On these surfaces long sequences of footsteps are difficult: walking on the spot is not really a very satisfactory way of creating footsteps, as it takes an expert to make them sound convincing. Walking along, the foot falls: heel, sole; heel, sole. When walking on the spot it is natural to put the foot down sole first, so a special heel-first technique must be worked out: a fairly crisp heel sound, followed by a soft sole sound. This is not at all easy to get just right. Even when the foot action is the same, the two are still readily distinguishable by the fact that the surface 'ring' sounds different on every step when walking along, but is monotonously the same when the foot keeps landing at the same place.

Walking or running up and down stairs are also better done with real steps: there is a very different quality for every step. But if the effect has to be simulated, the correct foot-fall is with the sole of the foot, going up, or sole-heel, going down. But again the result is difficult to get right unless there is some variety of surface.

Some attention should be given to footwear. Leather heels and soles are best for men's shoes. High-heeled women's shoes are usually noisy enough without special care in their choice. If only one person is available for footstep effects, remember that it is easier for a woman to simulate a man's footsteps than vice versa. It is not easy to get the right weight of step if shoes are held in the hand. The best way of getting two sets of footsteps is to have two people making the sound: except by pre-recording one person cannot conveniently do the job – particularly as the two are not generally in step. But two sets of footsteps will stand for three, and three will stand for a small crowd.

Gravel 'pit' for footstep effects
This one is built into the studio and concealed below the wooden floor.

Some surfaces are simulated. For footsteps in dry snow, cornflour (in American, corn starch) has been used, but this can be messy. A fine canvas bag of cornflour can be crunched between the hands – or try pressing a spoon or two down into a tray of the powder. For a frosty path try salt; for dry leaves or undergrowth, unrolled quarter-inch tape.

Telephone, bell and buzzer effects

Effects bell
This has two press-button controls.

For electric bells and buzzers it is useful to make up a small battery-operated box that can be used to give a variety of sounds (or make up several boxes, each giving a different quality). The mobility of such a box allows it to be placed at any appropriate distance, in any suitable acoustic, or against any surface that will act as a sounding board to it.

For telephones in radio, a standard hand-set can be adapted, so that the bell may be worked by a press-button. In order to suggest that the ring is stopped by picking up the hand-piece it is usual to finish up with half a ring. For example, the ringing tone in Britain being buzz-buzz ... buzz-buzz ..., the ideal amount of ring is one which goes buzz-buzz ... buzz-b ... followed quickly by the sound of the hand-piece being lifted. As this last sound is much quieter than the ringing, two things can be done to make it clear. First, the rattle of picking it up should be emphasized a little, and second, the telephone can be brought closer to the microphone at the end of the effect. The first ring is fairly distant (e.g. partly on the dead side of a directional microphone). This is doubly convenient in that it may be interrupting speech, and it is better not to have a foreground effect in this case. The second ring is closer, rather as though a camera were tracking in to the telephone, to become very close as it is picked up. The words following must also be in close perspective.

The action of replacing the receiver may be balanced at a greater distance: this is naturally rather louder than picking it up, and is an expected sound that fits neatly in the slight pause after the end of the conversation.

Dialling is more of a performance than being on the receiving end of a phone call, simply because most dialling codes are much too long: 10 or 11 digits are commonplace. If there is any way of cutting this sort of hold-up from the script it should usually be adopted, unless dialogue can continue over the dialling. Period rotary dials are a menace: codes were shorter, but even so, it is better to dial only a few numbers.

For television, telephones are used so often that it is worth making up 'specials' – a portable kit using real telephones, properly (and permanently) polarized so that they are fully practical, and linked together. This is necessary because good sound separation between the microphones picking up the two halves of the conversation may make it difficult for the parties to hear each other clearly without a telephone. Indeed, if one of the speakers does not appear in vision it is better to place him in an acoustically separate booth.

If both speakers do appear in vision at different times their voices are balanced on normal studio microphones, but that of the speaker who is at any particular time out-of-vision must have telephone quality distortion. Manual switching is risky, as the result of the slightest error may be very obvious. A better arrangement is therefore for the distortion circuit to be linked to the camera cut buttons so that the two ends of the conversation are automatically reversed on a cut. When there are more than two cameras (i.e. one at each end) it is necessary to make sure that *any* camera that can possibly be used in the sequence is linked in to the switching system.

In one kit, up to six telephones are connected through a box that is effectively a small telephone exchange that can produce manual or automatic ringing tones at a range of audio frequencies. Lights above each key show the condition of the associated line, showing 'called' or 'calling'. When the automatic button is pressed the selected telephone rings immediately, starting with the start of the first buzz: and it goes on automatically (using the proper ringing tone for the country represented) until either the operator's finger or the phone itself is lifted – when the ringing stops in a realistic manner. The use of such an automatic telephone also ensures that the bell itself is heard in the same acoustic and perspective as the actor.

Most 'far end of the telephone' dialling, ringing or engaged tone effects can be provided by the use of a buzzer, etc., on strong filter or by the use of a recording.

Control unit for telephones used in television plays
1, Input sockets for up to six lines.
2, Keys to select mode of operation (manual or automatic) for each line.
3, Line keys: down-select up-cancel.
4, Indicator lamps: 'line called'.
5, Indicator lamps: 'line calling'.
6, Ringing switch for use by operator: up – automatic ringing: down – manual ringing. 7, 'Auto-dial' switch. When 'on', this permits extensions having dialling facilities to ring each other automatically; in addition normal dialling tone is heard when a receiver is raised. In all other forms of operation the switch is left 'off'. 8, Ringing frequency selection. A simpler arrangement for interconnecting telephones is often adequate.

Personal action effects

There are many sounds that are so individual in their application that it would be almost ludicrous to use recordings: sounds of a newspaper being folded, a letter being opened, a parcel undone; or eating and drinking effects.

First, paper sounds. Newspaper makes a characteristic noise, but in a spot effect the action may be formalized somewhat, e.g. shake slightly, fold once, and smooth the page. Opening a letter should be similarly organized into two or three quick gestures, the last action, smoothing out the paper, being tucked under the words that follow. Work fairly close to the microphone. The paper should be thin but hard surfaced: thick brown paper crackles like pistol shots, and airmail paper sizzles like a soda siphon. Keep the actions simple but definite. For undoing a parcel: a cut, perhaps; two deft movements to smooth away the paper, and a final slower crinkling noise trailing under speech.

Many sounds, such as an orchestral conductor tapping his baton, can be made realistically (using a music stand); similarly for the chairman's gavel and block (but soften just a little to avoid sharp peaks, by binding a layer or two of electrician's tape round the face of the gavel). And the surface struck must have the proper resonance: it will not do to hold a small block of wood in the air; it should be on something solid.

Pouring a drink is also a spot effect. The sound of pouring may not in itself be sufficiently definite, and for all but the longest of drinks is over too quickly; so a clink of bottle on glass helps. But even without this clink it is important to use the right thickness of glass. Health salts do fine for fizzy drinks. A pop-gun may be a shade less acceptable as a champagne cork than the real thing, but is more predictable. If 'tea' or 'coffee' is to be poured into a cup, the effect can be confirmed by placing a spoon in the saucer afterwards – and again the right thickness of china is essential.

For a scene set at a meal table, the complete noise-making kit includes knife, fork, plate, cup, saucer and spoon. But do give the noise a rest from time to time – the right points to choose for this depend on mood and pace of the script. At any point where the tension rises, stop the effect; as the mood relaxes again, the knives and forks resume. It takes only a little of this noise to suggest a lot: listen to a recording of a tea-party for six and you will find that it sounds like crockery noises for twenty.

Gunplay

A rifle shot is one of the simplest sounds you can meet. It is an N-shaped wave like that from a supersonic aircraft: a sharp rise in pressure is followed by a relatively gradual drop to as much below normal pressure. Then there is a second sharp rise in pressure, this time returning it to normal, then nothing – of the original sound, at any rate. The length of the N (its duration in time) depends on how long the object takes to pass a single point. For a supersonic airliner this is short enough – for a bullet it is practically instantaneous. There is just amplitude and brevity. A rifle shot travels at something like twice the speed of sound, and that from a smaller gun at about the speed of sound. Whatever the speed, any loudspeaker would have a hard job to match the movement, however simple. It is fortunate that what happens after the double pulse is more complex.

In fact, the very size of the initial pulse is itself, in recording or broadcasting, unacceptable; everything must be done to discriminate against it. One way is to fire (or simulate) the shot on the dead side of the microphone; another is to take it outside the studio door – or into some small room built off the studio – and the door can then be used as a 'fader'. Very often a script calls for loud shouts and noisy action when shots are being fired, which clearly makes it a little easier to deal with the sheer loudness of the effect than if the microphone has to be faded right up for quiet conversation. The balance used varies from case to case, and experiment is always necessary to check it, so that the shot is as loud as the equipment can comfortably handle.

Gunshot effects generator
This may be contained in a portable box for use in the studio. Output is fed via a studio amplifier and loudspeaker.

Gunshot and other percussive effects
Above: Slapstick. *Centre:* Wooden board and hard floor surface. *Below:* Gun, modified to prevent firing of live rounds.

Experimentation is also necessary to select the appropriate acoustics: for the sound to be picked up is a picture of the acoustics in their purest form. Shots recorded out of doors are all bang and no die-away, which means that at the level at which they must be replayed they can sound like a fire-cracker. So always check the acoustic when balancing a studio shot, and watch out for colorations and flutter echos that may not be noticeable on a more commonplace sound.

A third element in the sound is (or may be) a ricochet – the whining noise made by the jagged, flattened, spinning scrap of metal as it flies off some solid object. And whereas the previous sound is just a noise of the purest type, the ricochet is a highly identifiable and characteristic sound. Even if in reality ricochets are rarer than they might appear from Westerns, they may be used in practically any outdoor context, i.e anywhere where there is space for a ricochet to take place.

Here are some of the techniques that have been used for gun effects:

● A real gun. A modified revolver is better than a starting pistol, which sometimes tends to sound a little apologetic. Legal regulations must be conscientiously observed, and, needless to say, care used in handling the gun. Even blanks can inflict burns. Guns for BBC radio are modified to allow the gas to escape from a point along the bottom of the barrel, and not from the end. The microphone balance adopted is generally that for a maximum ratio of indirect to direct sound.
● An electronic gunshot generator. This consists of a white noise generator which when triggered into action can produce 'bangs' with a variety of decay characteristics, ranging from staccato to reverberant. There will also be associated circuitry for producing a range of ricochets.
● A slap-stick. This is a flat stick with a hinged flap to clap against it. Depending on balance and acoustic conditions, it may sound like a gunshot (though not a good one), the crack of a whip, or just two flat pieces of wood being clapped together. Narrow strips of thick plywood are as good as anything for this.
● A long, fairly narrow piece of wood, with a string attached to one end. Put the wood on the floor with a foot placed heavily on one end, then lift the other end off the ground with the string and let go. The quality of the sound depends more on the floor surface than it does on the type of wood used. A solid floor is best – stone or concrete.
● Cane and chair seat. The sound quality again depends on the surface struck. Padded leather gives a good crisp sound.
● A piece of parchment on a frame, with an elastic band round it: the band can be snapped sharply against the resonant surface. This method is included as an example of the kind of improvization that the effects man might invent.
● Recordings of real bangs or of any of the above. The apparent size of any *explosion* can be increased by slowing down a recording. A sharp clap may become a big, rumbling explosion if dubbed down to an eighth or a sixteenth of the original speed.

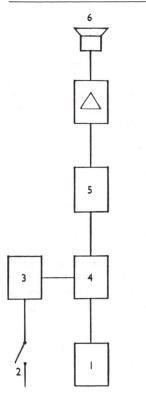

Gunshot effects generator: general principles
1, White noise generator. 2, Push button. 3, Impulser relay. 4, Gate. 5, Low pass filter. 6, Studio loudspeaker.

Creaks and squeaks
Methods using resin and string; or cork, resin and tile.

If the gun itself is used in radio there is no problem in taking it well away from the action. But for television this may not be possible: particularly if the gun is to be fired in vision, perhaps several times, with realistic flashes and smoke and perhaps with dialogue before, between and after. In this case two microphones are used. The close (dialogue) microphone has a limiter set for 2 dB below 100% modulation; the second is distant and is balanced for a high proportion of reverberation, and may also have reverberation added.

For television gunshots out-of-vision or for radio the BBC has used its own design of generator linked directly to a studio loudspeaker that is balanced acoustically. As they can hear the bang the actors are able to react satisfactorily. The basic sound is provided by a white noise generator. It passes through a gate (operated by the 'trigger', a push button) and a low-pass filter. The gating circuit allows the white noise passing through to rise sharply from zero to a predetermined peak, automatically followed by an exponential fall that eventually shuts off the white noise once again. A low-pass filter set for 5 kHz cut-off gives a quality simulating the close shots of a revolver or automatic; set to remove all but the lowest frequencies, it gives distant heavy gunfire – a 'cannon' quality.

A machine-gun effect can be obtained by using an impulser which operates the gate at regular intervals for as long as the button is pressed. In this case the exponential decay is made faster in order to keep the shots separate and distinct. Ricochets are produced by a separate pulsed oscillator, the output from which is rich in harmonics and falls in frequency with time. This oscillator works from the same initial trigger action, but delayed a little.

Creaks, squeaks, swishes, crashes and splashes

Creaks and squeaks offer rich opportunities for improvisation. Over a period of time the enterprising operator can assemble an array of squeaky and creaky junk just by noticing the noises when they happen and claiming the equipment that makes it. Wooden ironing boards, old shoes, metal paper punches, small cane baskets; anything goes in a collection of this sort.

But there are two rather more organized ways of producing creaks and squeaks:

● String, powered resin and cloth. The resin is spread in the cloth, which is then pulled along the taut string. If the string is attached to a resonant wooden panel the sound becomes a creaky hinge. Varying pressure varies the quality of sound.
● A cork, powdered resin and a hard surface, e.g. polished wood, a tile or a saucer. Place some resin on the surface and slowly grind the flat of the cork into it. Again, the squeak varies in quality with pressure and speed, and with the type of surface and resonator, if any.

The swish of a javelin or arrow may be simulated by swishing a light cane past the microphone. In real battles the arrows must have bounced off stone walls and parapets, making an undramatic clatter. An arrow landing in wood is, however, more satisfying aurally, so convention demands that all the best misses land in wooden panelling and tree-trunks. For this, throw a dart into a piece of wood close to the microphone. Proximity makes a sound grow bigger – so the twang of an elastic band may do well enough for the bow.

Crashes are another reason for hoarding junk – ranging from empty tin cans up to pretty well anything that makes a resonant noise when dropped. But glass crashes demand care. For these, spread out a heavy ground sheet and place a stout cardboard box in the middle. Sprinkle some broken glass in the bottom and place a sheet of glass across the open top of the box. Give this sheet a sharp tap with a hammer, and as it shatters it should all fall inside the box. Have a few odd pieces of glass in the free hand to drop in immediately afterwards. Not difficult – but it may still be just as well to pre-record such a crash, if only to save the worry of having broken glass around on the take (and the same goes for footsteps in broken glass). An additional reason for pre-recording is that successive crashes may not be predictable in quality and volume.

One of the fittings for a well-equipped drama studio is a water tank. If convincing water effects are to be obtained the tank must be nearly full, to avoid any tinny resonance of the space at the top. For any but the gentlest water effects it may be a good idea to protect the microphone – which may have to be close – by hanging a gauze in front of it, or putting it in a plastic bag.

Horses' hooves

Horses' hooves
Traditional method: a pair of half coconut shells.

Horses' hooves are included as an example of highly specialized spot-effects technique that dates back a long way but is still valid. Coconut shells really can be as good as the real thing, if not better – and certainly they are a great deal easier to fit to dialogue.

The secret of a good sound lies not so much in the shells (one shell sawn neatly across the middle) as in the surface. Start with a tray about 3 ft (1 m) across and several inches deep (e.g. a baker's tray). Fill the bottom with a hard core of stones and put gravel and a mixture of sand and gravel on top. For a soft surface place a piece of cloth or felt on top of this at one end of the tray, and bury a few half bricks at the other end to represent cobbles. This will give the qualities of surfaces ordinarily needed for horses' hooves. For horses' hooves in snow a tray of broken blocks of salt may be used.

There are three basic actions to simulate: walk, canter and gallop. The rhythm is different for each. When a horse is walking, the rhythm goes clip, clop . . . clip, clop . . . With a shell in each hand use them alternately, but with alternate steps from each shell move forward, digging in the 'toe' of the shell, then back, digging in the 'heel'. The sequence is left forward, right back (pause), left back, right forward (pause). ... In this way a four-legged rhythm is obtained. For the canter the same basic backward and forward action is used, but the rhythm is altered by cutting the pauses to give a more even clip, clop, clip, clop, clip, clop, clip, clop. In galloping, the rhythm alters to a rapid clipetty clop ... clipetty clop ... 1, 2, 3, 4, ... 1, 2, 3, 4, ... In this case there is not time for the forward and backward movements.

For jingling harness try a bunch of keys and a piece of elastic: the keys hang between the little finger and a solid support to the side of the tray, and jingle in rhythm with the horse's movement.

A special spot-effects microphone is needed for hooves; a hypercardioid will usually be selected. This should not be too close, and the acoustic should match that used for dialogue.

Obsolescent effects for weather and fire

At one time, rain effects were regularly produced by gently rolling lead shot (or dried peas) around in a bass drum; sea-wash was produced similarly, but rather more swooshily, while breakers could be suggested by throwing the shot into the air and catching it in the drum at the peak of the swoosh. But high-quality reproduction would reveal the device for what it is, so now a recording of the real thing is likely to be used. However, a bass drum remains a useful item in a spot-effects store – a noise produced by something placed on top of it is given a coloured resonance that may be particularly useful in overemphasizing the character of a sound for comic effect. For example, brontosaurus' footsteps can be made by grinding coconut shells into blocks of salt on a bass drum.

A bright, crackling, spitting fire may be simulated by gently rolling a ball of crinkled cellophane between the hands.

A theatrical wind machine consists of a weighted piece of heavy canvas hung over a rotating, slatted drum – but again it is better to use a record of wind, except where the noise is wanted for an orchestral work such as Ravel's *Daphnis and Chlöe*, where the ability to control pitch is valuable. But wind comes in many qualities, and its tone should suit the dramatic context. A good recorded effects library has gentle breezes, deep-toned wind, high-pitched wind, wind and rain, wind in trees, wind whistling through cracks, and blizzards for the top of Mount Everest or the Antarctic. A wind machine produces just one sound, wind; and the same applies to thunder sheets. They are not convincing except for comic effects.

The recorded sound picture

There are two distinct types of actuality sound recording: those that create a *sound picture* in themselves, and those that are selective of some particular *sound element* to the virtual exclusion of all others. These two general categories are not only different in content but are based on totally different concepts of sound use. The sound picture is the complete picture, a story in its own right. But the true recorded sound effect is, like the spot effect, heightened reality; it is the distilled, essential quality of the location or action. It is a simplified, conventionalized sound, the most significant single element of the complete picture.

A sound picture of a quiet summer countryside may have many elements, with plenty of birdsong, perhaps the lowing of cattle a field or two away, and in the distance a train whistle. A beach scene may have voices (shouts and laughter), the squawk of seagulls and distant rush of surf on shingle, and, beyond that, the popping exhaust of a motor boat. As pictures in their own right these may be delightfully evocative, but how would they fit into a play?

In either case it is easy to imagine dialogue to match the scene – but, significantly, it is the speech we have to fit to the sound effect, and not the other way round: in the countryside picture the cows might play havoc with an idyllic love scene, and the distant train whistle would add a distinctly false note to Dick Turpin's ride to York. Similarly, the beach scene is fine – if we happen to be on the sand and not in the boat. And suppose we want to move about within this field of action? With a sound picture we are severely restricted: the only view we ever have is that seen by the artist from the point at which he set up his canvas; we cannot stroll forward and examine one or another subject in greater detail unless he has chosen to do so for us. But if we ourselves are given paints and canvas we can recompose the scene as we wish. The true sound effect is the pigment, not the painting.

In a play or in any other programme where recorded effects are to be combined with speech, the final composition of the mixture will at many points be very different from that of a sound picture that has no speech. Consider the beach scene. The first sound to be heard might be a surge of surf on shingle, and, as this dies, the hoarse cry of a gull; then, more distantly, a shout and answering laughter. The second wave does not seem so close, and from behind it emerges the distant exhaust note of the motor boat for a few moments. Then the first words of scripted dialogue are superimposed at a low level and gradually faded up to full. Now the composition of the effects alters. Staccato effects, such as gulls and shouting, must be, in general, more distant, and timed with the speech, being brought up highest at points where the dialogue is at its lowest tension, or where characters move off-microphone, or where there are gaps in the dialogue (other than dramatic pauses). The motor boat, too, may go completely or be missing for long stretches, appearing only as a suggestion in such a gap. The individual elements can be peaked up much more readily than the waves themselves, which cannot be subjected to sudden fluctuations in volume without good reason.

When using effects in this way, avoid those that have become radio clichés. Because of 'Desert Island Discs', which starts with seagulls, sea-wash and a record of 'The Sleepy Lagoon', it has been difficult to use seagulls and sea-wash together on British radio. Another warning concerns authenticity in such matters as the location of birds and railways. There are so many experts on the sounds of these things that it is as well to try to get the region right – and in the case of birds the time of year as well.

The use of recorded effects

As the aims differ, when recording a sound picture or sound effects, so also must the recording methods. For effects, the backgrounds must not be obtrusive, nor must the acoustics be strongly assertive on effects that are of a continuous nature; the levels must not be subject to sudden high peaks, and there must be no unexpected changes of character.

A suitable microphone placing for an effects recording may not be the most obvious one. For example, a microphone placed inside a modern, closed car may not pick up any clearly recognizable car sound, so for a car recording – even for an interior effect – the microphone must be deliberately placed to emphasize the continuous engine noise. The mechanical noises associated with gear changes, on the other hand, should not be emphasized (except for starting and stopping), as their staccato quality may interfere with the flow of the script.

An effects sequence, like the composite sound picture, is best constructed from individual elements. Take, for example, a sequence with dialogue in which a car stops briefly, to pick up a passenger. The elements of this are as follows:

1 Car, constant running.
2 Car, slows to stop.
3 Tick over.
4 Door opens and shuts.
5 Car revs up and departs.

Of course, a recording might be found with all of these in the right timing and in the right order. But:

● If there is any considerable amount of dialogue before the cue to slow down, the timing is more flexible if the slowing is on a separate recording. The cue to mix from one to the other should be taken at a point in the dialogue some 10 seconds or so before the actual stop, or there will be too marked a change in engine quality. As the sound of a car stopping is lacking in identifiable character, a slight brake squeal or the crunch of tyre on gravel helps to clarify the mental image. The overall level of effects may well be lifted a little as the car comes to a halt.

- The tickover might possibly be part of the same recording as the stop. But the microphone balance for a good clear tickover is so different that it may be better to record it separately. Fade in the tickover just before the car comes to rest.
- Doors opening and slamming must be timed to fit in with dialogue. A real car door shutting is very loud compared with the door opening and the tickover, so it needs a separate recording and separate control.
- Departure could be on the same recording as the tickover, but again timing is more flexible if the two are separate and the departure can be cued precisely.
- Extraneous noises, such as shuffling on seats, footsteps, etc., may well be omitted.
- Additional sounds – e.g. another door, a motor horn or the setting of a mechanical taxi-meter – may be cued in at will.

This example is just one of a multitude of such sequences which must all be thought through as fully as this if the result is to blend into the landscape of sound.

The best way of putting together such sequences depends on the facilities available. Some may originate from effects records reproduced on a bank of turntables. Tape cassettes are used in a similar way. The flexibility of such a technique is beyond question, and it is necessary to pre-record only the most complex sequences. Tape cartridges tend to be used for shorter sounds, in some cases replacing spot effects, and besides their use as a vehicle for spot announcements, jingles and commercials are also employed for game shows where a limited number of stereotyped musical effects are punched in to signal success, failure or as 'stings' to mark a change of direction.

Effects can also be recorded digitally on to magnetic or recordable optical disc, in order to allow precise press-button replay, or cueing via a MIDI controller – a keyboard for greatest versatility, or (for single cues) a picture of a keyboard on a computer screen with a mouse to 'press' the keys. Cueing for these is described in Chapter 13. For the stereotyped effects, these may now replace banks of cartridge machines. And for those who believe in 'canned' applause, this provides the perfect can.

An alternative method, using two tape reproducers, is to pre-record into short tailor-made sequences and assemble these in the form of two insert tapes, which can then be used to mix from one to the other at the points where flexibility of timing or of the level of individual components is required. If only one tape reproducer is available, then all the tailoring must be done on a single tape by editing – trimming the tape to fit the dialogue – and it is best to record the effects sequence specially for the programme in which it is to be used.

Recorded effects in stereo

Ideally for stereo drama, it is best to use an effect actually recorded in stereo, steering it to its appropriate position and width. Where no suitable recording is available and it is impracticable to make one specially, it is often possible to adapt existing monophonic sounds.

With continuous noises such as rain or crowd effects a stereo picture can be created by using several recordings played out of sync (or one, passed through delay lines with different time delay settings) and steered to different positions within the scene. Rain or crowd scenes could also be fed through an electronic device called a *spreader*, which distributes a mono sound over the whole sound stage like wallpaper.

For a distant, moving sound in the open air (or one that passes quickly), panning a mono source is satisfactory. A larger object such as a troop of horses passing close would require a spread source to be panned across, with the spread broadened as the troop approaches and narrowed again as it goes. This effect could also be achieved by panning two mono troops across one after the other; while a coach and four might benefit by spread horses from one recording being followed closely by wheels from another.

In an example illustrating all of these, a horse and carriage may be heard to pan through a spread crowd, until a shot is heard from a particular point, its reverberation echoing over the whole of the sound stage. Note that in stereo radio drama there is a danger of lingering too long on movement that is meaningless in mono: for the sake of compatibility this should be avoided.

Changing the pitch of a recorded effect

The range of sound compositions available from a given set of effects recordings can be increased if their pitch can be changed. The simplest (and oldest) way of doing this was to change the speed of replay. The best sort of turntable for disc effects has a continuously variable wide range of speeds – including backwards, though this is more use for setting up than it is for generating weird effects.

Most high-quality domestic turntables have only certain specific speeds. The changes of pitch obtained by switching from $33\frac{1}{3}$ to 45, and again to 78 are musically meaningless, but do amount to a certain degree of versatility. Tape speeds, too, can usually be lifted or dropped an octave. Equipment designed to operate at given fixed speeds can sometimes be modified to provide speeds that are continuously variable between, say, half and double the nominal speed. This is done by powering the drive motor from a three-phase variable-frequency oscillator.

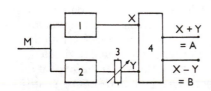

Spreader circuit
The monophonic input (M) is split equally between paths X and Y. 1 and 2, Delay networks producing differential phase shift. 3, Spread control fader. 4, Sum and difference network producing A and B components from the X and Y signals.

When the speed of an effect is changed the first difference in quality that presents itself is a change of *size:* slow down the two-stroke engine of a motor lawn-mower and it becomes the lazy chugging of a motor boat – or a small motor boat becomes a heavy canal barge. In the same way, water effects are increased in scale by slowing down the recording – just as filming a model boat in slow motion may simulate the real thing. As a by-product, it may sometimes happen that reducing the speed gets rid of some unwanted rumble or resonance in the bass.

Speeding up an effect produces the reverse change of quality, e.g. mice talking, and so on. It is worth remembering these techniques when planning unusual new recordings; they can sometimes save a lot of work.

A change of speed also changes the apparent size of the room in which an effect was recorded: the characteristic colorations go up or down in pitch with everything else. This means that recordings made out of doors may be more conveniently speed-changed than those with indoor acoustics attached. A good subject for drastic transformation is wind. Record a good deep-toned wind and double its speed to get a high, stormy wind; double again for a biting, screaming Antarctic blizzard.

It may be possible to vary the speed of effects while they are being played, but the results often sound weird. Nearly every sound contains some coloration that is essentially fixed in frequency. For example, footsteps should not be brought to a halt by slowing down the last couple of steps, as it will result in a sudden apparent enlargement of the feet and a deepening of the tone quality of the surface.

Pitch may be varied independently of duration by means of a *harmonizer* or by using a particular program with a hard disc editing system that does the same thing. Early versions were mechanically altered tape replay machines: the single head was replaced by a drum of heads, advancing or reversing against the movement of the tape – this offers an analogue for the way a modern pitch changer works. This digitally samples very short sections of sound and reads out the data at a different rate. Decrease of pitch creates overlaps, so that part of each repeated section has to be deleted; increase in pitch creates gaps, which are filled by repeating material. However, if a rapidly changing part of the sound, such as a transient, is repeated or deleted, the resulting glitch is untidy, detracting from the desired effect, and perhaps also distracting the listener. The best harmonizers are programmed to avoid such glitches if possible: this is one of those cases where the shift to digital methods provided vast extra benefits – including, because it uses similar programmes, more precisely controlled samplers. As suggested earlier, a MIDI keyboard provides a musically recognizable way of controlling pitch, and this can be combined with a fast cueing system.

It follows that duration can also be changed without change of pitch.

The effects library: contents

The contents of an effects library fall into a variety of general classifications.

First, and most important, there are *crowd* effects – a crowd being anything from about six upward. It may seem odd that the very first requirement among recorded effects should be the human voice in quantity, but the fact remains that the illusion of numbers, the cast of thousands, is often essential if a programme is to be taken out of the studio and set in the world at large. The classifications under this general heading are many and various: chatter, English and foreign, indoor and outdoor; the sounds of fear, apprehension, agitation, surprise, or terror; people shouting, whispering, rioting, laughing, booing, applauding, or just standing there and *listening* ... for the sound of a crowd that is silent except for the occasional rustle, shuffle or stifled cough may at times be as important as the roar of a football crowd. A subdivision of crowd effects is *footsteps* ... in this case a crowd being anything from two pairs of feet upward.

Then there are *transport* effects: cars, trains, aircraft, boats and all the others (such as horses), doing all the various starting, carrying on, and stopping things that are associated with them, and recorded both with the microphone stationary and moving with the vehicle, and with interior and exterior acoustics where these are different.

Another important group that has links with both of the above is *atmosphere of location* recordings. Streets and traffic (with and without motor horns), railway station, airport or harbour atmosphere of various types and places; all these are needed, as well as the specific identifying noises of London, Paris and New York; also the atmosphere of schoolroom, courtroom, museum, shop or concert hall, and the distinctive sounds of many other locations.

The *elements* – *wind, weather, water* and *fire* – provide another vital field. Then there are the sounds of all branches of human activity: *industrial* sounds. *sport, domestic* effects from the ticking of clocks to washing up, and the sounds of *war.* There are *animals, birds, babies;* and dozens of other classifications of sound, such as *bells, sirens, atmospherics, electrical* sounds, etc.

The effects library: disc or tape

An effects library may be built up using analogue disc or tape, digital tape, CD or recordable optical disc – or a mixture of these.

The disc effects library. One or more copies of a wide variety of records is kept (these may be purchased from a variety of commercial sources, or, as at the BBC, specially produced). The records from such a library get a great deal of use in the process of selection and rehearsal, and analogue discs must be replaced frequently. In a large library it is easy to sift through a dozen different recordings that are nominally suitable to find out which is best and which will not do at all. It is possible to go to the studio with two or three specimens of each sound: one the likely best bet, and others sufficiently different in character to provide alternatives in case, on adding the sound to dialogue, the first turns out to be wrong in mood, level or timing. A different disc can be readily substituted at any point in a production. But some skill in disc-playing is needed for programmes with complex effects. The recordings on effects discs are laid out in bands which may be preceded by a burst of tone, growing in volume, then cutting off sharply. Accurate cueing is made possible by the quick-start techniques described in Chapter 13.

The tape effects library. When tape is used there are two conflicting requirements. One is for the storage of a wide selection of effects; the other is for immediate accessibility, both for sampling and for rapid setting up in the studio. To store a very large number of effects, small reel-to-reel spools may be used, with many recordings each identified by a separate spacer or some other marking on the tape. Such a store is, however, inconvenient for operational use and is likely to hinder a thorough search for the exact sound.

Tape cassettes and cartridges are more satisfactory, in that no time is spent on lacing up and they can be automatically reset to the start. But for a store of substantial size a very large number is required. This can be condensed by digital storage, for the most rapid access on magnetic disc or recorded optical disc. They can be located fast and cued almost instantaneously – but from the start of the recordings, not at an internal cue, although plainly that can be arranged by copying from the internal point to make it a new start.

Accurate cueing from an internal point can be arranged by using a medium and replay equipment that uses time code.

Acoustic foldback
1, Reproducers, which are normally fed direct to the mixer via switch contact (2), may be switched (3) to provide an acoustic feed to the studio loudspeaker (4).

Acoustic foldback

In certain cases, usually where lively acoustics are being used, it is possible to route pre-recorded sound effects via a loudspeaker and back through a studio microphone. It is of greater advantage to do this when the pre-recording is very dry – for an outdoor recording, perhaps, or one made with a very close balance. Passing the sound through the studio adds the same recognizable acoustic qualities that are heard on the voices. In addition, the actors have something to work to, and can pitch their voices in a realistic way. The main danger with this technique is that the loudspeaker/ microphone combination can all too easily introduce excessive coloration or even distortion.

The volume of the studio loudspeaker amplifier should be set at such a level that most of the range of the 'send' control on the desk is in use. These two controls are used in combination to set the sound level in the studio. A little experiment is necessary to find the optimum settings for the various elements in the chain – if a satisfactory result can be obtained at all. This technique will not improve the match if the effect is already brighter than it should be and the studio is dead.

Sound effects for television and film

All the effects that have been described here work well out-of-vision on television. Recorded effects are likely to be preferred to spot effects, which in turn are more often recorded in advance and played in from a sound tape: this helps to limit the number of open microphones and produces more predictable results when the director has many more elements to watch and listen to at the same time. However, in-vision sound effects are still produced by some practical props. These need the same care in balancing that they would have in radio, but here in more adverse conditions.

Often an effect that can be picked up satisfactorily on a speech microphone sounds much better if given its own close effects microphone. For example, a studio discussion about radioactivity using a geiger counter could have a small electret hidden near the loudspeaker of the counter. For this, a recorded effect could not be used successfully because the rates would not vary realistically; because the people present would not be able to react properly to the rates; and because a meter may be shown in close-up, working synchronously with the clicks. The effects microphone makes it possible to vary not just the volume but also the quality of the sound at appropriate points in the discussion, so that the clicks could be soft and unobtrusive at points where they are not central to the discussion, and hard and strong when they become the centre of interest. This is just one of many cases where clear, crisp action noise enhances an otherwise quiet sequence.

As in radio, recorded effects are used to suggest atmosphere. Effects plus a camera track in over the heads of six extras may convince the viewer that he has seen a shot of a crowded bar; or film may establish a street scene and the sound may continue over a shot in the studio set showing a small part of the street.

Television introduces an extra dimension to sound effects that radio has never had, the ability to jump easily between different viewpoints – say, between interior and exterior of a car or train. For fast-cut sequences it is convenient to use a twin-track effects tape with interior on one track and exterior on the other. The operator simply fades quickly from one to the other on each of the appropriate cuts or uses a device that automatically links sound with vision cuts. For an effect that needs to be in stereo, spreaders or stereo reverberation may be added, or a four-track tape used.

It is in any case useful to pre-record effects sequences for complex scenes in television, for which little rehearsal time is available in the studio itself.

Footsteps in vision

Where footsteps are to be heard in the studio, the television or film designer must help sound by providing suitable surfaces. Rostra, in particular, cause problems. Stock rostra are generally made with a folding frame supporting a wooden surface which gives a characteristic hollow wooden sound to footsteps. An action area on rostra should therefore be treated to reduce this to reasonable proportions: a suitable treatment would be to 'felt and clad', i.e. to surface the rostra with a layer of thick felt with hardboard on top. On this, footsteps are audible but no longer have the objectionable drumming sound. If little or no footsteps are to be heard, a further soft layer of rug or carpeting is needed.

For 'outdoor' areas the use of a sufficiently large area of peat not only deadens the surface but may also deaden the acoustic locally. But certain materials commonly found out of doors cannot be used in television studios. *Sand* cannot be used, as it spreads too easily, gets in technical equipment, and makes an unpleasant noise when a thin layer is ground under the feet. Sawdust is a convenient substitute. For similar technical reasons salt and certain types of artifical snow cannot be used: again substitutes may be found that do not cause sound problems.

Rocks are rarely used in studios, because they are too heavy and inconvenient. Wooden frames covered with wire netting and surfaced with painted hessian have been used, as have fibreglass mouldings or expanded polystyrene carved into suitable shapes. Unfortunately, none of these has a satisfactory sound quality, so any climbing, rubbing or bumping must be done with the utmost care by the performers – and then preferably only at places that have been specially arranged not to resonate, perhaps by using peat or sawdust either as a surface layer or packed in sandbags. Falling 'rocks' of expanded polystyrene must have recorded sound effects exactly synchronized – which is easier if the action is seen in big close-up, with the rocks falling through frame, rather than wide-angle. Expanded polystyrene can sometimes squeak when trodden or sat on, and particularly when small pieces are ground against the hard studio floor. In such cases, post-synchronization is safer, and filming or pre-recording and track-laying appropriate sounds avoids it completely.

Sidewalk *paving stones* are often simulated by fibreglass mouldings, as these are light and convenient to hold in stock. When laid in the studio they should stand on cloth or hessian to stop any rocking and to damp the resonance. If, in addition, performers are wearing rubber-soled shoes and there is recorded atmosphere the results should be satisfactory. But if clear synchronous footsteps are to be heard there is no satisfactory alternative to real paving stones, despite the inconvenience and weight.

Stone floors can be simulated by the use of a special $\frac{1}{2}$ in (12 mm) simulated stone composition sheeting laid on underfelt. But 'stone' steps that are actually constructed of wood are very likely to cause trouble if the sound effect is required (and even more if it is not). If a close effects microphone with good separation can be used for the footsteps there is some small chance of improvement, as heavy bass cut can then also be used.

A play series about a prison in a castle had trouble throughout a television season with its simulated cobbles and fibreglass doorsteps around the courtyard. For a second run the film-set was constructed using real cobbles and stone – which, to everyone's surprise, turned out better not only for sound, but also for the programme budget.

Wind, rain and fire in vision

The sound effects for wind, rain and fire create no problem: recordings are used as in radio. But when they are combined with visual effects the device that makes the right picture sometimes makes the wrong noise, which must therefore be suppressed as much as possible, to be replaced by the proper recorded effect.

Wind is generated by machines varying in size from the very large aircraft-propeller-type fans about 8 or 9 ft (2.5 m) in diameter, through medium sizes 3 or 4 ft (1 m) in diameter and which can be moved round on a trolley, down to small machines about 18 in (45 cm) in diameter, which can be set on a lamp stand and used for relatively localized effects. The large and medium sizes are very noisy and push out a large volume of air over much of the studio (if they are used indoors). Windshields are needed for any boom or other microphones within range. The largest machines may be used for the controlled single shots of a film, but are inconvenient in the television studio, where the smallest size is best. When run at a relatively low speed, the noise produced actually sounds like wind (at higher speeds blade hum is heard). The visual effect is localized and controllable, so that microphones may be avoided.

Rain, when produced in the television studio, may be far too noisy, so it should be arranged for the water to fall on soft material in a tray with good drainage. Also, if it is being used in the foreground of the shot the feed pipe may be close to the ideal microphone position, making balance on a boom difficult. Another problem is that it takes a little while to get going and even longer to stop completely, to the last drip. This may be overcome either by slow mixes to and from effects (including recorded effects) accompanying the transition to and from the rain scene or by discontinuous recording and subsequent videotape editing.

Fire – including coal and log fires – in television studios is usually provided by gas (with visual flicker effects added by lighting). Unfortunately, gas does not burn with a sound appropriate to other flames, so again recorded effects have to be used (though usually at a low level). Any hiss of the gas tap can be prevented by leaving the tap fully open at the burner, controlling the flow instead from the tap at the studio wall.

Sound equipment in vision

When radios, television sets, record players and tape machines are used as props by actors in television programmes they are rarely practical – obviously, the sound gallery wants to retain control so that the most suitable sound balance can be maintained at all times, and also so that there is perfect sound cueing in and out. This means that careful rehearsal is necessary for exact synchronization of action.

Occasionally such material is produced live from a microphone in a distant part of the studio, but pre-recording is generally preferred, because control of sound quality (by filtering) is easier. Where it is necessary for cueing the actors, the sound is fed at low level to a studio loudspeaker.

Pre-recordings using the studio itself are sometimes used for out-of-vision sound. This may allow the correct perspective to be obtained or, more often, it is used simply to fix one element in an otherwise all-too-flexible medium. Music is often pre-recorded where complicated song and dance numbers are to be performed – and frequently in a different acoustic and with a close balance, both of which create matching problems.

Television sports and events

Sports and events create sounds that must be captured on television, preferably in stereo. There are two main types:

● Arena events. For these, crowd atmosphere and reaction is taken in stereo and spread over the full width available. Action noise, such as bat on ball, is taken on more highly directional microphones and usually placed in the centre. Occasionally an additional stereo pair is used for action, but this is rarely crucial.

● Processional events, where cameras and microphones are set up at a sequence of points along a route. Microphones are usually set to follow the action, using stereo according to the same conventions as those used for picture. For television, it is not always necessary to follow action precisely, provided that left and right match the picture: it is better to concentrate on dramatic effect. A racing-car approaching makes little sound, but its departure is loud. In mono the tail of the previous sound was often allowed to tail over the start of the next (approaching) shot, but in stereo a cleaner cut, and near-silence (apart from stereo atmosphere) on the approach, has been found more effective.

Sound effects on film

In film, effects are added in post production to maintain a continuous background of sound, or to define a particular quality of near silence. But in addition, effects of the 'spot' type are made easier by the process of synthesis that is used to create films from many individually directed short scenes, within each of which the sound element may be given greater care than in the continuous take of the television studio. A live action effect recorded synchronously as a scene is filmed may be enhanced in a number of ways:

- it may be changed in volume
- a second copy of the effect may be laid alongside on another track, so that it can be changed in quality and volume to reinforce the original sound
- the timing may be altered, e.g. for an explosion filmed from a distance the noise can be resynchronized visually, when the original recording is out of synchronization because of the time taken for the sound to reach the microphone
- the original effect may be replaced or overlaid by a totally different recording, again in exact synchronization with picture.

In the film industry, the provision of footsteps and other spot effects may be a separate stage in post-production, and is the province of spot-effects specialists, called foley artists in the USA. Their skill permits the film director to concentrate on picture and action in locations where there is extraneous noise from traffic, aircraft, production machinery or direction; or where the acoustics or flooring are inappropriate. The tracks are recorded or laid to picture and can be adjusted, treated and pre-mixed as required, before the final re-recording.

Even if a particular staccato sound works well on the speech microphone, it may be better if recorded wild-track (i.e. without picture) and given its own special balance. But when neither works satisfactorily, some other spot sound simulating the proper noise may be tried.

For example, take the case of a bulldozer knocking down a street sign. In the main synchronous recording the dominant noise is the engine and machinery noise of the bulldozer: the impact of blade on metal is hardly heard. So a second version of the noise of impact is simulated, e.g. by using a brick to strike the metal, recorded without any bulldozer engine noise. This is added on another track and controlled separately. If the bulldozer is knocking down a wall the sound is complicated still further, and the dramatic importance of three sounds may be the inverse of their actual volumes: impact noise may once again be the most important; then the sound of falling masonry; then the bulldozer engine and mechanical noise. In this case several recordings of the principal event could be made with different microphone positions (on either side of the wall), perhaps with additional sounds recorded separately to be incorporated at the dubbing stage.

Commercials and documentary features sometimes make use of punctuating sound effects. An item of equipment built up stage by stage by stop-action techniques, for example, has a series of parts suddenly appearing in vision: each time this occurs there may be a staccato noise in the sound that might be a recorded 'plug-in', or something tapped on a metal surface, or indeed any other sound that may be felt to be relevant. Cash-register noises accompanying messages about the amazingly low price of a product have been overdone: such music may delight those who pay for the commercial more than the customer whose money is taken. But there is a vast potential for non-representational, comic or cod effects in commercials.

Chapter 16

Shaping sound

One aim of audio engineering has been to approach ever closer to the ideal of perfect fidelity in sound recording and reproduction. But as our technical grasp of the means of eliminating distortion in all its forms has been extended, so also has our ability to use and control such changes for creative purposes. We have seen that, today, distortion of the sound as picked up by the microphone is often deliberately introduced.

An alternative is to create the audio signal (or some part of it) electronically: today's *synthesized sound* has been developed from electronic music. Similarly, *sampled sound* began life as musique concrète. At the BBC, those early methods were also used to create *radiophonic music and effects*, but have largely been superseded by digital techniques, which continue to evolve in flexibility and speed of operation so fast that commercially available devices appear and then become outmoded at a bewildering rate. Since they mostly mimic earlier forms, this chapter takes a historical approach to their development.

From the start it was clear that anyone with a microphone, recorder and razor blade can demonstrate that the formal organization of any sequence of sounds has some resemblance to music, although its quality depends on the musical sense of the person making the arrangement.

Sound conventions

Practically every sound we hear reproduced – however high the quality – is a rearrangement of reality according to certain conventions. For example, monophonic sound is distorted sound in

Classification of sound

The dynamics or envelopes of sounds: early classifications devised by musique concrète composers and engineers as a first step towards a system of notation. The terms shown here could represent either the quality of an original sound or the way in which it might be transformed.

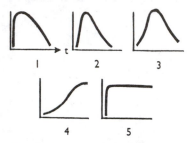

Attack. 1, Shock – bump, clink, etc. 2, Percussive – tapping glasses, etc. 3, Explosive – giving a blasting effect. 4, Progressive – a gradual rise. 5, Level – full intensity from start.

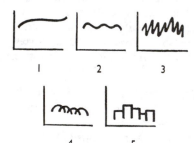

Internal dynamics. 1, Steady – an even quality. 2, Vibratory or puffing. 3, Scraping – or tremulous. 4, Pulsing – overlapping repetitions blending together. 5, Clustered – subjects placed end to end.

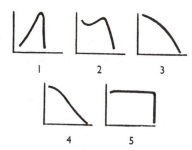

Decay. 1, Reversed shock. 2, Accented. 3, Deadened. 4, Progressive. 5, Flat – cut off sharply.

every case except where the effective point source of the loudspeaker is being used to reproduce a sound that is itself confined to a similarly small source. Every balance – every carefully contrived mixture of direct and indirect sound – is a deliberate distortion of what you would hear if you listened to the same sound live. This is accepted so easily as the conventional form of the medium that very few people are even aware of it. Perhaps distortion would be better defined as 'unwanted changes of quality', rather as noise is sometimes defined as 'unwanted sound'. Sound is continuously being shaped in some degree, and the listener is constantly being required to accept one convention or another for its use.

Note, however, that good listening conditions are even more important when we listen to 'new sounds' than they are for conventional music. When a listener with good musical imagination hears instruments he knows and musical forms he understands, he is able to judge the qualities of writing and performance in spite of bad recording and bad reproduction. Although they may reduce his enjoyment, noise and distortion are automatically disregarded. The musical elements are identified according to the listener's previous knowledge and experience: imagination reconstitutes the original musical sound.

Plainly, this does not happen on the first few hearings of music in which the new sound is produced by what, in other situations, would be regarded as distortion. Thus we reach the novel situation in which any imperfections produced by the equipment are, for the listener, indistinguishable from the process of musical composition, and indeed become part of the music.

The traditional composer commands a large number of independent variables: each instrument and the way in which it is to be played; the pitch of each note and its duration; the relationship in time and pitch between successive notes and between different instruments. Even so, however much is specified, there is always room left for interpretation by the performer.

This leads to the first and most obvious difference between conventional music and synthesized or sampled recordings, for an electronic or concrete composition or its radiophonic successor exists only in its final state, as a painting or carving does. Unless it is combined with conventionally performed music or generated in real time, for example from a MIDI keyboard, it is not performed, but merely replayed. In this one respect, at least, musique concrète is like electronic music, from which it otherwise has certain fundamental differences.

Musique concrète was made by sampling the ready-made timbres of some particular group of sound sources, transforming them in various ways, and then cutting and assembling them in a montage. Electronic or synthesized music began, as might be expected, with electronically generated signals such as pure tones, harmonic series or coloured noise (noise contained within some frequency curve). This basic divergence in technique produced results that are characteristically different – although in theory the concrete sounds could be imitated by electronic synthesis, and, indeed, an oscillator is a valid 'concrete' sound source.

One characteristic quality of sampled sound is that the pitch relationship between the fundamental and the upper partials of each constituent remains unaltered by simple transformation unless any frequency band is filtered out completely. This is both a strength and a restriction: it helps to give unity to a work derived from a single sound or range of sounds, but severely hampers development.

In electronic music, on the other hand, where every sound is created individually, the harmonic structure of each new element of a work is, in principle, completely free. Where musique concrète has restrictions, electronic music offers new freedom – an essential difference between the two. But together, these techniques and their successors have a proven value in extending the range of sound comment – that is, in providing radiophonic music and effects (to use the BBC term).

This is the field in which sound effects have taken on formal musical qualities in structure and arrangement; and musical elements (whether from conventional or electronic sources) complemented them by taking on the qualities of sound effects. Radiophonics has never attempted to assert itself as an art form in its own right; it has always been an element in a larger picture, and indeed rarely even moves into the foreground of the audience's attention. It is still best to avoid latching too firmly on to any particular musical style. One extremely valuable attribute for anyone working in this field is, I would say, a strongly self-critical sense of humour.

The construction of musique concrète

Musique concrète is built up in three distinct phases – selection, treatment and montage.

The first characteristic of the form lies in the choice of basic material: a sampled sound which, by the choice of beginning and end, becomes complete in itself. In the early days of musique concrète the emotional associations of the original sound were incorporated into the music. There was a later reaction away from this idea, for the lack of an immediate mental association can lend power to sounds. The basic materials may include tin cans, fragments of human speech, a cough, canal boats chugging or snatches of Tibetan chant (all these are in an early work called *Etude Pathétique*). Musical instruments are not taboo; for instance, one piece uses a flute – both played and struck. And there are other things besides the natural sound qualities of an object that may affect a recording: such things as differences in the balance or 'playing' of a sound help to extend the range of sound materials enormously.

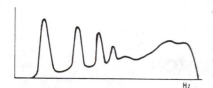

Hz

Line and band spectra
Instantaneous content: this sound is composed of individual tones superimposed on a broad spectrum band of noise.

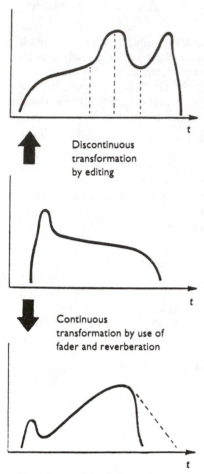

Discontinuous transformation by editing

Continuous transformation by use of fader and reverberation

Transformation of dynamics
(Original sound in centre)

Samples integrated into popular music have included political statements, news actuality and fragments of other people's compositions.

This preliminary recording may be considered analytically in terms of a variety of qualities:

● the instantaneous content of the sound, its frequency spectrum or timbre (this may contain separate harmonics, bands of noise or a mixture of the two)
● the melodic sequence of such sound structures
● the dynamics or envelope of the sound (the way in which the sound intensity varies in time).

The second stage in building musique concrète is treatment of the sound materials to provide a series of what may be termed 'sound-subjects', the bricks from which the work is to be constructed. A wide range of technical operations is now available, and in France the *Groupe de Recherche de Musique Concrète* distinguished between various types of manipulation:

● *Transmutation of the structure of sound* – changing the instantaneous sound content in a way that affects its melodic and harmonic qualities, but not its dynamics. This includes transpositions of pitch (which may be continuous or discontinuous) and filtering (to vary the harmonic structure or coloration).
● *Discontinuous transformation* of the constituent parts of a sound by editing. An individual sound element may be dissected into attack, body and decay; and particular sections subjected to reversal, contraction, permutation or substitution. This form of manipulation varies the dynamics of the sound, though at any instant the sound content is not altered.
● *Continuous transformation* in which the envelope of a sound is varied without editing, by the use of faders, artificial reverberation and related techniques.

The third phase in the manufacture of musique concrète is that of construction; the sound-subjects are put together piece by piece, like the shots of a film. Several techniques have been used for construction. One is to edit a single tape by cutting and joining pieces together (montage); another is dubbing and mixing in the same way that sound-effects sequences are put together. More satisfactory, however, is to lay tracks or their equivalents in digital memory, and to mix them down.

Transposition of pitch

When a sound is originally being made, its creator generally has some control over its pitch, timbre, duration and envelope (the way in which the attack and decay are related in intensity to the main body of the sound). The independent control of these variables is easier if a

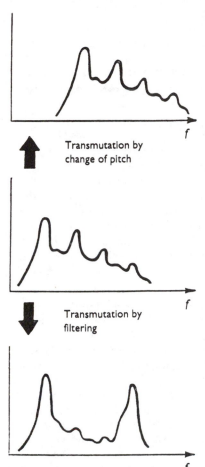

Transmutation by
change of pitch

Transmutation by
filtering

Transmutation of frequency
(Original sound in centre)

musical instrument rather than some 'concrete' sound source is chosen. But once the original recording has been made, all of the formerly independent variables are locked together. For example, speed up replay to increase the pitch by two octaves, and the duration comes down to a quarter, the attack and decay characteristics change so that the sound is a great deal more percussive and dry than before, the timbre changes to that of an instrument a quarter of the size, and so on. Everything is changed together.

One way of modifying the result is to alter the quality of the original before recording, but this is usually less effective than one would wish: speed transformations easily swamp all but the most emphatic differences in the raw material. Not surprisingly, some of the most successful results are obtained by listening to the raw material together with its most effective-sounding transformations and allowing these to suggest their own arrangement.

When treating actuality sound, e.g. a foghorn in a busy harbour, any background recorded with it that cannot be filtered out is treated as well. The background may be hardly noticed with the ordinary sound, because it is natural, but treatment may make it vividly apparent. Close microphone techniques, providing good separation, are desirable.

A simple example of pitch change is that of a double speed piano mixed with normal piano. On the first recording the piano is played with deliberation, to avoid the appearance of inhuman virtuosity when speeded up. This is then played back at double-speed, and a conventional accompaniment added. The double-speed piano sound is no mere variation on this instrument's ordinary quality: it sounds like a new instrument, and must be treated as such in the musical arrangement.

When a singer's voice is subjected to a speed change the result can sound particularly odd. This is due to vibrato, the cyclic variation in pitch that gives colour and warmth to a singer's sustained notes – a technique that is copied on orchestral instruments such as the violin. Analysis of voices that are generally considered beautiful suggests that pitch variations of as much as 8% may not be excessive (a semitone is about 6%). But the really critical quality of vibrato is the number of these wobbles per second. The acceptable range is narrow: it seems that to create an attractive sound the vibrato rate must fall somewhere between about 5 and 8 Hz, and is independent of the pitch of the note sung.

Outside the limits of 5–8 Hz the effect resembles wow or flutter: the vibrato range can be taken as dividing the two. Plainly, only small speed changes are possible without taking the vibrato outside its normal range. At worst, the result is to make the vibrato both obvious and ugly. At best, doubling the speed produces the effect of a small, vibrant-voiced animal.

Tremolo, a similar variation on the volume of a note, is also dramatically affected by speed changes; so also is *portamento*, an initial slide into a note or from one note to another which, if slowed down, changes its character, or if speeded up, almost disappears.

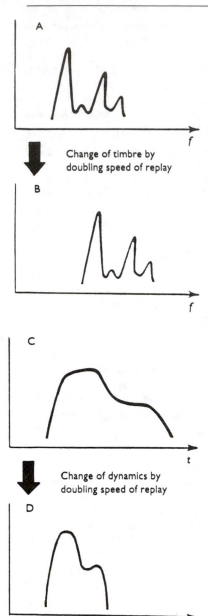

A

Change of timbre by doubling speed of replay

B

C

Change of dynamics by doubling speed of replay

D

Transposition in pitch
Replay at twice the original speed changes both timbre (A and B) and dynamics (C and D) of a sound.

Whereas a male and female voice may differ in fundamental by an octave, the *formant* ranges (i.e. those parts of the vocal sound emphasized by the throat, mouth and nasal cavities and used to give the main characteristics of speech) are, on average, only a semi-tone or so apart, and vibrato ranges are also little separated. Small changes of pitch therefore tend not to change the apparent sex of a voice, but rather to de-sex it.

It is possible to control pitch independently of duration by using a device such as a *harmonizer* (described in Chapter 15). Within a hard disc editing system (Chapter 18), *timeflex* can be used to change duration independently of pitch. Experiment will show the limits within which the results are satisfactory: it may be plus or minus 20%. Another use may be provided by a program which slightly spreads the nominal pitch as an anti-feedback measure: if the treated signal is returned through a studio by an acoustic feedback or public address circuit, the loudspeaker level can be raised substantially. Unlike the frequency-shifted feedback sometimes used for public address, it can be fed to musicians, too, as the spread is centred on the original frequency.

An older and quite different device for changing pitch independently of duration was the *ring modulator*. It allowed a fixed number of cycles per second to be added to (or subtracted from) all frequencies present (e.g. components f_1, f_2, f_3, become $x + f_1$, $x + f_2$, $x + f_3$; whereas simple alteration of speed multiplies all frequencies by a common factor). Alternatively, frequencies could be inverted ($x - f_1$, $x - f_2$, and so on) so that high frequencies become low ones and vice versa. By means of this type of transmutation (by whatever means) certain curious qualities of consonance and dissonance can be obtained. These are of a type that do not occur naturally, i.e. in conventional music, but which have been used in some electronic music. However, with the development of fully digital techniques, the ring modulator fell into disuse. Its function should be fairly easy to reproduce, but has been absent from early digital equipment.

Modulators have been used in telephony, so that narrow-frequency bands containing the main characteristics of speech can be stacked one above the other on a line capable of carrying a broad band. In their modulated form some of the speech bands may be transmitted in an inverted form – and it is in fact possible to bring them back down to normal frequencies but still have them inverted. For example, a band of speech might be inverted so that what was 2500 Hz becomes zero, and vice versa. The most important effect of such a manipulation is to invert the three most important bands of formants, so that the speech is translated into a sort of upside-down jabberwocky language – which curiously enough it is actually possible to learn.

More complex effects can be achieved. For example, in the field of sound effects, a filtered human voice can be used to modulate a train siren to make it talk. The effect of an exaggerated machine or computer voice can be obtained by modulating an actor's deliberately mechanical-sounding speech with a tone of 10–20 Hz. In electronic music, two relatively simple sound structures can be combined to

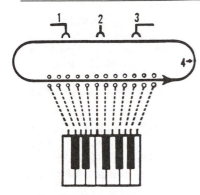

Chromatically controlled tape machine
This is a mechanical precursor of the MIDI (electronic) controller. There are 12 keys, 12 spindles and a 2-speed motor. 1, Recording head. 2, Erase head. 3, Reproducing head. 4, Tape loop.

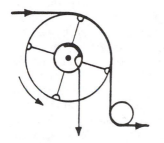

Tempophon
This is a mechanical precursor to the harmonizer. Changes duration and pitch independently. Head assembly can rotate either way: *If sense is opposite to that of tape, pitch increases; if heads chase tape, it decreases.* Signal is fed from whichever head is in contact with tape, via commutator and brushes, to a second recorder.

make a much more complex form. After use of a modulator it may be necessary to filter to get rid of unwanted products.

The harmonizer, too, can produce a useful hollow-sounding alien voice effect when feedback (without delay) is used in conjunction with pitch change.

The *Vocoder* is an older device which offers changes of a more dramatic nature. It is designed to encode the basic parameters of speech, including the tonal and fricative qualities, so that they can be imposed on any electronic, or other, source and thereby impart to it the characteristics of a voice, including intelligible speech and song. This extends the range of technique available to television programmes that specialize in alien voices and mechanical monsters, which had previously depended heavily on ring modulators and swept filters (the latter usually wobbles a filter up and down in frequency). The Vocoder also offers possibilities for humanizing electronic music by encoding some of the characteristic 'imperfections' of musical instruments and operating with them on electronic (or other) source material. Used in conjunction with the independent change of pitch and duration, the opportunities are vast.

When there is picture as well as sound, it is possible to link the two directly, deriving or modulating an audio signal from some characteristic of the (electronic) video signal; or alternatively, allowing the audio to modulate video or to control the generation or selection of picture elements.

Pure tones and white noise

In creating *electronic music*, the composer has, in principle, complete control of resources, synthesizing every detail of the sound. With no performer to get between him and the final product, he has the whole responsibility for the work, and more freedom than any other composer – subject only to the limitations of time, patience, and his equipment.

In practice, synthesizers offer a range of programs – short-cuts to particular effects – which at best simplify choice; at worst channel and so limit it.

The basic equipment for producing electronic sound (now part of any synthesizer) includes generators for producing pure sine tones (single frequencies) and combinations of tones, perhaps sawtooth or square-wave generators. Precise calibration is essential: if musical instruments are off-pitch, it is only one performance that is affected, and the score remains unaltered, but in electronic music the first 'realization' is likely to be the only evidence of the composer's intentions. Some early synthesizers were notoriously unstable in frequency, but this should be no problem today.

A characteristic sound in early electronic works was short bursts of tone at various frequencies, sounding like a series of pips, beeps and bloops. If a work was assembled from a series of short lengths of tape to give this effect, variations in attack and decay could be introduced by varying the angle at which the tape is cut. A 90° cut gives the strongest effect of a click at beginning or end; other angles of cut soften it.

Today, however, a synthesizer would be used to provide not only the tone, together with conventional harmonic coloration, if required, but also a range of different qualities of attack and decay. These are often described by relating them to some musical instrument of similar attack and tone quality.

White (and *coloured*) *noise* is another type of sound that is peculiar to electronic music and radiophonic sound, as this incorporates electronic techniques. It can be produced in several ways – originally by amplifying the output from a noisy vacuum tube. Special tubes were made for the purpose. Another source is the amplified output of a VHF or UHF receiver tuned off-station.

White noise is in itself not particularly useful; but in combination with a versatile filter it can be used to produce 'colours' that can be of indeterminate pitch, or related to any desired musical scale by placing peaks at particular frequencies and troughs at others, or by using bands of particular widths. Electronic wind effects can also be manufactured in this way. Another effect is achieved by imposing a sinusoidal envelope, or any other waveform, on the noise so that it becomes a series of puffs of sound. If this is speeded up sufficiently it is transformed into a tone that can be used musically – then on slowing down again the granular structure of the original sound may be observed. The principle is similar to that of the siren.

Conventional harmonic structures can be derived from sawtooth or square-waves. The saw-shaped wave contains a full harmonic series, while the square shape includes only the odd harmonics, like those of a clarinet. These are subsequently treated by *subtractive synthesis* (filtering, etc.), which in appropriate cases is less time-consuming than an additive process.

Many *electronic organs* have been used in the past, with names like Monochord, Melochord and Spinetta, besides the more conventional Compton and Hammond organs. The Trautonium was one of a group of instruments that could be used (and played, live) to provide notes of any pitch, not just those corresponding to a keyboard. Organs, as one class, shade gradually into another, the *synthesizers*, which includes instruments with manufacturers' names such as Arp, Moog, Yamaha and Fairlight. Most of the electronics of early synthesizers has been 'analogue' in form in that the instruments copied or reconstructed the required waveforms or noises by means that provided a continuous shape, which could also be continuously modified from the start. The next stage of development was to add digital circuitry – for example by using a digital control device to programme musical sequences of treated sounds. This was only the beginning of the digital invasion, and synthesizer design has developed as fast as that of computers: by the time a choice has been made and the equipment installed, it may already be outmoded. In practice it is necessary to set reliability against range, for a given cost, size (or compactness) and ease of performance.

In its time the Moog (pronounced to rhyme with vogue) was a revolutionary design of analogue synthesizer relying on compact multi-function voltage-controlled oscillators and amplifiers making possible a generous capability for the modification of sound structures in interesting ways. Increased versatility has been one line of development, but this has been paralleled by a feeling that that which can do anything does nothing in particular: in the workshop a vast available range of functions may for the most part rarely be used, with the composer seeking out a synthesizer mainly for its more unusual qualities. More recently there has been a tendency toward design for specialized purposes; for electronic instruments that can be used with conventional orchestral instruments in performance, or perhaps for re-creating classical sounds such as that of the baroque organ.

It remains legitimate to use the simplest means that offer the desired effect, so many workshops have also used such sources as the electric guitar and autoharp, zither, harmonium, and piano, among other musical instruments, conventional and rare.

Radiophonics

Where a work is being synthesized from basic sources it is obviously a major task to treat and assemble them. A radiophonic studio will, of course, have a good range of conventional equipment, including filters, compressors and limiters, versatile reverberation equipment and advanced (ideally, fully automated) control desks with all their other usual facilities. To this is added a range of more specialized items, which have advanced even faster and more radically than in other sound departments.

Since it began in 1958, the BBC's Radiophonic Workshop has updated its equipment many times. Successive generations seemed to last about 10 years at first, but speeded up in the late 1980s until it was dominated by digital techniques.

Such a studio will have a computer capable of running a substantial range of software packages, including several used as composition, performance and sound-design aids. There will be a variety of synthesizers and a routing matrix. Some analogue recording machines may still be found (mainly for transfer purposes), but digital (and optical) recorders have largely replaced analogue multitrack machines – initially with a smaller number of channels, but expanding as the technology develops.

The most significant change has been the adoption of MIDI keyboard techniques, which are particularly suited to this application – and, in

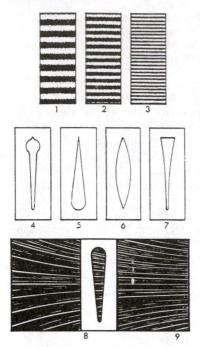

Amimated sound created using optical film track
1–3, Cards from a set giving different pitches. More complex patterns of density give richer tone colours. 4–8, Masks (to be superimposed on tone cards) giving dynamic envelopes of a particular duration. Long durations are obtained by adding together cards that are open at the ends. 9, A card giving a range of pitches of a particular tone colour. A technique developed when optical tracks were more widely used: the 'sounds' were all photographed directly on to it. A wide range of similar techniques has been employed, using both variable density and variable area.

the view of BBC radiophonic composers, indispensable. They can be used either in the 'conventional' keyboard mode for controlling pitch or with each key designated separately for some other control function. It follows that unconventional scales or progressions can also be constructed and routed through it. It is used in conjunction with its own MIDI routing matrix.

The range of conventional synthesis is extended further by the digital envelope shaper. An envelope is drawn by light-pen on to a display tube, and then converted by a processor to digital form.

Radiophonics demands an unusual combination of skills, both operational and creative, and it is difficult to predict whether any particular individual will have either the aptitude or temperament. It is best to create ways for aspirants to try their hand before any commitment is entered into. At the BBC (as distinct from some electronic music studios) the product of its Radiophonic Workshop is applied to the needs of general radio and television programming; it is not a 'pure' experimental music laboratory. Where the latter exists there has been a tendency for them to produce extreme avant-garde or computer music, much of which deserves to be played only to other computers.

New laws for music

Most radiophonics is based on the conventional musical scale in some form: this is directly related to human powers of perception and appreciation, which are attuned to harmonic structures that arise naturally. But by starting from pure electronic sources the musical 'laws of nature' can be relaxed somewhat. In view of the way in which human minds work, whether or not this is much use depends on what is offered in specific cases. Unfortunately there is little incentive to other than the more remote experimental musicians to work in this field, so little has been done to establish its broader value.

But before we can begin to examine the composer's new freedom to create new harmonic structures and scales, consider first the concept of scale as we know it, and the restrictions that this imposes. All scales ever devised for conventional music have one thing in common – the interval of an octave, in which the frequency of the upper note is exactly twice that of the lower. If the two notes are sounded together on orchestral instruments the combination sounds pleasant because of the concord between the fundamental of the higher note and the first harmonic of the lower; and there are also many other harmonics in common. If, however, one of the notes is shifted slightly off-pitch these concords are lost – they are present only when the frequency of the two fundamentals can be expressed as a ratio of small whole numbers.

Note that without the harmonics there would be no sense of concord or discord at all, unless the fundamentals were themselves fairly close

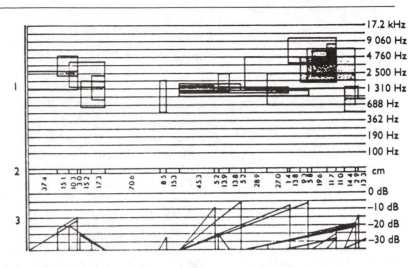

Electronic music
Score of the type devised by Stockhausen to show the structure of his 'Study II'. 1, Pitch: showing the groups of tone mixtures used. Each block represents a group of five tones (see opposite). 2, Intervals of tape measured in cm. Tape speed 30 i/s (76 cm/s). 3, Volume: dynamics of the tone mixture.

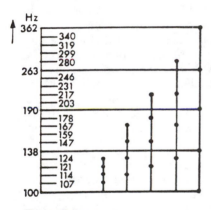

Tone mixtures
Five typical mixtures used in Stockhausen's 'Study II': 193 were used in all.

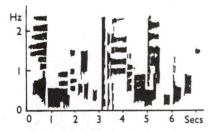

Electronic music and reverberation
Spectrogram of part of Stockhausen's 'Study II', showing the softening effect of reverberation.

together. If we start instead with two pure tones at the same frequency, and then increase the pitch of one of them, the first effect that we hear is a beat between them. As the difference becomes greater than 15 Hz (twice or three times this at high frequencies), the beat is replaced by the sensation we call dissonance. This increases to a maximum and then falls away again, until as the frequency ratio approaches 5:6 it is no longer apparent. For pure tones, dissonance occurs only within this rather narrow range. There is just one proviso here: that the tones are not loud. If they are, the ear begins to generate its own harmonics.

Some experimenters with electronic music, noting all this, have concluded that our conventional concept of scale is merely a convenient special case, and that if only we could create harmonic structures that were not based on the conventional 1, 2, 3, 4 . . . series, completely new scales could be devised – as, indeed, they have.

One early example of electronic music based on a totally new concept of scale has (exceptionally) been published as a complete score, showing everything that had gone into the making of a particular work: Karlheinz Stockhausen's 'Study II'. When the tape of this is played to a listener with perfect pitch, or even good relative pitch, it seems immediately that 'something is wrong' – or at the very least, that something is different. And indeed, by all conventional musical standards, something is; its arrangement of musical intervals could never have been heard before the introduction of electronic sound synthesis.

A conventional scale accommodates as many small-whole-number relationships as possible by dividing the octave into 12 equal, or roughly equal, parts. Each note in the chromatic scale is roughly 6% higher in frequency than the one before, so that the product of a dozen of these ratios is two – the octave. But Stockhausen, in his 'Study II', dispenses with the octave completely, and takes instead a completely new scale based on an interval of two octaves and a third, which is the interval between a note and its 'normal' fifth harmonic, or, to put it another way, the interval between any two notes whose frequency ratio is five. This large interval he subdivided into 25 equal small intervals, which means that each successive note on his scale is about 7% higher than its predecessors. So it is only to be expected that very few of the intervals based on this scale correspond to anything in the previous musical experience of the listener.

If music written to this scale were played on almost any conventional musical instrument, almost any combination of notes attempting either harmony or melody would be dissonant because of the harmonics present. Violins and trombones are examples of instruments that *could* be immediately adapted to playing in it; others, such as trumpets and woodwind instruments, would have to be specially made. The only feasible form of construction was to use completely synthesized sound derived from electronic sources.

Stockhausen, having devised his new scale, proceeded to construct new timbres in which the partials would all fit in with each other on the new scale. He limited himself to five basic sound groups, each composed of five tones. The most compact of these contains a group of five successive frequencies or 'notes' from the new scale. The quality of this first group could be described as astringent, as it consists of a series of dissonant pairs. The next group contains members spaced two notes apart, and the internal dissonance has now almost gone; in the other groups, whose members are spaced at three-, four-, and five-note intervals, it has gone completely, and the difference in character between them depends solely on the width of their spectrum. Before use, each of the 193 basic groups were replayed through a reverberation chamber (modern electronic devices were not then available) and re-recorded, in order to blend the sound.

This single example is offered as a stimulus to the reader's imagination. Plainly, the opportunities for shaping sound go far beyond what is conventionally achieved.

Chapter 17

Editing

Editing can mean various things: physically cutting and rejoining a tape or, alternatively, reorganizing and recombining signals within computer memory. It may involve *copy editing*, in which selected sections from originals are copied in a required sequence on to an assembly tape; *transfer* between recording media or *loading* or reading the material onto hard disc; or *mixing* (or *film dubbing*) where several magnetic tracks and other sound sources are combined in a final mix. *Rough editing* is assembling the main body of a programme or sequence in the right order and taking out the longer stretches of unwanted material. *Fine editing* is tightening up this assembled material to provide for the best continuity of action and argument, at the same time deleting where possible the hesitations, errors and irrelevancies that distract attention and reduce intelligibility.

These processes may employ recording media (usually magnetic) such as audiotape, videotape, film separate magnetic track ('sepmag'), and increasingly digital storage systems (random access memory, RAM) are used as an intermediate stage.

Editing is generally undertaken for one of five reasons:

1 *Programme timing:* adjusting overall duration to fit the scheduled space. In radio or television this often overrides all others, whether or not editing improves the programme.
2 *Shaping* the programme: giving it a beginning, a middle and an end, and ensuring that the pace and tension varies, and does not drop too low. Awkward parentheses, repetitions, phrases, whole sentences and even paragraphs have to go if they obstruct flow.
3 *Inserting retakes.* This restores or improves a predetermined structure, especially in music and drama.
4 *Cutting 'fluffs',* etc. Minor faults can lend character to speech, but too many make it difficult to follow. Mistakes in reading a script do not often sound like natural conversational slips.

5 For *convenience of assembly:* to combine film or material recorded on location with studio links, to allow material of different types to be rehearsed and recorded separately, or to pre-record awkward sections.

Radio and television techniques that aim to avoid editing presuppose either that the broadcast has a well-rehearsed and perfectly timed script or that those taking part are capable, in an unscripted programme, of presenting their material logically, concisely and coherently, or are under the guidance of a presenter who will ensure that they do. Such self-control under the stress of a recording session is rare except among experienced broadcasters. Even when a sound recording or film is made in familiar surroundings, the presence of the microphone, even more than the camera, produces stilted and uncharacteristic reactions. One purpose of editing is to restore the illusion of reality, generally in a heightened form.

This process of editorial selection generally carries a responsibility to avoid distorting the character and intentions of the speaker. On hearing an edited playback, any contributor should recognize the essence of what was said – within the limits of what was needed for the programme, which should have been made clear in advance – but it is also easy to use editing to make content conform more closely to the purposes or preconceptions of those in control of the process. At best, this may be to mould it to some openly-declared format. At another *extreme*, it may completely subvert the intentions of contributors. Plainly, editing can be a source of controversy, and may lead to debate on ethics – which may overshadow the process, even as it is being carried out. But the purpose here is to describe what can be done, and how. Fortunately, in most cases, the purposes are both clear and desirable.

Retakes and wildtracks

Nearly all recordings benefit from, at the very least, a little tidying up, but it will be easier if, right from the start of the first recording, you are mentally checking what might be needed later. The things to watch for depend on the type of programme. For example, if a 'fluff' occurs in speech, will it be desirable or even possible to edit, or would a retake be advisable?

In film the error may occur at a point when a cutaway (in vision) is intended, or can easily be arranged. In this case a repetition may be left for the editor to remove, or if this is awkward a retake may be taken in sound only (*wildtrack*). If a retake in vision is necessary the director has to decide whether to go again from the start of the scene, whether to vary the framing of the picture (usually, if possible, to a closer shot or to a different angle), or whether the same framing can be continued after an earlier cutaway. The sound recordist should also (and generally without being asked) supply background atmosphere at the same recorded level in sufficient quantity to cover any gaps that may be introduced by editing or where there are more pictures than synchronous sound. This saves time and effort later, at the film dubbing session.

In a radio discussion between several people, the director should note whether the various voices have identified themselves, each other, and the things they are going to talk about sufficiently clearly. If not, he might record a round of names and subject pointers in each voice so that these can be cut in where necessary. Watch for continuity of mood: if the point you are likely to join up to ends with a laugh, try to start the new take with something to laugh about (which can be cut out). Discourage phrases like 'as I've said', and retake if they are not clearly editable.

When editing radio programmes, 'pauses' may be required. So, here again record 15 seconds of atmosphere (although atmosphere for pauses can often be picked up from hesitations between words at other points on the tape, it is better to be safe). Do this while the speakers are still there: after they have gone the sound may be slightly different. If much editing is likely to be required, be cautious in the control of levels; over-control can make difficulties with the background atmosphere. If the background is heavy, record a spare track of 20 seconds or more.

At a recording session for classical music it is again useful to record a few seconds of atmosphere at the end of each item in order to have something to join to the start of the next piece. Music retakes should be recorded as soon as possible after the original, so that the performers can remember their intonations at the link point, and also so that the physical conditions of the studio are close to the original. An orchestra quite literally warms up; its tone changes appreciably in the earlier part of each session in a studio.

In describing the techniques used I shall deal first with editing on quarter-inch analogue tape. The same general techniques apply to other editing processes. For hard disc digital editing, the display on the screen mimics tapes passing replay heads, with the edits clearly visible. Digital editing involves the use of time code, which was originally developed to control video recordings and is described in Chapter 18.

Tape editing: preparation

Electronic editing can be used with quarter-inch (6.25 mm) sound tapes (as with videotape), but physically cutting the tape is more convenient except when the levels have to be adjusted. So this account of the techniques begins with traditional tape splicing. Here we shall deal only with techniques where there is no tape overlap, so that no tape is lost when a joint is made. In the event of an unsatisfactory edit it should always be possible to reconstitute and try again. This encourages experiment and, in general, a more adventurous approach to editing. For programmes in which most of the material

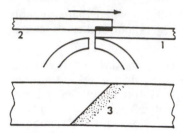

Lapped joint

These were originally used for editing tape. These so-called 'permanent' joints were sealed with a special adhesive. The overlap was arranged with the first part (1) of the tape or film past the head on top of the second part (2), so that contact is quickly regained as the second part drops on to the head. If the join is angled (3), the temporary loss of contact is less important as its effect is spread over a length of tape.

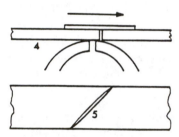

Butt join

With this (4) there is no loss of contact at any time unless the butt is imperfect (5). Again, the effect of any slight gap or overlap is minimized by angling the join.

can be edited sequentially, it remains a rapid and easy method – and at lower capital cost than a hard disc editing system.

Many freelance contributors to radio programmes prefer to do their own editing before submitting their tapes. They, too, benefit by using a method that is technically simple. For speed and accuracy, the following tools are all that are required:

● A simple block with a channel to hold the tape. The channel is lapped slightly so that tape smoothed into it with a fingertip is gripped firmly. The block also has an angled cutting groove that can be used for mono but a groove at 90° may be necessary for stereo. It should be fixed firmly to the near edge of the tape deck.

● A fairly sharp stainless steel razor blade. Use the one-sided type to avoid cut fingertips, and only fairly sharp for the same reason. Change the blade as soon as it begins to get blunt, however, because it then drags on the tape and may prevent butt ends matching correctly.

● A soft wax pencil (yellow or white) or felt-tipped pen for marking edit points on the tape backing.

● A roll of jointing tape slightly narrower than the recording tape to allow for a little inaccuracy in use without overlapping the edges of the tape.

● As an optional extra: leader and spacer tapes. It is convenient to standardize: white for leaders, yellow for spacers and red for trailers. Green is also used to mark the head of a tape.

Non-magnetic scissors can be used instead of a razor blade but rarely are. Similarly, the accidental magnetization of non-stainless-steel blades is an uncommon event. A dusting powder such as French chalk is sometimes recommended but should not be necessary if genuine jointing tape is used, except in hot climates, where the adhesive may ooze a little. It is, nevertheless, important that tape should not be sticky, or layers will pull as they come off the spool, drag on tape guides and clog the heads.

When preparing to make a cut, first play the tape to the point at which the cut is to be made, and then switch from replay to a condition in which the tape can be spooled by hand with the tape still in contact with the head. The appropriate method for this varies from machine to machine. In some cases there is a 'pause' control, in others it is 'replay/off', and so on. The exact point can be checked by hand, pulling the signal back and forth over the replay head. Mark the tape on the backing. When in doubt about a mark, set up the tape with the mark at the replay head, switch to 'replay', and run. If the machine has

Equipment for joining tape

1, Splicing block. 2, Blade. 3, Leader tape. 4, Wax pencil. 5, Jointing tape. 6, Spare spool.

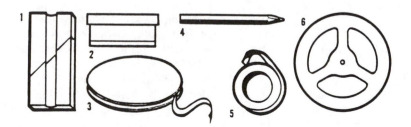

Tape editing
1, Move the tape backward and forward over the heads, by hand, in order to find the exact point. 2, Mark the tape with a soft wax pencil. 3, Cut the tape diagonally using the groove in the block. 4, Butt the two wanted sections together in the block and join with special adhesive tape.

a quick start it should be clear whether the mark is correct. Alternatively, pull quickly up to speed by hand.

Always mark the cut in the same way, e.g. along the lower edge of the tape. This ensures that if a scrap of tape is reinstated, it is put in the right way round.

A simple edit is made like this:

1 Locate and mark the joint, as already indicated.
2 Switch the recorder to whatever condition allows the tape to be lifted from the heads, and place the tape coated side down in the editing block.
3 Cut. The razor blade should be sharp enough not to drag when drawn lightly along its guide channel. It should not be necessary to apply pressure.
4 If a fairly long piece of tape is to be taken out, replace the take-up spool by a 'scrap' spool and wind off to the new 'in' point. Mark this, cut again, and swap the take-up spools.
5 Place the two butt ends of the tape coated side down together in the block (and a little to one side of the cutting groove) so that they just touch each other.
6 Cut off an inch or so of jointing tape and lay it over the butt, using one edge of the channel to steady and guide it into place.
7 Run a finger along the tape to press the joint firm. If it seems advisable, dust with French chalk by dipping a finger in the powder and running it along the tape again.

After a joint has been made, place the tape back on the heads, and spool back a foot or so (again, this can usually be done by hand against the braking effect of the clutch plates), then play through the joint checking that:
● it is properly made
● the edited tape makes sense – and that you have in fact edited at the point intended
● the timing is right: that the words after the joint do not follow too quickly or too slowly on those before
● the speaker does not take two breaths
● neither the perspective nor volume of voice takes an unnatural jump
● there is no impossible vocal contortion implied
● the tone of voice does not change abruptly, e.g. from an audible smile to earnest seriousness
● the background does not abruptly change in quality; nor is any 'effect' or murmured response sliced in two
● (for stereo edits) no 'flicker' – a momentary sideways jump – has been introduced by an imperfect butt.

It is worth thinking of these possible errors in advance, too, both when recording and when selecting suitable places to cut. But a virtue of this kind of editing is that it is very easy to reinstate material if the cut does not work, or to change it by cutting in a different place.

You may need to cut down a gap that is too long; or insert a pause or breath or even an 'um' or 'er' to cover an abrupt change of subject or mood; or you may have to restore part of what was cut. After marking any new point at which you propose to cut, undo the old joint. To do this turn it over and break it; grip one of the points of the butt and pull it away from the jointing tape, then the other. After remaking the cut, reseal the joint with a strip of new jointing tape.

It is always recommended that monophonic tape (including monophonic contributions to a stereo mix) should be cut at an angle – the exact angle is not vital. A 90° cut has the disadvantage that any sound on the tape, even atmosphere, starts absolutely 'square' at the point of cut: and any sound that cuts in at full volume seems to start with a click. Examples are time signal pips or 'beep' tones: the very words 'pip' or 'beep' suggest the effect that is heard. To demonstrate this, record a pure tone and cut in a short length of spacer at 90°, then try the same thing with angled cuts.

It can be demonstrated experimentally that if a fade in or out exceeds 10 ms in duration there is no click. In fact this is over 20 times more than is actually needed in most circumstances. This duration corresponds to 0.15 in of 15 i/s tape (i.e. 0.38 cm of 38 cm/s tape). As a 45° cut on full track tape extends nearly twice this length along the tape, there will therefore be no click. Similarly, 10 ms corresponds to 0.075 in of 7½ i/s tape. A half-track recording fades in or out over 0.1 in for a 45° cut, so again there is no click.

The standard angle of cut provided on some joiners used for magnetic film sound is 10°. A recording on separate magnetic film track is 0.2 in (5 mm) wide: a cut therefore fades in over about 0.02 in (0.5 mm); for 16 mm film the corresponding duration would be 2.5 ms, and this too should be quite long enough to avoid a noticeable click.

Tape recorded at 7½ i/s (19 cm/s) is the most convenient for editing when using cine spools. Lower speeds give reduced quality and less precision in marking, and higher speeds make it more difficult to pull the tape over the heads by manually rotating the spools against the clutch or brakes. Using the larger NAB spools, 15 i/s (38 cm/s) is a convenient recording speed for editing.

For stereo, an angled cut should only be used when editing between identical sounds or in a pause, otherwise a momentary change of position (flicker) will occur. When a 90° cut is also unsatisfactory (for the reasons already given), mixing techniques must be employed.

Rough editing

In rough editing the objective is much the same in radio, television and film: it is simply to assemble the material in its intended order, so that the effect, logical or artistic, can begin to be seen. The simplest and most flexible is the case where sound only has to be considered – typically, for radio. In this application, you may perhaps start by putting in spacers to indicate where there will be a studio link. Next, work on structure.

Remember that you can *transpose* material – especially when several unrelated subjects have arisen in an unscripted discussion. The factors that governed the order of recording may not hold for the finished programme. You might want to start with a bang, with something that may not be of vital importance, but which engages the listener's attention and involvement. The middle should, while progressing logically, have light and shade: with variety of pace, of speech lengths and rhythm and of mood. And the tension should build to an effective 'curtain': for example something that contains in a few lines the essence of all that has preceded it, or will offset and illuminate it.

As you select material, feed it on to your make-up tape in programme order, with transpositions and retakes in their proper place. At this stage try to cut at 'paragraph pauses'. Rejoin the unwanted material after each cut, by feeding it on to a reject spool which should be kept until the programme is complete.

Even at this stage exercise care: do not cut in the middle of a breath, or too close to the first word when there is a heavy background. If you are joining different sections by the same speaker, check that voice quality, mood and balance are sufficiently similar: it is all too easy for the listener to get the impression that someone new has entered the conversation and to wonder who it is instead of listening to what is being said.

When an additional replay machine is available the rough editing stage can be done by copy editing. In this way the original tapes can be kept intact, which will be of advantage if you want to retain them for re-use. It also allows for levels to be matched and provides insurance against mishaps. The selected programme material is re-recorded to a make-up tape in the appropriate order, perhaps taking a little more at the beginning and end of each insert than will be required. The inserts are then cut together on the make-up tape in the normal way.

If subsequent mixing is envisaged, the material will be cut or copied to 'A' and 'B' reels, or possibly laid down on multitrack tape. Mixing between two or more tracks is easy in hard disc editing.

Fine editing for radio

After your rough editing give thought to the shape of the programme as a whole, to the intelligibility and conciseness of individual contributions, and also to whether the character and personality of the speakers is adequately represented. The first of these considerations may mean that you have to make severe internal cuts in one section and not in another; you may have to cut out repetitions or verbal embroidery. Try to keep the theme of what remains tight and unified.

Here are some other things you may wish to cut:

- Heavy coughs, etc. These hold up the action and may cause the listener to lose the thread. But if the speaker has a frog in the throat (and this section of the recording cannot be cut) it does no harm to leave in the cough that clears it – just for the psychological satisfaction of the listener!
- Excessive 'ums' and 'ers' (or any other repetitious mannerism) should be cut if this improves intelligibility – which it often does if they appear in the middle of a sentence. But some actually *improve* intelligibility by breaking the flow at a point where it needs to be broken. So be careful: there are people who use these noises to support or modify meaning; to add emphasis or emotional expression, and while the result may not read well, it sounds right. Others seem to express character in these punctuating sounds. In these cases, do not cut. An 'er' moulded into the preceding or following word often cannot be cut, anyway.
- Excessive pauses. In real life pauses can often afford to be relatively long, because we can watch the speaker's face. On audiotape, a pause is just a gap, unless it lends real dramatic emphasis to its context. But pauses should not be cut down to nothing if the effect is of impossible vocal contortions. Try it: record the sentence 'You (pause) are (pause) a (pause) good (pause) editor' and then cut the pauses out completely. The result sounds like a case of hiccups. The minimum gap depends on how much the mouth would have to change in shape during the pause. This applies to 'ers', too.
- Superimpositions. Two people talking over each other are irritating. But edit with care: it is generally necessary to leave a partial overlap, or to take out part of the sentences before and after.
- 'Fluffs', verbal errors, where the speaker has gone back over the words to correct them. Again an edit generally improves intelligibility. But take care here too: the first word of the repeat is often overemphasized.

Each different speaker presents a new set of problems and decisions on whether to cut or not. For example, an interview may begin hesitantly and gather speed and interest as it progresses. This is wrong for the start of a new voice in a programme tape, where the listener's attention and interest must be caught in the first few words. If you cannot cut this warm-up period it may be necessary to tighten up the opening sentence: this matters more than hesitations at a later stage, when the voice and personality are accepted by the listener.

In certain cases, particularly where the recording is of importance as a document, no editing should be done at all (with the possible exception of cutting down really excessive pauses), and the greatest care should be exercised if any condensing is to be attempted. If in doubt, resort to a studio link.

In any case, do not over-cut. It is useful to have good ideas about where to cut: but it is just as important to know where *not* to cut.

Finding the exact point

There are two places to cut at in a 'sentence' pause. One is after the word (and its associated reverberation) has finished and before the breath; the other is after the breath and before the next word. For most purposes it is best, and safer, to choose the latter. You retain the full natural pause that the speaker allowed, and you can cut as close as you like to the new word which helps to mask any slight change in atmosphere.

Cutting between the words of a sentence is trickier. In addition, the words themselves sound very different when they are wound through slowly by hand, so that a suitable point is more difficult to locate. But certain characteristics of speech can soon be recognized, e.g. the 's' and 'f' sounds. The explosive consonants 'p' and 'b', as well as the stopped 'k', 't', 'g' and 'd', are easy to pick out (though not always as easy to distinguish from each other) because of the slight break in sound that precedes them on the tape. In a spoken sentence the words may be run together to such an extent that the only break may be before such a letter – perhaps in the middle of a word.

Do not assume that because a letter should be there it will be. Complete vowel sounds may prove to be absent. The personal pronoun 'I' is typical of many such sounds. In the sentence 'Well I'm going now' it may be missing completely. If you try to cut off the 'well' you may even find the 'm' was a vocal illusion, no more that a slight distortion of the end of the 'I' sound. Similarly, in 'it's a hot day', spoken quickly, there is probably no complete word 'hot'. Nor is it possible to isolate the word from 'hot tin roof', as the first 't' will be missing and the one that is there will be joined to the following vowel sound. Before attempting fine editing it is worth checking what the beginnings and ends of words sound like at slow speed, and where words blend into each other and where they do not.

In cutting between words, cut as late in any pause as possible. When searching for the right pause on a tape do not be misled by the tiny break in sound that often precedes the letter 't' and similar consonants. You can often cut *in* to a continuous sound, for example where several words have been run together, provided that (for physical editing) the cut is at an angle, but do not attempt to cut *out* of one, unless you join straight on to a sound of equal value (one for which the mouth would have to be formed in the same shape).

It is very difficult to insert missing 'a's' and 'the's'; these are nearly always closely tied with the following word, so that unless the speaker is enunciating each word clearly and separately, it is necessary to take this in combination with the sound that follows. The letter 's', however, is somewhat easier to insert or (sometimes) remove.

These principles apply equally to hard disc (digital), videotape and film sound editing: though in the latter case the cut is normally made at discrete points defined by the frame interval; on hard disc the minimum separation may be shorter.

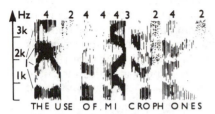

THE USE OF MI CROPH ONES

Human speech
Analysed to show formant ranges. 1, Resonance bands. 2, Unvoiced speech. 3, Vocal stop before hard 'c'. There would be a similar break before a plosive 'p'. 4, Voiced speech. These formants are unrelated to the fundamentals and harmonics, which would be shown only by an analysis into much finer frequency ranges than have been used here. This example is adapted from a voiceprint of the author's voice. Each individual has distinct characteristics.

Editing sound effects

Editing staccato effects and absolutely regular continuous effects presents no problems. But effects that are continuous yet progressively changing in quality can be difficult to edit, as can speech accompanied by heavy but uneven effects or speech in heavily reverberant acoustics.

Rhythmic effects on their own are fairly simple, as they can be treated rather like music: and so, possibly, in its different way can reverberant speech. But with more uneven noises it may be difficult to find two points sufficiently similar in quality, and in this case it is again better to mix. Editing applause is typical of the type of problem that seems as though it ought to be simple, but when physically cutting audio tape, it is not. A burst of applause lasting, say, 10 seconds is difficult to cut to five without a slight blip at the join – but to do this with two copies of an effects recording or on hard disc, it is easy.

Sometimes the change of quality behind a speech edit can be covered by mixing in some similar, but heavier, effects afterwards, using an effects disc or tape, or a loop. Even so, a change of quality can often show through unless the added effects are substantially heavier than the earlier backing, or are peaked at the point of the join.

There are certain cases where the untidy effect of a succession of sharp cuts in background noise can be tolerated. It may be accepted as a convention that street interviews (*vox pop*) are cut together by simple editing, with breaks in the background appearing as frank evidence of the method of assembly.

Editing music

There are various reasons why we may wish to edit music: to cut out a repeat, or one verse or chorus; to insert a retake; to delete everything up to a certain point so that we hit a particular note 'on the nose'; to condense or extend mood music to match action, and so on. Most of the skill lies in marking the right part of the right note, and this requires practice. Remember also when handling music tapes that damage due to spillage or stretching, etc., is more noticeable than on speech.

The important thing in music editing is not that the next note after each of the two points marked should be the same, but that the previous one should be – that is to say, when cutting between two notes it is the quality of the reverberation that must be preserved. Any retake should start before the point at which the cut is to be made: there is nothing so painfully obvious as a music edit where a 'cold' retake has been edited in, resulting in a clip on the reverberation to the previous note. It might be supposed that if the new sound is loud enough the loss may be disguised, but this is not so unless it is cut so tight as to trim away part of the start of the note. This attack transient is particularly important to the character of a sound.

So unless the cut is made in a pause that is longer than the reverberation time as recorded on tape, the first requirement for a successful joint is that the volume and timbre preceding both the 'out' and 'in' points on the original tape should be identical.

It may be possible to make an exception to this rule when cutting mood music to fit behind speech and effects – if the level of the music can drop low enough. For preference, choose a rest to cut out on and something rather vague to come back to, and if necessary dip the music level at the edit point. However, no such liberties may be taken with foreground music. For example, when linking music is being trimmed to length, any internal cuts must be made with the same care that would be given to featured music. And speech that has been recorded with music in the background is often practically impossible to edit, because an untidy join in the music may coincide with a slight gap in the speech.

When everything up to a certain point in a piece of music is cut, in order to create a new starting-point – e.g. to come in on a theme – the effect of the tail of reverberation under the new first note is often unimportant. Except in the case when this tail is loud, and contrasts strongly in quality with the first wanted sound, it will be accepted by the ear as a part of the timbre of the note. Even so, in such a case a sharp manual fade-in often sounds better than a cut. For getting out of a piece of music at a point before the end (where there is no pause) there is no alternative to a fade. On a hard disc or video, or any other system where events are controlled by the use of time code, a manual fade can be replaced by a programmed fade or *ramp*.

When the cut is to be made on tape at a place that is not clearly defined by the music itself it is often possible to identify and match the cutting points by: (a) measuring off the distance on the tape from a more easily located point, or (b) marking out the rhythmic structure of the music with a wax pencil or felt pen on the back of the tape. Of the two methods, the second is possibly the better guide, as the easily located note may not have been played exactly on the beat.

Hard disc audio editing

Digital editing, using specialized hard disc equipment and also the time code which is used to control it, is described in more detail in the next chapter, as the opportunities it offers have been more fully exploited where sound is allied to picture. For sound-only editing it comes into its own where there are many layers of speech, music and effects, and for this some of the early systems that had only eight tracks (possibly arranged as four stereo pairs) were too limited to be used except as just part of a mixture of assembly techniques. More recent developments that have many more tracks are more versatile, but at a capital cost that demands high use and a marked need for the advantages offered. Most versatile of all (and most expensive) are fully digital combinations of mixer desk and editing system.

The editing system will accept either analogue or digital sound, but there are great advantages to using digital audiotape (with time code) as the original recording medium, so that tape and hard disc system can have a common set of numbers by which to locate material. DAT is also used to load down from the system after editing, and if intermediate versions or sequences are kept, a large number of tapes can eventually be generated, even by a single programme. Digital identification numbers recorded on the tape and corresponding labels on the tapes and their boxes are essential. Event lists (information on the current program of operations required to edit the material) are recorded on specially formatted floppy discs.

The great advantage of fully digital editing is the minimal loss of quality over many generations of recordings, and the precision with which many layers of sound can be assembled, together with all the details of volume, cross-fades, etc. Adjustments within the stereo image are simple, as are overlapped edits – so that, for example, an intake of breath can be put in its most natural position, whether or not the gap left for it is long enough. Curiously, a single track appears to hold both parts of an overlap, but needless to say, it does not: the audio for overlapped material is held elsewhere on hard disc.

A disadvantage is that where so many changes can be made in rapid succession, while the last two are easily retained in a 'quick save' facility – necessary, as complex computer systems have been known to crash – any change of mind to go back to an earlier version can be time-consuming. Also, it is possible to make changes on one track or a stereo pair that loses synchronization downstream. Further, any technical problems encountered when learning such new techniques can be frustrating and also expensive in studio time, if this is locked into the use of the digital equipment. Loading and downloading (if this is in real time) must also be allowed for.

As with so much in digital technology, the main benefits have not been in cost or time reduction, but in permitting more ambitious and sophisticated results, while maintaining technical quality. Where appropriate, the simpler, cheaper – and often quicker – techniques should be retained.

Chapter 18

Film and video sound

This chapter offers a compact account of the principles involved in recording and editing sound with picture in a range of formats – beginning with film, for which the principles were initially developed, and also relating them to current television practice.

Although the standard professional film gauge is 35 mm, where film is still used in television 16 mm is most common – with negative and print stock used for applications where there is time for the extra process. Quarter-inch analogue audio tape (full track for mono or twin track for stereo) or digital audio tape (DAT) are used for recording the sound. Although A and B stereo is acceptable, M and S is preferred for some television applications. For traditional film editing the sound is transferred to a 16 mm separate magnetic track.

Videotape standards (within television) have also progressed to narrower gauges, initially from 2 in to 1 in (retained now mainly for archive material which has not been re-recorded) or, for some purposes, $\frac{3}{4}$ in; then again to $\frac{1}{2}$ in metal tape, which can be used for analogue or digital recording – the latter being less subject to drop-out and more suitable for on-line editing, since there is no significant loss of quality over the several generations of copying that are necessary in videotape editing. The number of channels of sound has increased from one, then two, to four or so; and while four is too limited for sound dubbing it can be generous for much location recording. It may, however, be useful to have one for a microphone on the camera itself (helpful, at the very least, for reference purposes) as well as a pair for stereo – again with M and S preferred by some television users.

Formats for specialized use include $\frac{1}{4}$ in metal tape (analogue) and plug-in ultra high density hard disc (digital). The latter may employ compression to increase recording duration.

Where television programmes are still shot and edited on film, they are now usually transferred to a video format in the final stages of post-production, and are transmitted on video, often from digitally recorded tape.

Digital video will continue to progress, and fully digital editing systems are increasingly used, so no account of the use of sound alongside vision can remain unchanged for long. Even so, tape still has some disadvantages as a storage medium and even greater deficiencies when used for editing, so systems in which picture and sound are transferred to and processed in computer memory are better for all but the simplest editing.

Many of the techniques now used were originally developed for use with film and still apply to the other media as well as for the films that are still shot – not least for the cinema film industry, which shows little sign of making a switch, though even here computer memory is increasingly used as an intermediate stage for editing.

For film itself, most of what follows applies equally to 16 mm and 35 mm, but where slight differences exist the method for 16 mm is described. In editing, archaic methods of joining film which involve the overlap of part of a frame (or on 35 mm, the frame bar) are disregarded, as are the noisy editing and viewing machines based on intermittent action. Film (and particularly film sound) is most conveniently edited using guillotine tape joiners and viewed on equipment that allows continuous movement of picture, stabilizing the image by the use of rotating prisms. This allows good-quality sound to be heard without loud machine noise.

But this account must begin at the place where the combination of picture and sound first appears. Since good pictures usually take precedence, this is often far from the studio where sound could (in principle) be controlled with relative ease, and instead at some location where the achievement of good sound may begin with an exercise in damage control.

Choice of location

When a location is chosen for film or video shooting, it is essential to check the sound qualities and environment. At worst, it may be necessary to replace dialogue and effects completely: one branch of the motion picture industry specializes in such techniques. But plainly it is simpler to record and use the original location sound if this is possible, and for documentaries this is almost always preferred – not least for reasons of cost.

Location surveys should establish that aircraft are not likely to interrupt sound recordings. Direct observation is a good start, but in addition a glance at an air traffic control map will show what type of disturbance there might be – jet airliners along the main airways and in the terminal control zones, and military or light aircraft elsewhere. The survey should also ensure that road traffic noise is not obtrusive (or, if traffic is accepted as a visual element, that it is not excessive), and that machinery noise from factories, building sites or road works will not create difficulties – and so on. If the investment in good sound is high, a visit to the site at the same time of day in the appropriate part of the week is a useful precaution.

In many cases where people unconnected with the production are found to be making noises that mar a performance, a courteous request to stop often produces results, in which case courtesy also demands that the noisemaker should be told when filming is complete or when there are long gaps. It is, of course, unreasonable to expect people to stop earning their living in order that you may earn yours, so in such cases (or where it is the only way of gaining silence) payment may have to be offered.

For interiors, acoustics as well as potential noise should be surveyed. On this will depend the choice of microphones, type of balance, remedial measures against deficiencies in the acoustic environment, debate (and often compromise) with camera, lights and design and, where necessary, recommendations to the director on how sound coverage could be improved.

Film and video sound recording

A video or film unit includes a director, a cameraman, a sound recordist and an electrician (for anything more than minimal lighting), plus as many assistants in each of these departments as is necessary for efficient operation. There are few occasions when a sound recordist actually *needs* an assistant for television location work, except as a mobile microphone stand for certain types of dramatic subject.

Video recordings are made directly on the same tape that takes the picture, and synchronization is automatic. Some have four sound recording tracks, of which more later. What follows here applies mainly to film.

The film sound recordist generally has a high-quality lightweight quarter-inch (6.25 mm) analogue tape recorder or DAT recorder. Either can be used to record both the output from the microphones and a reference signal related to camera speed. In countries where the electricity supply frequency is 50 Hz, 25 frames per second has been adopted as the standard television film speed, and the pilot tone is 50 Hz; but where the mains frequency is 60 Hz the film industry standard of 24 frames per second has been retained for television work, and a pilot tone based on 60 Hz is used.

One commercially well-established sound recording system was originally developed to employ the full width of the tape for mono, but with the pilot tone superimposed down the centre. There are, in fact, two narrow centre tracks each of 0.01 in (0.25 mm) with a similar gap between them. Pilot tones on the two tracks are recorded out-of-phase, so that on full-track replay of the recorded sound the two cancel and are not heard. For stereo, if the outer tracks carry the A and B signals, full-track replay will produce an A + B, i.e. mono, signal. For much television work, MS is preferred, but the signal from MS microphones can usually be converted to A and B (marked on some control boxes as X and Y) and the same layout.

In each case the corresponding system must be available for transfer of the tape (conventionally to magnetic film stock) for editing: it is this transfer equipment which matches the speed of tape and magnetic film by scanning the pilot tone, so that the picture and sound tracks that are presented to the editor are perfectly in register, sprocket hole to sprocket hole, whatever the duration of the take.

To make this work, the camera and tape must be controlled to a common standard speed. In older film equipment (as for video camcorders) this was achieved by running a cable (a *sync lead*) between camera and recorder: the tone signal is generated by the camera and transmitted to the recorder. For film, this has largely been replaced by systems in which the camera and recorder are controlled by a matched pair of crystals, vibrating at the same frequency. The camera motor has its speed accurately controlled, and the tape records the corresponding pilot tone. Note that if the crystal control on the camera is switched off (if, for example, the mains frequency is used in order to synchronize with the frame speed of a television monitor) the sound synchronization may be lost – but can be regained by a return to the sync lead. In cases where the monitor is running at a frame speed which is radically different, and the camera speed is reset to avoid flicker, picture and sound will have to be considered separately.

Filming and videorecording routine

Consider first the case where a film director has control over the action before the camera.

If there is any rehearsal, this allows the recordist to obtain a microphone balance and set the recording level. In a full rehearsal the action, camera movement, and lighting are close to those to be adopted on the actual take, so the recordist has a final chance to see how close the microphone can be placed without it appearing in vision or casting a shadow and can then judge the ambient and camera noise levels in realistic conditions.

When the rehearsal is complete and the performers are ready, a film camera is normally set up for some form of identification of the shot, and the lights (where they are used) are switched to full power. While this is going on, the recordist listens for any rise in noise level due to traffic or aircraft, or unwanted background chatter, and will warn the director if necessary. When the director has checked with each department – often very informally – the camera assistant (in the conventional start to each take that is shot on film) holds up a clapperboard to the camera, and the sound recordist or assistant directs a microphone towards it.

On the director's instruction 'turn over', the sound recording tape is run: note that on DAT, in particular, a few seconds run up is required before the first usable sound in order to lay down time code for subsequent replay. Otherwise (for analogue sound) tape and film are run together, and the recordist and cameraman then both check that their equipment has run up to speed, so that they have crystal 'lock'. Alternatively, if a sync cable is used they check that the sync pulse is reaching the recorder and that both camera and recorder have reached stable speeds. When they have, the camera assistant is told to 'mark it'. Note that although other systems are available for synchronization of picture and sound, for example, one which puts light on a number of frames of film and a tone (or silence) on the corresponding duration of sound, the traditional sound clapper board is still preferred by many film-makers. It uses more film but saves on editing by providing ready identification of both picture and sound.

Marking the shot (*boarding* or *slating* it) consists of stating clearly the shot and take numbers, e.g. 'one–two–seven take three', while holding the board up to the camera with the bar open. The board is then clapped, the assistant and the camera operator both taking care that the action of the board closing is clearly in vision, and the recordist noting that the clap is identifiable and distinct. If any fault is observed, a second board can be called for. The camera assistant puts the board back in shot, this time saying something like: 'one–two–seven take three, sync to second clap', then clears the shot quickly but quietly, while the cameraman checks the framing and (if necessary) focus, and the recordist checks the microphone position and gives a final thought to noise problems – as the director may also do at this time.

The director checks that the performers are absolutely ready, waits for a signal that the camera is running at its proper speed, notes that there are no new shadows or unwanted sound, then calls 'action'.

At the end of the take the director calls 'cut': the camera stops running first, and as soon as it has, the recordist stops too – or, if there is some continuing noise, he may signal for everyone to hold positions while extra sound is recorded for a fade or mix; the director's cue might later be deleted in the editing.

Mute (in American, *MOS*) shots, i.e. picture filmed without sound, are marked with an open board (perhaps held in such a way that it could not be shut without chopping off a few fingers).

Wildtracks are recordings made without a picture. Identify them by relating them to a particular shot number, e.g., 'Wildtrack extra background of car interior with gear change for slate one–two–eight' or 'Wildtrack commentary for one–three–two onwards'. Alternatively, give them letters: 'Wildtrack B: atmosphere quiet classroom'.

Atmosphere or *'buzz' tracks*: the recordist should make a separate recording of the characteristic continuing sound of each different location – atmosphere ('atmos') or 'room tone' – unless specifically told it will not be wanted, and may also wish to re-record important effects that can be improved by a balance in the absence of the camera, together with any additional noises the director wants.

In the case where the director does not have complete control over the action, it may still be predictable to the degree that a front-board is possible, or failing that, an end-board, which is inserted and identified with the words 'board on end'. The director (if any) may give instructions to turn over (start the camera) and cut, as before, in which case the recordist takes cues in the normal way. (The recordist should call for the camera to keep running if the director calls 'cut' before the shot has been slated.) Alternatively, the director may instruct the cameraman to take cues from the action itself: in this case the recordist follows the lead of the camera-man. If the synchronizing mark is missed, it is often simplest to run the shot again and board it properly, rather than leave the editor to find it. This depends on the time available and the ease of repeating the shot.

Where the scene that is being filmed would be unduly disturbed by the standard procedure for marking shots, it can be simplified by starting each sync take with the recordist speaking the shot number and tapping a microphone in vision. The recordist can help the editor by leaving fingers touching the microphone for a fraction of a second and some confirmation of the number may be provided by its being spoken in vision. In such cases, keep the numbering simple: there should be no 'take two'. A hand clap, pencil tap on a table or any similar action can also be used for rapid synchroniza-tion.

Where informal action is covered by a video camera with time code (see p. 290), and particularly where non-linear (digital) editing is envisaged, conventional film shot-numbering may be omitted – though will often be replaced by a comprehensive, time-coded log, generally made (or checked) using viewing tapes, if possible before editing begins.

Film and video documentation

The director's assistant, the (film) camera assistant and the recor-dist each have their own documentation to complete. The main (director's) shot list should include the following:

1 Film roll or video cassette number.
2 Board and take number of each film shot. For video, detailed time codes.
3 On film, some indication, such as a suffix 's' or 'm' to the shot number to indicate sync or mute; also 'eb' for end-board if appropriate (this use of suffixes is not general, but is convenient for quick reference).

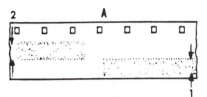

16 mm film: separate magnetic sound
1, Edge track used in US and Canada. 2, Centre track used in Europe. The full width of the film is coated and can therefore be used for either system.

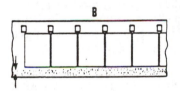

16 mm film: combined magnetic sound
The narrow magnetic stripe occupies the space formerly used for optical sound. Sometimes the coating is applied to the surface of the film (in which case a narrow strip is also applied to the edge close to the sprocket holes in order to ensure even winding); sometimes it goes into a groove in the film, so that the surface of film and magnetic track is flat. Picture and sound are displaced, to allow for the two sets of equipment required for synchronous reproduction. Also, so that some projectors can have heads permanently in place for both optical and magnetic-stripe tracks, their displacement from picture is different. The older, optical recordings are closer to picture: 25½ or 26 frames ahead of 16 mm picture and 19½ or 20 frames for 35 mm. Magnetic stripe is recorded 28 frames ahead on 16 mm film.

4 Shot description: opening frame (on start of action); action within the shot (including camera action; pans, tracks or zooms, timed if possible); continuity points; faults or errors, including sound problems; and the overall duration of significant action.

5 Duration or footage of film used for each take (optional).

6 Conventionally, for film 'P' ('print'), 'OK', or a tick for a shot to be printed; 'NG' for a shot that is not worth printing. Note, however, that where 16 mm film is printed to a low-cost cutting copy it may be as cheap to print everything. In this case the 'P' 'OK' or 'NG' serves as an editing guide.

7 Film sound roll numbers as new rolls are started.

8 Wildtrack details (inserted in the shot list between the sync takes where the sound itself is to be found).

The film sound recordist's own report should include as many technical details of recordings as may be useful later, including frames per second and frequency of pilot tone used (24 or 25 fps and 60 or 50 Hz) when the film is shot in a country where the standard may be different from that in which the film is to be edited. The format should be stated: analogue systems may have been controlled by time code or f.m. pulse, while DAT has a single worldwide standard. Each shot for which sound is recorded will be listed, followed by a list of takes. A way of indicating the 'good' takes is to put rings round those but not round the others. But again the simplest course may be to transfer all sound. Wildtracks are included on the sound report in the order in which they were recorded, and notes on quality may be added.

In addition, the recordist identifies each new roll of (separately recorded) tape over the microphone, with a standard list of details that might include his name, that of the film, any costing code or job number, date, location, recorder type and number, and (in case of doubt) the camera speed. This may be followed by a 10-second burst of reference tone at an agreed (or stated) standard volume such as 'minus 8 dB' (i.e. 8 dB below his nominal maximum volume).

The film sound recordist is responsible for getting the tapes back for transfer within the shortest convenient time unless otherwise agreed: the simplest arrangement is for them to be despatched with the film (which is often sent for overnight processing) and then forwarded with the rushes for transfer while the picture is being viewed without sound. If so required, the synchronized rushes could then be viewed later the same day. This schedule assumes a fast, routine operation; but in a slower schedule the rush print would still be viewed for technical faults as soon as possible, though the sound might not be synchronized with it until later.

Video sound recording differs from the systems used for film in that a single tape is generally used for both picture and sound. Betacam SP, widely used for documentary and news, has four audio channels, but only two are linear (i.e. along the length of the track and independent of picture), while two are coded into, and so combined with, the picture recordings. Stereo recordings are generally taken on linear tracks (1 and 2) and copied to the coded tracks 3 and 4, but other formats are possible and should be indicated in the documentation. The digital D3 system is more versatile in that all four of its audio tracks are independent of picture: there are less problems when the same tapes are used for editing. These and other professional video recording systems also have time code recordings.

Time code

A time code, as applied to video, is an eight-character number indicating hours, minutes, seconds and frames. It can be linked to a 24-hour clock, and so reset automatically to zero at midnight. Alternatively it can be started from an arbitrary positive number (such as 01 00 00 00, which means 'one hour exactly') after which the numbers will continue to rise in real time as the machine is running but will hold while it is switched off, then continue in sequence when it is run again; this indicates programme or recording (and so replay) time. If the tape starts at a number sufficiently far above zero, it can park at a point before the nominal start and be ready to run up without ever encountering negative numbers.

The first form of time code was *linear* or *longitudinal* (LTC). This is recorded continuously (originally on an audio cue track, where it made a high pitched buzz). It can be 'read' when a tape is played or spooled, but this breaks down at low speeds. So on videotape LTC is generally supplemented by VITC, *vertical interval time code*, which is recorded intermittently in the video signal just above the top of the picture, and replayed at low speeds or when the tape has stopped. Either can be used to generate characters that feed number displays on the equipment itself or superimposed on the picture as BITC, *burned-in time code*. A video system with time code has separate feeds, with and without BITC. If it is re-recorded on a domestic VCR, BITC permanently overlays part of the picture.

The digital numbers that contain the time code have some spare capacity, called *user bits*, that can be used to assign tape numbers.

Different time codes are used for film (24 fps), European television (25 fps), and American television (30 fps and 30 fps dropframe).

On videotape, time code provides a method of individually marking every frame. A tape can be so marked in real time, so that sequences separately recorded over a long period are marked with the actual time of recording and so can be identified simply by consulting a programme log, or it can be built up, so that as new material is edited in sequence the timing is cumulative.

It is often arranged that at the press of a button the real-time flow of numbers can be stopped, so that the time of a particular event

BITC
Burned-in time code offers eight numbers for time and eight as 'user bits' (sometimes available for a second time display). The position on screen is arbitrary, and is chosen from options offered on replay machine. On professional equipment the data is kept separate and combined, on demand, only on the screen display. The combined signal can be fed to domestic equipment, where it will then be permanently superimposed.

can be logged; when the flow is restarted it returns to the advancing real time display. In this way logged material can be readily re-located. More importantly, however, it also means that computer-controlled operations can be initiated and related to any given frame on a tape. In most editing systems using time code any required number can be keyed into the memory and the control system set to search for that address. Once found, the tape automatically sets itself back and parks, ready to run up and display the required material.

For time-coded tape editing to work, times must be recorded on both replay and recording tapes, which are addressed separately. When 'out' and 'in' points have been selected and checked by replay the tapes are commanded to reset, to go automatically to their starting positions. On the replay command, the tapes start up one after the other, lock, and the edit is rehearsed. When a satisfactory edit has been demonstrated, the tapes are reset once again and the edit command is given. Auxiliary registers can be used to switch the sound at a different point from picture or to control external equipment, such as audiotape or a digitally recorded audio file. Three-machine editing can similarly be controlled, and this allows both picture and sound to be mixed instead of cut. The mixes themselves may be set for automatic or manual control.

When these automation systems were introduced they were based on editing techniques that restricted the points at which an automatic cut could be made – particularly in the case of the PAL system. Further, there have been anomalies between vision cuts made in the studio, which change the picture between frames, and those made in editing where the switching may have been designed to occur between the two fields of a single frame. This might lead to the loss of small amounts of material – sometimes with knock-on effects in the sound. However, where sound cuts are required at neither frame nor field change but at some intermediate place, it is better to offload the sound either to audiotape or (preferably, given suitable equipment) a digital system within which the appropriate adjustments can be made.

'Off-line' editing systems, of which more later, all lead to the generation of strings of time codes called edit decision lists (EDLs) which can be used to control the original sound and picture tapes in a final 'on-line' conforming process. For the computer control of digital sound (and also picture), time code is no longer simply an advance on manual techniques, but an essential starting point for identifying information within computer memory.

Time code can also be used for time-cued playback in automated broadcasting systems. It synchronizes video with multitrack tape or digitally recorded sound files in order to locate separate sound tracks (e.g. the commentaries made for different users of a common picture of some event or sporting fixture); television with stereo sound that is broadcast simultaneously by radio, and cinema film with optical discs carrying sound for additional loudspeakers. It also provides an accurate measure of duration.

Sound transfer

When separately recorded audiotape is transferred to separate magnetic film, the recording speed is controlled by the sync pulse or time code on the original tape. The original recording is kept as a master, and can be used for re-transfer of worn or damaged sections, or for finding extra sound for transfer to digital sound file or for direct, unsynchronized replay to fill gaps at a dubbing session.

It is simplest to transfer stereo – MS or AB – in the form received, unless otherwise specified; but MS will be converted to AB when sound is being prepared for the dubbing (re-recording) session. Where videotapes are copied for off-line editing using non-broadcast quality videotape (linear off-line editing, see later), no more than two audio tracks can be copied, usually M and S or A and B as received (again, the dub will require A and B). In order that the tapes can be controlled and edit-decision lists generated, a digital time code (VITC) is transferred; while for visual reference, 'burned-in' time code and user bits (BITC) are superimposed on picture.

For the transcript of unscripted dialogue (useful in editing), audio-cassette copies can be made at the same time.

Videocassette sound may also be transferred to a range of other formats at this or later stages, for example to multi-track U-matic tape. Note also that where U-matic tapes are used for off-line video-editing, the sound they carry may be edited on-line – that is, it may be used directly as a stage in the editing process.

Prior to off-line editing using a computer hard disc system (non-linear editing, see later), there is no separate sound transfer: sound and picture are loaded into the system together.

Film and equipment in the cutting room

Cutting rooms or editing suites are specialized, each dedicated to a particular technical process. Again, this account begins with film, which served as the first major industry standard for many years, and in the operation of which many of today's principles for cutting picture and sound were first developed.

For safety, the original film camera negative (or reversal master) may be stored at the labs where it was processed. The film editor gets:

- film positive, 'rush prints'
- the original sound tapes
- sound transfers on the same gauge as the original film
- documentation about film and sound, provided by cameraman and sound recordist; and perhaps also laboratory and viewing reports on the rushes

Flat-bed film viewer

A simple example of the most convenient equipment for viewing 16 mm film to give normal playing speeds and picture or sound of medium quality (i.e. the best, short of projection in a theatre). 1 and 2, Picture feed and take-up plates. 3 and 4, Sound feed and take-up plates. Lacing is simple, with drive spindles on either side of the sound and picture heads (which are in the centre). The sound path has additional, tensioning rollers. It is not necessary to have rolls of film as shown here: the machine requires only enough head or tail of film to be held by the drive rollers. Typically the film can be run at normal and 'fast' speeds, and either forward or backward. In addition to the heads shown, there is normally a further group to the right of the picture column, to replay combined magnetic or optical sound. For editing there will be an extra sound path, so that the machine has six plates (which may be of different sizes, to allow for the thicker base used for picture), and a slightly different pathway for the magnetic film. Further variants allow for stereo sound and other features.

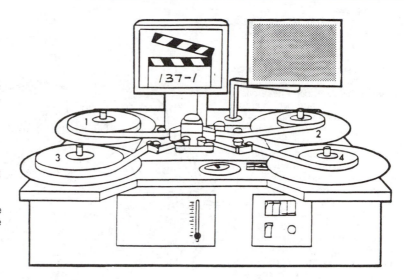

● a shot list
● a script, or sometimes a transcript of the sound
● alternatively or additionally, a cutting order, which may have shot descriptions taken from the shot list and dialogue copied from the transcript.

To this list may be added: the frequent presence of the director. The creative talent of the film editor should be subordinate to that of the director. It follows that creative editing devised in the cutting room takes the form of suggestions to the director, perhaps in the form of a trial cut.

The cutting room has a flat-bed viewing and editing machine, tape-joiner, wax pencils, trim bin, blank spacer film, leader film and a waste bin.

There is also a film synchronizer (with its own small amplifier and loudspeaker) on an assembly bench. This has two soft-fabric-lined bins into which picture and film sound can fall and between them a flat platform wide enough to hold the synchronizer. The platform

Editing bench

1, Picture synchronizer. 2, Amplifier and small loudspeaker. 3, Horse, supporting rolls of film and spacer. 4, Winding spindle with 'split' reels fitted. 5, Bins, lined with a soft but tough fabric, to hold excess film that has not been wound up. 6, Racks for small rolls of film awaiting selection of material.

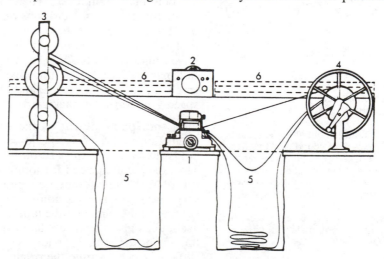

Synchronizer

Picture synchronizer with simple loudspeaker. 1, Track one (picture) passes below lamp (2) which projects a picture via a system of rotating prisms on to the screen (3). 4 and 5, Two of the three sound paths on this example. These pass over the sound heads, which are on the centre line behind the frame through which the lamp is projected. Each path passes over a 1 ft diameter wheel with (for 16 mm film) 40 sprockets. 6, The wheels are fixed to a common axle turned by hand at the front. Some models have a motor and clutch plate, to enable edits to be checked by running them at normal speed. A disc attached to the axle indicates the frame within each foot, and a nearby digital counter (7) shows feet or time from when the counter was set at zero. The reproduced sound signal is fed through a simple low-level mixer (8) to the loudspeaker input (9). Power is fed to the projector lamp through a transformer (10). Synchronizers may have two picture heads or more sound paths and a more sophisticated sound system.

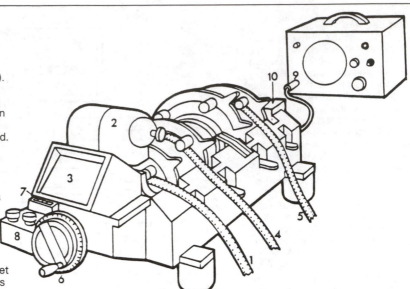

Looping back

Sound and picture are synchronized on both sides of the picture synchronizer; but at the join there is an excess of sound, and this has been taken up in a loop between two of the sound paths. Picture and sound can now be rolled backward or forward together until suitable points for cutting the sound are found. Similarly, excess picture may be removed to match sound.

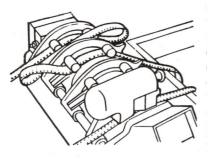

may also have an illuminated plate of frosted glass against which film may be seen and identified. On either side of the bench and outside the bins are supports for the rolls of film and sound to allow them to pass across the bench from left to right. On the left-hand side is a 'horse', several uprights with holes, so that rolls of film can be supported freely on spindles passed through the centre of the plastic cores on which film is generally wound. On the right is a take-up spindle to hold a number of split spools, with a winding handle to take up film.

The synchronizer has sprocketed wheels, all rigidly mounted on the same axle so that they are always mechanically synchronized. Each wheel is one foot in circumference, so that for each revolution 1 ft (16 frames of 35 mm or 40 frames of 16 mm) of film passes through the synchronizer. There is a counter showing feet of film. Sometimes this uses the 35 mm measure for 16 mm film: i.e. the counter is calibrated in '35-mm feet', which means that 16 frames equals one digit on the counter. '35-mm feet' are a finer measure in dubbing than 16-mm feet. There is also a disc that gives a number to the separate frame positions on the circumference of the wheels: on 16 mm equipment these run 1–40. Picture is carried on the front film path and sometimes on the second as well. The other paths have sound replay heads. The synchronizers can be turned by hand but may also have a motor which can be engaged by means of a small clutch so that picture and sound can be run at their normal speed – forwards or backwards.

Film and sound are cut by the stainless steel (and therefore non-magnetic) blades of a tape joiner (guillotine splicer), which then holds the two ends in position while tape is stuck across the join. The tape recommended for joining film is similar in appearance to transparent plastic domestic tape but is of polyester; it is very thin, very strong, and has a non-oozing adhesive. As there is no overlapping of the film, and the tape is much more flexible than the film itself, joins held slackly are liable to appear to 'break' to an angle at the line of the join. Tape joins can be undone easily, for remaking edits or to reconstitute the original, without loss of frames.

Wax pencils for writing on film 'cel' are available in colours such as white, yellow, red or black.

A trim bin is used for hanging up sections of picture and sound that have been cut. Sound trims may be needed later for overlapping sound in order to cross-fade between scenes.

Blank spacer film is used for insertion in the sound assembly where no sound is available or in the picture where a shot is missing. Spacer is also used to protect the ends of rolls of assembled film. Note that spacer is made opaque by coating one side of the film support. On the sound head some types can cause unnecessary wear, so it is usual to cut spacer into sound with the coated side away from the head. Rough and smooth sides can most easily be distinguished by touching them with the tip of the tongue.

Leader film for picture and sound is printed with start marks and 35 mm footages counting down from 12 to three, after which the print is black to the point at which the film starts (which is where zero would be on this count). Sometimes a frame of tone (a 'sync plop') at 1000 Hz (for 16 mm) or 2000 Hz (35 mm) is cut in on the sound leader at '3', or in BBC practice at '4' to allow greater latitude in fading up sound at the start of a programme. As it passes through the projector or telecine prior to this being switched, the presence of the tone confirms that sound is live on the replay itself.

The waste bin, designed specifically for inflammable film waste, should be made of metal and have a lid that will exclude the air in case of a fire in the bin. As a further safety measure, all technical equipment should be on a single circuit with a readily accessible isolator switch.

In addition to all of these, there will be a supply of plastic cores and split spools; perhaps a fast rewind bench; spare lamps, and so on.

It is not the purpose of this book to describe picture editing, and the principles of cutting separate magnetic film sound are mostly similar to those used for audiotape, though a few extra problems (and possibilities) emerge from the combination of sound with picture.

Tape joiner
The film (picture or magnetic track) to be cut is placed on the lower plate (1) and located precisely over the sprockets which stand along one edge. The knife is brought down on the film. The cut end is then moved to the centre of the plate and the film to be joined with it butted to it. Again, the sprockets hold the two ends precisely in position. A strip of adhesive tape is drawn from the roll (3) and placed across the butt and stuck also to the bars (4) on either side of the base plate. The upper head (5) is then brought down on to the film and the arm (6) pressed down. This causes two guillotines to cut the tape on either side of the base plate, and a punch to perforate the adhesive tape at the three sprocket holes nearest the centre of the plate. For simplicity a second, angled knife used for mono sound cuts is omitted. It is located to the right-hand side of the main knife.

Picture and sound in the cutting room

Film picture available to the editor includes:

● film shot with synchronized sound:
   (a) lip-sync sound
   (b) synchronized effects or music
   (c) material where picture and sound are related, having been shot at the same time; but for which there is no need to maintain perfect sync

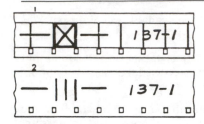

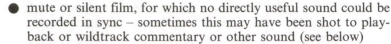

Synchronization marks

Marks on picture (1) and separate magnetic sound (2). In the conventional clapper board system these are marked at the picture and sound of the board closing, and in the case of electronic synchronization, with the flash on the picture and the corresponding buzz on sound.

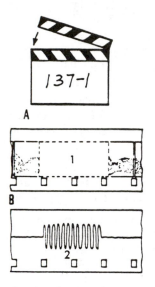

Synchronization of separately recorded sound

A, Clapperboard. In addition to the sound and take number there is other information on the board. B, Simple 'blip sync'. In this case after the camera and sound have both started to run (so that the camera has a picture) a light is flashed to blank out the picture (1) for the same duration as the blip (2) is recorded on sound. This lacks the precision of identification given by the clapperboard, but is compensated by increased operational flexibility and reduction in film stock used.

- mute or silent film, for which no directly useful sound could be recorded in sync – sometimes this may have been shot to play-back or wildtrack commentary or other sound (see below)
- library film for which suitable sound may or may not exist, or only in a technically imperfect form (e.g. optical track).

Film sound available to the editor includes:

- sync material as in the first item above
- wildtrack commentary or dialogue that will generally be played only with pictures that do not show the speaker's mouth
- wildtrack effects that will be synchronized to action in mute film (e.g. a gunshot) or sync film for which the effects recorded at the time were inadequate (e.g. too distant)
- location atmosphere tracks
- library sound effects, atmosphere or music recorded originally on disc or tape but selected and transferred to magnetic film stock.

Synchronizing picture and sound

The film editor's first responsibility, in practice usually undertaken by an assistant, is to synchronize the rushes or, if this has already been done, to check sync. To synchronize film, run the picture and find the frame where the clapperboard has just closed. With a wax pencil, mark this frame with a boxed X. On the roll of transferred sound, mark the corresponding frame with three bars, III, on the backing with a thick felt marker pen (wax could stick to the magnetic coating of the next wind out). Then write the shot and take number ahead of these marks in both picture and sound. For convenience of continuous replay the spacing is then adjusted so that this picture and sound follow in the same synchronization as that of the preceding shot on the rolls of rushes. If preferred, intervening mute film shots and wildtrack sound may be cut out at this stage and wound on to separate rolls: all of the good takes of sync material can then be seen in the sequence in which they were shot.

In order to be able to identify trims of sound and picture at all later stages of editing, the paired rolls are then sent off to be run through a machine which inks corresponding serial numbers on the edges of both film and sound backings, so that they fall between sprocket holes, every 16 frames (40 in older systems).

While it is in this form, the film may be viewed by the director and editor together, so that the content, intention and cutting style can be discussed. Meanwhile the assistant logs the material (see below) and breaks down the rolls that have been viewed into single shots or related groups of shots and rolls them up with picture and sound, if any, together. These are kept in large film cans: each can contains a substantial number of small rolls each of which is labelled with its shot number. The can itself is labelled on the edge with the shot numbers it contains. The assistant looks after this rough filing

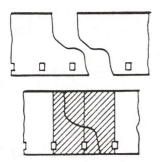

Repairing broken film with a tape joiner
A simple butt and rejoin. If time permits, the original sound is re-transferred, and the damaged track replaced.

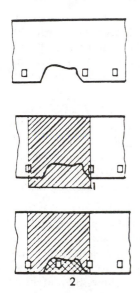

Temporary repair
Torn sprocket holes with film missing. For picture the damaged section is simply taped on both sides, turned over again and the sprocket holes punched through. For sound film the tape is stuck to the upper (uncoated) surface only, but is cut with a razor-blade or scissors a little way from the edge of the film (1) before the guillotine is brought down. The loose edge is then doubled under, on to the coated side of the film (2), but will not reach as far as the recorded track. This will serve during editing until the damaged section can be re-transferred.

system from which material can be taken, restored, and retrieved as necessary.

Note that the film (only) is already edge-numbered by its manufacturer, every 20 frames. These numbers, optically exposed on to the film, are later used by the negative (or camera master) cutters when the edited film is ready to be printed, as those on the master will necessarily correspond to those printed from them on the rushes. Only in exceptional circumstances do the neg-cutters ever have to look at the picture to 'eye-match' original and print.

Up to this stage much of the work can be done on a synchronizer. Most of the subsequent rough and fine editing will be done on the flat-bed viewing machine, with the synchronizer used mainly for finding additional shots and their corresponding sound.

Post-synchronization

Post-synchronization (automatic dialogue replacement, ADR) may be employed in order to re-record voices that have been inadequately covered or were overlaid by inappropriate background noises; or because a director must give priority to a costly setting or unrepeatable event. It may employ an improved version of the original speaker's voice or substitute a more appropriate accent, vocal quality or performance. It may replace one language by another. In any of these cases it is a specialized job that must be done well if it is to be really effective.

Good post-synchronization has to be done phrase by phrase using words that give reasonable approximations to the original mouth movements. A loop of film or its video or digital equivalent is repeated over until the best match is obtained; then the next phrase is taken with the next loop. Technically, there have been several variations on the 'loop' idea, including systems for high-speed roll-back that are very similar in effect to a loop. Any final trimming or tracklaying to match lip movements perfectly can often be done later. Near-perfect results can be achieved by the use of digital editing techniques (see later).

Post-synchronization of spot effects ('foley') is – for a few talented individuals – a well-paid job in the film industry. These experts are capable of seeing through a reel of film a couple of times, then collecting together a pile of odds and ends which they manipulate in time to movements both in and out of picture, in a convincing simulation of the casual sounds of live action. In a low-budget production this may be sufficient, but if the schedule permits, the sounds will be refined by further editing, tracklaying and treatment.

A complementary technique is shooting picture to playback of pre-recorded sound. This is regularly done for musical production numbers (miming), where the quality and the integrity of the sound is of primary importance. It also permits action (such as dancing) to be combined with song in a way that would be difficult if not impossible in real life. In miming, the illusion is, however,

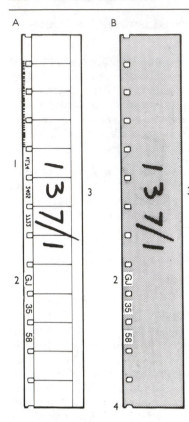

Identification marks on film trims
A, Picture trim: 1, Edge number (and bar-code on extreme edge above it) on picture only, printed through from negative and used by neg-cutters working with a cutting copy to identify the original. B, Magnetic sound trim: 2, Code number (rubber number) printed on edge of both sound and picture after the rushes have been synchronised. They are repeated every 16 frames. By this means, sound and picture can always be re-synchronised.
3. Original shot number, written on picture and sound trims by hand for quick identification. Not necessary if trims are rubber-numbered.
4. Stereo sound is cut square, in order to avoid wiggle.

easily shattered by the sudden change of acoustic between dialogue and song, emphasizing and making awkward a transition that should be smooth.

Rough and fine cutting

In the cutting room the editor usually works with the picture and just one sound track until the picture cut is complete.

The editor begins by making a rough-cut – almost invariably at this stage cutting sound and picture parallel, making a 'straight' or editorial cut. The inked-in edge numbers will usually identify the sound and picture trims, but some sections will be too short to carry a number, so the editor will mark these with a nearby code plus the number of frames to it.

In the rolls of synchronized rushes, spacers will have been cut in, so that in sections where there is only picture, the sound has a spacer inserted into it, and at some places where the sound has run on, there is spacing instead of picture. Even after the rough cut, some of these spacers will still be left in place.

Where picture and its related sound have been shot (or recorded) separately, and so may differ in length, matching is sometimes just a matter of making the longer one equal to the shorter with a parallel cut. Sometimes picture is longer than the sound intended to cover it: in this case the picture can usually be allowed to run its proper length, and the sound will later be filled out with matching atmosphere or effects.

Cuts may be made by removing lengths of sound or picture, including spacing that has been inserted to maintain sync, and, having found another suitable length from the other track, simply measuring one against the other by hand, marking and cutting. It is easy to check that the lengths are the same, frame for frame, by matching the sprocket holes by eye.

Note that a short cut in picture can often be accommodated by making several shorter cuts in sound, perhaps hesitations or in-essential words, that add up to the same length. Picture can also be opened out for sound by the use of freeze frame or slow motion. Opening out sound, in particular, wildtrack commentary, to match picture is a common and often desirable device, because it is better to have more picture than words; but lengthening picture to match sound is an exceptional measure – and one that should have a better reason than solving an editing problem.

In all cases, the timing of speech – both in its spoken rhythm and in its cueing – should remain comfortable and natural. One of the objectives of the fine cut will be to restore these qualities if they have been lost or diminished in the rough assembly.

At the fine cut, as further sound and picture is trimmed away, sound overlays (where the sound from one shot runs over the

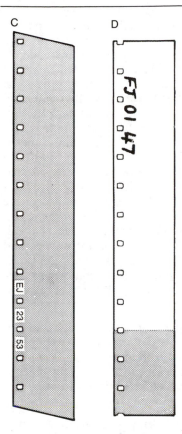

Magnetic film trims
C. Mono sound (cut at an angle) clearly identified by its rubber number. D. A stereo sound trim with no rubber number printed on this short segment. The spacing attached to it shows where the nearest number would be: this is written in by hand.

next) may be allowed, but with film the cut in picture is marked on the sound track. Before any complex non-parallel cut is made, the editor ensures that sync can easily be found again both before and after the region of the cut.

Much of the technique for the detailed editing of sound on film is identical to that described earlier for audio tape: the mechanics of choosing the place are just the same; joining the magnetic film is governed largely by the same principles; and the things to listen for in checking a join are precisely the same. Below are a few additional points to note.

Words, syllables or pauses can often be transferred from rejected takes, replacing the same number of frames as are inserted. If this results momentarily in an apparently inappropriate mouth movement it is still worth considering whether this may be preferable to inappropriate words.

A change of sound that happens with a cut is much more natural and acceptable than an unexplained change at another point.

Short gaps in background atmosphere created by lengthening the picture (e.g. with a cutaway that needs to be seen for longer than the master action it replaces) can sometimes be filled with closely matching sound. This is better than trying to find similar effects from library sources, as it takes a lot to obliterate a drop-out of atmosphere, particularly where there is a continuous steady noise such as machinery hum or the whine of jet aircraft.

Guillotine tape joiners have two blades, one cutting at a right angle to the edge of the film, so that the picture can be cut between successive frames, and an angled blade that is used for mono sound. The angle gives a quicker fade up or down than that used when editing audiotape, but is just sufficient to avoid the effect of a click on the end of material containing continuous tones. However, it does not avoid the effect of sound drop-out unless a correspondingly abrupt picture change explains the difference. The angled blade can be retained for some stereo cuts, but must be replaced by a 90° cut if there is any danger of wiggle, a momentary displacement of the image.

Videotape editing

A variety of techniques has been used for editing picture and sound recorded on videotape. One major division is between *off-line* and *on-line* techniques. The latter is essentially a computer-controlled copy-editing system.

Off-line is further divided into *linear* and *non-linear* techniques. The term linear means that everything downstream of a cut has to be recopied, as in any copy-editing system. In non-linear editing a change can be made in the middle of a section without affecting material on either side. Film editing is a non-linear process.

Off-line linear editing (which came first) was developed when there were severe losses, particularly of picture quality, after tapes had been copied through only a few generations. Copies were made, so that most editing decisions could be worked out and demonstrated without degrading the originals. A further advantage was that such off-line tape editing equipment could be provided at low capital cost and for this reason the technique has continued in use. Where these employ domestic-quality tape, a disadvantage is that the cheaper copies degrade over even fewer generations, so that it soon becomes difficult to see (and to show others) what has been done. Worse, the sound on some off-line viewing machines is copied only very roughly, with fades in and out that can extend over a whole (wanted or unwanted) word, so that precise sound editing cannot be done at this stage, but has to be delayed until the return 'on-line', to the original materials or good-quality copies of them. Three-quarter inch off-line is better than $\frac{1}{2}$ inch in this respect, and the sound recorded on $\frac{3}{4}$ inch U-matic tape is sometimes treated as 'on-line' even though the off-line picture will be conformed later and the sound copied to it. However, in all but the simplest cases, a full sound dub is desirable.

Off-line tape editing rooms generally contain two machines, so that only picture cuts can be seen; dissolves (and the corresponding sound mixes) are made later. On-line editing may also begin in a two-machine suite, progressing to three-or-more replay machines at a final (and more expensive) on-line picture editing session.

In any linear editing system flexibility is severely limited, either to few editing stages or in complexity. In particular, it becomes increasingly time-consuming to make fine adjustments as longer and longer sections for which no changes have to made must be recopied. It is possible that some time can be saved by editing in shorter segments and assembling the final cut only at a late stage, but after that point the same problem returns. Film, in contrast, gets easier to recut in the later stages, so the process of refinement is not halted by diminishing returns for the effort and time expended.

Given its greater flexibility, film editing was unlikely to be fully superseded by videotape editing: indeed, sometimes video recordings have been transferred to film for off-line editing in order to take advantage of the non-linear editing process. However, as off-line videotape editing gives way to digital processing this advantage of film is lost. Digital systems are non-linear, just as film is, but faster, which makes it even easier to try out different ideas for both picture and sound, and also speeds consultation between editor and director. As a result, digital off-line video editing is increasingly used in the film industry as well as in television.

Digital video editing

Digital video editing requires a substantial investment in dedicated hardware and software – carrying the associated bugs and glitches

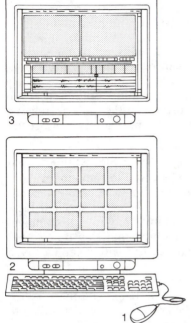

Avid hard disc editing
Adapted from a Macintosh computer
this has: 1, a keyboard and mouse
to control a range of display
windows on two screens. 2, Picture
monitor. 3, Edit monitor, here
operating in 'trim' mode, showing
pictures before and after a cut, with
a bin of clips and the picture and
sound timeline below.

Avid 'manual user interface'
This is one of a range of control
options.

that inevitably seem to emerge in such a rapidly developing field in which frequent upgrades are offered. Initially, the most severe limitation was in memory for picture, but systems offering many gigabytes of hard disc storage (with rewritable optical disc as backup) are now available. For example, about ten gigabytes would offer satisfactory picture quality for editing most television programmes of an hour or so. It helps if the rushes are loaded selectively and unwanted material is eliminated for more to go in its place.

Partly as a result of this, but also because sound and picture have to be loaded in real time, it is best, if time permits, to shotlist the available material with time codes throughout. The production team can work with domestic-quality viewing copies that have BITC (time code in vision) and also time-coded transcripts of dialogue. From these the director can search out the preferred scenes, takes, segments of interview or other selections. The in and out time codes are then entered into a program that can be used to copy from the original tapes directly into the memory of the digital editing system without the intervention of the editor other than to feed in the tapes on demand. As the selections are loaded, this is a good opportunity for the director and editor to view the material together and discuss their objectives.

Alternatively, if there has been no time for preparation, the time codes may be typed in manually as selections from the rushes are reviewed. If other material – additional scenes or alternative takes – is required at a later stage, it is easy enough to add, provided that a replay machine of the appropriate format is available. For picture, the editor selects a compression factor that offers an acceptable compromise between the detail required for editing and viewing and the memory available (which affects the cost of the system). In fact, the actual compression may change automatically from shot to shot, depending on the amount of detail within it.

Audio tracks are loaded in a digital format. If the sound is sampled at 32 kHz or higher (e.g. 44.1 kHz for CD or 48 kHz for DAT) it will remain at broadcast quality. Lower sampling rates are acceptable for editing where sound as well as picture will be recopied (conformed) later to match the editing decisions. On-line-quality sound can be cut with either edit-quality or on-line-quality picture.

Picture and sound are loaded together, if this is how the material was shot, or separately in the case of mute picture or wild sound. It is helpful if the system also records and holds a few seconds more sound at either end of the selected picture in and out points: this will allow for mixes or 'soft' cuts without having to take extra care, except where there is unwanted sound very close to a cut that will become a mix. During replay, a marker travels along time-line viewstrips which look like film and audio tracks laid out across the screen one above the other. These are also used to show cuts and fades. A simple display might have the picture and one (mono) or two (stereo A and B) tracks, but there is provision for many more, of which perhaps eight may be seen at once.

To make an edit, a shot (or some other length of picture or sound) is marked, then some chosen operation carried out. In this way complex edits can be undertaken, and if they do not work little time is lost, as almost any change can be reversed by an 'undo' command, and the choice of system will dictate how many stages it will be possible to go back in this way. (Even so, if a long sequence of operations is envisaged, the edits up to the starting point should be saved so that they can be reloaded, if necessary.) This, together with the drag and paste facilities that are widely available on computers, makes digital editing both easier and faster than conventional film editing: here the user is relieved from the feeling that new developments must always be accompanied by new restrictions.

One of the few disadvantages of digital over film editing is that computer programs can crash, so it is essential to back up regularly to ensure that a recent edit and perhaps also some earlier versions are always available. Another is that the increased speed reduces thinking time. A documentary, for example, may require commentary, and if this is written in the editing suite it may slow down other work, so it may be helpful to divide the editing with a gap of, say, a week or more in the middle. But there is a cost to this: if the equipment is used for another job in the intervening period, the edit may have to be consolidated and the remaining memory (or part of it) cleared. Time must then be allowed for 'lost' material to be reloaded before editing can be resumed.

Editing sound to picture is more complicated than sound on its own in that, as the picture is opened out or condensed, all the tracks of sound must have a corresponding length (an integral number of frames) inserted or deleted. Shots or sequences can be inserted, pushing the exisiting sound or picture apart, or replaced, overlaying it. Keeping it all in sync requires care.

Digital editing systems must operate in a range of modes, which include initial loading (digitizing), then editing and, after that, tracklaying. As with film, it is best to do most of the editing with just one track or a stereo pair, at first mostly with parallel cuts, then at a later stage putting in simple overlays. When the picture edit is virtually complete, tracklaying begins. Even then it is still possible to recut the pictures, although more laboriously and with an increased possibility of confusion over sync.

It is helpful if complex picture dissolves can be made and viewed immediately: this is a great improvement over film, not least because the interaction between picture and sound can then be judged. Otherwise, most of the principles developed for film editing apply with little change, although plainly the mechanics are different – to a degree that depends on the digital system employed. All require computer-operational skills or a capacity to acquire them quickly. An early leader in the field, called Avid, was a direct progression from video editing operations that are controlled by arrays of buttons and screen ikons. Another, Lightworks, was designed to mimic film cutting room techniques. In practice, these (and many others) have tended to converge, so the facilities offered now depend more on the grade (and therefore cost) of the system.

Lightworks employ a single, large screen to display pictures, tracks, and a wide range of filing systems, each with its own separate

Lightworks editing system
1, Editing screen, with layout chosen by editor – here including two 'viewers' (windows displaying pictures running in sync); a 'rack' containing the material and information on selected groups of shots; a 'gallery', here of 8 shots; and a viewstrip with a segment of the timeline of picture and two audio tracks. 2. Keyboard. 3. Mouse. 4, Flatbed variable-speed control. At low speed the sound is heard at proportionally reduced pitch, as with film, helping the editor to locate sound cutting points precisely.

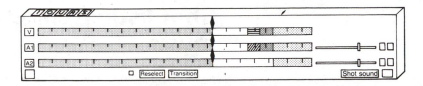

Lightworks viewstrip

V, Segments of picture, each represented by a different block of colour. A1 and A2, the associated sound, here as two tracks with, *right*, faders which can be clicked and dragged to new settings. The upper and lower parts of the fader 'knob' can be moved separately to set volumes before and after a fade, the ends of which are marked on the associated strip.

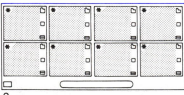

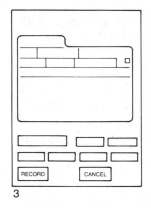

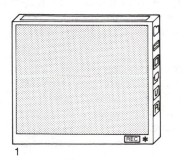

Lightworks display elements

1, Viewer: a simulated 3D box with the picture on the front and control ikons and data displays on two sides and plinth. 2, A gallery, in this case of 8 selected shots, with the names or numbers that have been allocated to them and ikons which allow the editor to identify, select and use them. 3, Part of a recording panel, with controls and a file card.

window which stands out like a three-dimensional block seen in perspective, with control ikons that may be ranged along the top and down one side. There is also a keyboard with a range of function buttons and an external VDU which displays data. Useful, also, are an external monitor to show the output picture and a conventional PPM.

The main screen window areas include one or more 'viewers' for displaying selected individual pictures, 'racks', or filing cabinets of time-coded and annotated lists of shots and 'galleries' of scenes (banks of tiny initial frames with their designated titles below them). For sound, the most important window is the stripview with a scale of seconds along it. In its simplest form the strip corresponds to ten seconds or so of picture with the sound below it and it is possible to 'zoom' out for a broader view or in for detail. Cuts are indicated by changes of colour and cues by displacement of the strip.

From the start, Lightworks had a large film-flatbed type of lever control to run picture and sound at variable speed back or forward in sync, in order to be able to stop precisely at the point required for a perfect sound cut, even though this might not be on a junction between two frames. Gaps, sound effects or changes in background or voice quality can be quickly located by listening for them, as with film.

The Avid system has two screens, one showing pictures where the effects of small changes can be seen; the other offering information (timeline, access menus, bins of clips, etc.) and some of the control ikons required for editing. To see where the sound is on the audio timelines and as an assistance to sub-frame editing, Avid users are offered a simplified waveform (essentially indicating momentary sound volume). For editors with film experience an early limitation was that replay was designed to run at a fixed forward speed, but now flatbed variable-speed replay is available at extra cost among a range of control options – an example of the convergence in product development.

A picture compression ratio of 8:1 is barely noticeable to most television viewers: there is some loss of fine detail and smoothness of rapid motion but the quality is still a little better than from a domestic VCR. So for some current affairs programmes 8:1 has been used both to edit and transmit. In a further development, this makes it possible to use a camcorder that has two slot-in ultra high density hard discs that employ the same compression to increase recording duration. The same discs, inserted directly into an editing machine, avoid any delay due to copying, and can be rapidly edited and then broadcast in the same format. Here the compression cannot, of course, be reversed but this may be regarded as acceptable in high-speed operations such as news-gathering.

Shot listing, commentary and music

A shot list giving times or film footages is usually required for a variety of reasons: for documentaries this includes commentary writing. Narration will generally have been drafted earlier and rewritten or modified as editing progressed, but it is better to cut the film completely so that the exact length and value of each picture is known before the final version of the words is set.

The shot list can conveniently be combined with a script of the dialogue as cut and notes of other important sound: in this case it can be laid out to look like the standard television script with picture on the left of the page and the corresponding sound on the right.

Times should be noted not only at cuts in the picture but also at important points within the action of a shot, e.g. when a detail of importance first appears in vision, and also at the start and finish of dialogue, important effects or dominant music.

Commentary can then be accurately written to picture: for a fast-speaking narrator allow about three words per second (or, on film, two per 35-mm foot or five per 16-mm foot). Where a speaker is slower or more expression is wanted, these figures must be reduced. It is better to write too little than too much and have to rush the words; and in any case strong pictures are often better without words.

The audience's primary need from a commentary is that it should interpret the picture: this is a more effective use of the medium than when the reverse is attempted. The worst possible case is when pictures and words do not agree, either because the commentator wants to talk about something else or to continue about something that is no longer in the picture. In these cases an even averagely interesting picture demolishes the commentary entirely: the words are not heard.

There is a particular danger when a presenter records part of a commentary in vision, either in the studio or on location, and another part later in the confined space of a commentary box. Even if the voice or microphone quality can be matched by equalization, there may be abrupt changes of acoustic. Short links should if possible be planned in advance, and recorded at the location; a variety of versions might be recorded.

Nothing is more clumsy than a succession of different acoustics on the same voice for no apparent reason. At best it is mildly distracting (and certainly destructive of artistic unity); at worst it is actively confusing, leaving the audience unsure whether the new sound is the same voice or a new and unexplained character. On the other hand, well-chosen words, written precisely to fit and complement the pictures that have been shot, may add far more than is lost by the mismatch in sound. If so, one compromise is to allow a period of time, a music link or burst of some sound effect to intervene: the change will then be less noticeable.

If it is obvious from the picture that the dubbed voice is not that filmed on location, the change in quality may be an advantage: the commentator finishes speaking in vision, turns and walks into the action; and the dubbed voice takes over: here a clear convention has been adopted, and the change of acoustic helps it.

Specially composed film, television and video music is often written at a late stage. The composer requires a copy of edited picture and sound that is as complete as can be arranged, a cue list with time codes or footages and also a script (for documentary, a shot list with narration that is as advanced as possible). However, the editor will find that the picture, words and music all interact: after it has been recorded, the music may suggest changes to the other components, so it is wise to allow time for these. Final changes in the picture may in turn require changes to the music, so as this is written and recorded generally to a tight deadline, it is worth checking that there is sufficient flexibility.

Music from existing sources can be selected at an earlier stage of editing in order to allow more time for the pictures to be accommodated to it.

Tracklaying

Tracklaying is the first stage of preparation for dubbing (making the final sound mix). Here again the standard techniques were pioneered with and are still used for film, and are functionally almost identical when used with digital editing. The editor beings by 'splitting the tracks'. In videotape editing the process is similar but the mechanics are a little more awkward, as in order to extend it the sound has to be recopied from the original, generally onto multitrack tape or to a digital sound-editing system. In all cases, when the fine cut is complete, the constituent parts of the cut sound are separated on to several tracks or to time-coded tape or computer memory, or a mixture of these.

For as long as the sound continues at the same level and quality, whether of the main or background sound, it is retained on the same roll; but at points where these change, the sound is switched to another roll. If by simple adjustments of level and equalizer settings it will be possible for the sounds to be matched, a parallel shift from one roll to the next is all that is necessary. But if there is a more drastic change that is not immediately explained by an equally strong change of picture, then extra sound must be found for the dubbing mixer to fade on. These fades are generally made quickly, but it is better to have too much than too little to fade on, or the sound may be heard cutting in or out at low level. Such chopped-off sounds appear unnatural when they do not occur at a cut.

For film, extensions are found from the trim bin, generally by an assistant film editor. On digital systems they can be easier to find: in

Lightworks, for example, overlaps reappear automatically as the sound is split onto neighbouring tracks.

During the editing, extra tracks that had to be synchronized with action may not have been cut in because of other synchronous material already present: on film, these were marked and put to one side. When tracklaying, they go in on their appropriate rolls.

Music may also be laid: where it is dominant it will already have been arranged for the picture cuts to be related to it, or alternatively for specially composed music to match the picture. The precise manner in which picture and music should be coordinated is a complex study: but a strong rhythm demands that some, but not too many, of the cuts are related to the beat; similarly, strong notes may cue a cut.

As many of the sound effects as possible are accurately laid by the editor; this includes both library effects, which should be selected

Dubbing cue sheet

The original on which this was based has provision for more sound rolls. Here 35 mm film footages are given (i.e. 16 frames = 1 ft) but timings may also be used. The action column does not list every picture cut, but only sequences within which sound should remain roughly the same. In some cases sound has been laid to several feet beyond a picture change, but will be faded out quickly on the cut (this is represented by the shallow V). Fine points are discussed or decided by consultation on the spot. For example in this case it was not made clear on the chart, but the choir was taken behind the boys' chatter between 3 and 17 and was faded quickly after 17, although it was originally laid as far as 32. The rugby football sound (at 16) preceded the cut at 17 and, in effect, killed the sound on tracks 2 and 3. 'V/O' means 'voice over'. At 80–84 the teacher's voice was to be heard over background chatter and effects and from 84–105; this in turn was taken to low level while Richard's voice was heard (out of vision) talking about what is happening at this point in the story. 'Hum' and 'buzz' are terms used here to indicate 'low-level background sound'. The commentator's voice has been recorded in advance and has also been laid (on track 4). The final words of each commentary section may be marked in this column.

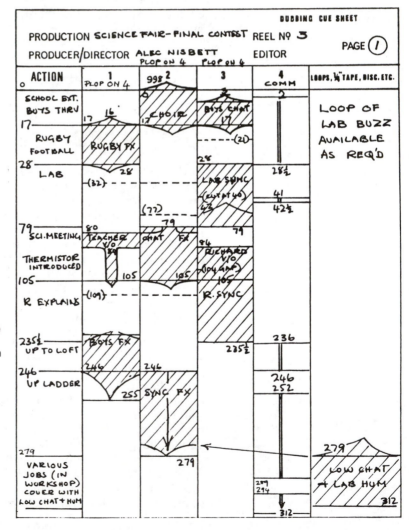

and transferred for the purpose, and wildtracks which were specially shot.

Some continuing sounds may be formed as loops: formerly this was done by making up lengths of recorded track with the ends joined back to the beginning. Now this can be replaced by a digital 'loop', using material selected and repeated from a sampler. Loops with strong distinguishing features must be long, so that the regular appearance of such a feature does not become apparent. If it is likely to, it could perhaps be trimmed out, though this would shorten the loop still further. Also, in joining the ends of loops together (or selecting the in and out points of a sample) care must be taken to find points where volume and quality are matched.

The editor or assistant prepares a chart with a column showing the main points of the action, and opposite this the sound on each roll, or source, showing the cuts in and out of sound with symbolic indications of fades and mixes and clearly marked footages or time codes, together with notes on loops and tape and disc effects.

Commentary that has been pre-recorded is also laid.

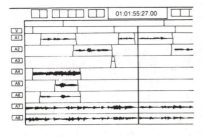

Avid sound editing and tracklaying displays
Picture and tracks are lined up in sync, as for film. The 'waveforms' representing sound help the editor to find cutting points to within a small fraction of a frame.

Conforming sound and picture

When an off-line video tape or digital edit is complete, the picture must be conformed (matched and copied) from the originals: this is equivalent to the negative-cut and showprinting of film.

When, in order to keep costs down, the off-line picture has been accompanied by on-line-quality sound, this must be transferred back in sync with the picture, where necesssary making any final adjustments such as volume correction, mixes, overlays, and adding music, effects, commentary, etc. However, if there is to be much of this it is uneconomic to attempt it in a suite designed primarily for on-line picture editing.

In the more general case, the sound (as well as picture) will be conformed, which means that it must be separately recopied from the original recordings. Tracks which have been laid off-line as a provisional set of editing decisions are used to create a corresponding set of full-quality audio tracks in a format which will serve as sources in a dubbing theatre. For this, there are two main options. One is to copy to multitrack tape and use this to provide the sources in the dubbing theatre; the other is to use a digital recording system.

Even if the previous tracklaying has been comprehensive, this can be a time-consuming process, which may require equipment for the replay of sound in a range of different formats. From being one of the more neglected components of video editing, sound preparation ('prep') has developed to the stage where it may occupy a room in its own right, in effect a small studio. Depending on the facilities available, some of the tracklaying may be left until this stage; some of the sounds may be processed here; ADR and spot effects ('foley') laid and edited; and some sequences might be pre-mixed prior to the main dub. In such a case, the preparation may be undertaken by a specialist sound editor who carries this to a stage where the final mix becomes almost a formality – except that the dubbing theatre also serves as a meeting place where production can hear what is offered and either accept or change it on the spot.

Before any of this can begin, in order to conform either picture or sound all of the selected in and out points on the original tapes must be located and logged. A cheap but laborious way of doing this is to read them from burned-in time codes on a recording of the off-line edit, or possibly even from a number display on the replay machine. Far more convenient, however, is to use a system that produces an edit decision list (EDL) for picture and audio tracks on a floppy disk that can in turn be used to control the conforming process automatically.

To do this, the EDL finds and displays the first time code, including the user bits which identify the source: in effect, it 'asks for' a tape. When this has been inserted, it will then cue and transfer all the material on that tape, leaving gaps for material that will be inserted from other tapes; then it will request a second tape and so on. Picture is generally conformed to a single tape, and dissolves are put in manually later. In contrast, sound is transferred with the tracks alternated in a chequer-board arrangement, so that they can be separately mixed and controlled.

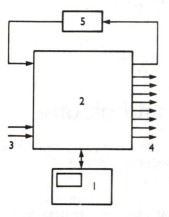

AudioFile editing equipment
1, Controls ('control surface') and VDU. 2, Equipment rack. 3, Audio inputs. 4, Eight independent audio outputs (allowing system to operate like a cartridge machine). 5, DAT recorder.

Where television which is to be transmitted with stereo sound has been edited using equipment on which the editor, for simplicity and to minimise the danger of losing sync, has been working in mono, a clearer idea of what is available in stereo sound may emerge only at this stage. Any stereo in the MS format is converted to A and B. Note that for the benefit of television viewers who do not have stereo decoders, television sound is still also transmitted in a form in which the A and B signals have been recombined to form a 'compatible' M signal. Stereo should not be made so extreme that mono listeners cannot hear all the information they need to make good sense of and enjoy the programmes.

AudioFile editing system: control surface
1, 'Soft key' function switches.
2, Soft key status displays on VDU.
3, Selected page of information: for example, see opposite.
4, QWERTY keyboard. 5, Numeric keyboard. 6, Trigger keys. 7, Floppy disc drive. 8, Transport, nudge keys, 'soft wheels' and LED display for locating and shifting material.

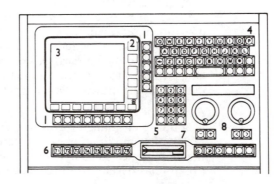

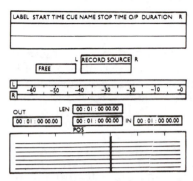

AudioFile control screen
'Audio record': a typical page shown on VDU. Many have the eight tracks of sound, which 'play' across the screen from right to left, past the vertical line which represents 'now'. There may be many more tracks which are not shown. Positions are based on time codes, but can be measured to greater precision. 'Audio record' is the only page which has a meter display.

Complex sound preparation, as an extended, separate stage of editing, is now generally undertaken using specialized digital equipment, of which an early and now well-developed example is AudioFile. This employs hard disk for storage and instant access from a 'control surface' VDU set in a small desk with keys and buttons, some of which are located below and alongside the screen. The display has switchable 'pages' which allow the editor to operate in a range of different modes.

The sound can be loaded (digitized) as usual under the guidance of an EDL floppy disk. This will have, in addition to the time codes, notes or remarks indicating where extensions may be required for fades and mixes, to extend atmosphere or overlay speech, to replace missing or unsuitable tracks, and so on. Each cue is automatically allocated a number; if time permits, short identifying names can be typed in, too.

If the sound comes from a source that has no original time code, this must be added, in order to locate and control it. Note that in the subsequent editing, so long as the original time code and reel numbers are retained, the original can easily be located and reconstituted in order to try out alternatives.

In such a system precision and repeatability of the programmed levels, cuts and mixes are incomparably greater than can be expected from manual control. Timings can be marked to one hundredth of a second, and levels matched to a quarter of a dB; while cuts may be made hard or soft, or fades or cross-fades varied in length or internal dynamics. All of these can be treated far more analytically, and the editor's selection retained on floppy disk for subsequent replay. A small mixer desk can be useful for rehearsal.

The most user-friendly element of the information shows 16 of the tracks (of up to 24 in memory) in a display that looks like a multi-track tape marching from left to right across screen, past replay heads in a vertical line down the centre. Each track is identified by source and edit number, and a suffix, 1 or 2, indicates the left or right track of stereo. A cut in or out is shown by a vertical bar across the track, which travels with it; a fade (here also called a *ramp*) by a diagonal. Sound can be bumped from one track to another instantaneously. Just as easy, but less obvious once it has been carried out, is a shift or *slip* to the side, moving it backward or forward in time. Incidentally, note one other item of jargon: here *balance* means 'stereo offset'.

Amongst other modes of operation is one for automatic dialogue replacement (ADR). Since it is digital, this is ideal for the purpose, and a *timeflex* facility allows the internal duration of segments to be changed independently of pitch, by up to a nominal doubling or halving of the length – far more than would be used in practice, other than to produce deliberately odd effects. The recording disc an also be used as a *cart machine*, replaying recordings in random order, rather like the kind of reproducer which carries a number of tape cartridges, all ready for immediate replay.

The dubbing theatre

A dubbing theatre has equipment to display picture and run several synchronized sound tracks, plus a recorder. There is often a recording studio if only for a single commentary voice, and provision for playing records, tapes, cassettes, digital audio files, etc., together with additional facilities on the control desk that can make operations simpler and more flexible.

At one stage, videotape sound re-recording seemed to be developing a separate set of techniques, but where digital processes are brought into the dubbing theatre (rather than used separately at the preparation stage) they have allowed the methods to converge and become less dependent on the recording medium. The equipment can therefore be described primarily as laid out for film dubbing.

Picture display. So long as a picture is plainly visible to operators and commentator, and there is sufficient detail to see all significant action, high-quality film projection requiring specialist staff is un-necessary. For film, the cutting copy, or a low-quality 'dupe' of it, is often used. Alternatives are a very simple optical projection system, or a low-cost telecine and monitor or both; a monitor provides a picture for the commentator if the main screen is out of sight. A monitor is, of course, always used where the source is a video signal.

Counter. A counter showing timings or time codes in minutes and seconds or footages (i.e. with 16 or 40 frames to each digit) as preferred should be displayed close to or superimposed on to the picture, perhaps with repeaters on the desk and in the studio. A commentator or musical director works to these numbers in conjunction with other cues.

Sound replay and recording equipment. This is available in various layouts, with (for film) a number of machines for the replay of laid sound tracks, and a recorder. The machines are under local control while they are being loaded, but are then switched to 'remote' operation so that synchronous replay of picture and all the replay

Re-recording studio (dubbing theatre)
1, Mixer at mixing desk.
2, Assistant, with tape, cassette decks, turntables, etc. 3, Film projector and footage counter (here with video cameras to relay picture and footage as required).
4. Magnetic film sound reproducers and recorder. 5, Loudpeakers.
6, Screen and footage monitor.
7, Narrator with desk, microphone, cue light and monitors for picture and footage. 8, Acoustic treatment.

and recording machines are controlled from the desk. There is provision for discontinuous recording, sometimes called 'rock and roll'. The mix continues until errors are made, when the sound and pictures are stopped and run backward until well before the point at which re-recording must restart. A counter runs back in synchronization with sound and picture. Then the whole system is run forward once again, and the mixer checks levels by switching the monitor rapidly between the input and the previously recorded signals, adjusting faders to match them; then switches back to record, and from that point the new mix replaces the old. The switch should not operate instantaneously, of course, or a click might be introduced.

Digital replay. Alternatively or additionally, tracks may be stored (and treated) digitally. This offers greater ease of adjustment of the position, duration or pitch of individual components.

Ancillary replay equipment. There must be the normal provision for replay of tape (for original recordings, etc.) and disc (for music and effects records). Cassettes may also be used for a library of standard effects, while cartridges may replay continuous or staccato effects, including recorded spot effects that must appear accurately on the footage cue. Cartridges with loops of various durations are available. There should also be provision for the replay of sampled sounds as loops and a range of floor surfaces for spot effects, perhaps in an acoustically separate booth.

Prereader. For mixing voices (especially commentary or narration) with music and effects it is convenient to have visual cues to show where the gaps are. A prereader is an extra replay head that picks up sound several seconds before it appears at the sync replay head. Its output is fed through a gating amplifier which converts it to a signal that is on or off according to whether there is speech on the track or not. When there, it lights an indicator bulb on the desk for a fraction of a second, then another to its right and so on, so that the signal is passed along a row of indicators arriving at the last at the same times as the signal arrives at the replay head. The display appears as a flow of light and dark passing from left to right, providing the cues necessary to control. But note that a short, unexpected burst could be a cough or some other extraneous noise, for which the cue must be taken as a signal to keep it faded out.

Monitoring equipment. This must be of high quality, and is usually set loud so that imperfections in the original material or of the mix itself can readily be heard; but as usual, the sound should also occasionally be switched to near-field (lower quality) loudspeakers to check what most of the audience will hear. This is particularly important when mixing dialogue or commentary with music or effects for television. Any sound recordist should listen to replay in an acoustically separate cubicle, because there is a delay for film typically of five frames (one fifth of a second) between the recording and replay heads.

Commentary box. A commentary is usually recorded in a dead acoustic. This ensures that it does not carry with it an ambience

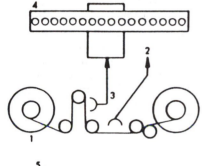

Prereader
1, Deck for magnetic film track.
2, Output to mixer. 3, Prereader feeding separate amplifier-oscillator which acts as a gate to detect presence or absence of signal.
4, Display on desk: a signal lights the first indicator in turn. The final indicator is lit as the corresponding signal reaches the main reproducing head. 5, Signal has reached prereader head. 6, Signal is still being registered at prereader head.
7, Signal has ceased: an illuminated strip is passing to the right, indicating arrival of signal at main output head.

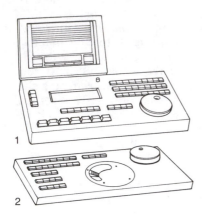

Augan multi-track recorder and editor

This system has re-usable optical disc for storage and hard disc for rapid access. 1, Eight of the available tracks are displayed, with time codes and a wide range of operational data. The control panel has a jog-and-shuttle facility. 2, Flatbed control option.

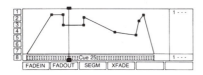

Augan editing

In one of the many facilities offered, the eight tracks are replaced by a display for one of them which allows a complex volume envelope to be set.

that is alien to the picture it accompanies, for example, by overlaying a live, indoor acoustic on to an exterior scene. It also seems to help the voice to cut through background effects or music, perhaps, in part at least, because a higher level for the voice can be set. If time permits, the commentary should be recorded in advance of the main dub and taken back to the cutting room to adjust each cue precisely to the frame at which it has the best effect. Sometimes fine trims of the picture may be removed or restored at this stage: occasionally the strength of the word enhances a picture to an unexpected degree and demands that it remain on the screen a little longer. The commentary track is then returned to the dubbing theatre along with the other tracks, although its addition is often the last stage of the mixing process.

Cueing systems. There will be provision for the commentator to be cued from the desk by the director, editor or assistant editor, or the dubbing mixer. The usual system is 'green for go'; the cue light should be close to the reader's sight line. The times corresponding to cues are usually marked on the commentary script, which omits most of the picture details and carries only the most significant details of other sounds, such as the in- and out-cues of dialogue already on film, or featured sound effects. The cues have to be given a little early to allow the speaker time to respond. The commentator will often require a headphone feed of music, dialogue and effects, with or without voice feedback (whichever is preferred): before the rehearsal begins check that the sound level fed to the commentator's headphones is comfortable. There is also keyed talkback between desk and studio.

The desk. This has the usual fader channels, each with its standard facilities for equalization and feeds to graphic filters, artifical reverberation, etc. Compressors, limiters and expanders are available, and also a noise gate. There will be a full range of controls for stereo, including panpots and spreaders.

Documentaries are often mixed in stages, with a so-called music and effects (M and E) mix being made first (in fact, this would also include any sync speech, and there might be no music). The final stage is to add the commentary; but the M and E track should be retained against the possibility of changing the commentary later, or making versions in other languages, or for use when extracts are taken for other programmes.

The techniques (as well as much of the audio equipment) used by specialist dubbing mixers are very similar to those of other sound studios, and so are much as are described in earlier chapters.

Differences reappear as the sound leaves the dubbing suite, and are mainly concerned with ensuring and checking synchronization in the various formats of film and both analogue and digital videotape.

Chapter 19

Planning and routine

Before entering the studio, some planning is likely to be necessary – if only to ensure that all the necessary technical facilities are available and ready for use. If a script is available in advance, whether or not you read it through in detail, do note and roughly mark up those parts that directly affect you. In television there may be a planning meeting to coordinate the expected contributions from the various technical and design departments. And in the studio itself there are routines such as line-up, cueing, checking timings and preparing documentation, all of which may be essential to its efficient operation.

The preparations and procedures described in this chapter apply to radio or television – often both – and also (perhaps with a few minor changes in terminology) to record company recording sessions and to the film and video sound dubs (re-recording sessions) that have been described in Chapter 18.

Planning: radio or recordings

For radio or other sound recordings, when there is a range of studios available the request for a booking is normally accompanied by a brief description of its use, e.g. the number of speakers or musicians, and if necessary a brief indication of the type of balance: 'discussion – 6 people' or 'novelty sextet with vocalist'.

Except for the simplest programmes, the production office prepares a running order or script which (except for very topical material) is sent to the programme operations staff in advance. Many productions follow an established routine such that the minimum of discussion is necessary before arriving in the studio; but if there is any doubt, check in advance. For example, before a recording session involving an unfamiliar musical group, the balancer should find out what instruments are expected and what sort of balance the producer wants to hear: if, for example, it were required to sound like an existing record the balancer would have to listen to it. The balancer would also want to know whether the music is to be featured, or for background use: whether it is to be brilliant and arresting or soft and undisturbing; whether it is to be heard with announcements at certain points within it, or applause (which would thicken the texture of the sound).

The recording arrangements are also discussed: the probable duration, whether it is to be stop-start or single take, the medium and number of copies; and arrangements for and time of delivery.

Any unorthodox or experimental layout has to be discussed with the musical director in sufficient time for studio staff to be given any plan for the provision of chairs, music stands, acoustic screens and other furniture; and studio instruments such as the piano. The balancer also sets up microphones (or checks their layout) and arranges their output on his desk, working out and plugging up a suitable grouping of channels, and checks that equalizers, compressors and limiters are ready for use; that reverberation and other devices are plugged to the appropriate points; that talkback is working; and that foldback or public address levels are set.

Television planning

For a television programme of some complexity the sound crew may be six or seven in number: a BBC team might include an operator for the tape and disc replay decks, two boom operators, and a floor assistant (who may track the boom, reset microphones, etc.). Second in command and the most versatile in duties is the sound assistant: in some programmes the assistant leads the sound department's forces on the studio floor, supervising the rig for musical programmes, operating the main boom on a complex play, or sitting alongside the supervisor and handling the audience microphone pre-mix, public address and foldback in a comedy show.

Representing the facilities that the team can offer, the sound supervisor attends the planning meeting (several if necessary) called by production to coordinate design and operational plans. For a simple programme, this may involve the director, the designer, the lighting supervisor, and the technical manager (who is in direct overall charge of all technical services other than lighting and sound). More complex programmes may also require the presence of the floor manager, senior cameraman, costume and make-up supervisors, and others dealing with house and studio services such as audience management, the assessment of fire risks, allocation of dressing rooms, and dealing with any special problems in catering caused by breaks at unusual times or the arrival of distinguished guests.

From this planning meeting, and from subsequent contact with the director and the production team – and often also from direct observation of outside rehearsals – it is the sound supervisor's responsibility to ensure that all non-standard equipment and facilities are booked and available, and that the microphones and other equipment are set ready for rehearsal on schedule.

But the planning meeting also serves a secondary purpose: it is a forum within which ideas can be proposed and discussed and then either adopted or abandoned. There may be conflicts of interest that must be resolved in compromise and cooperation. The director listens to suggestions and generally arbitrates rapidly between conflicting aims. Sometimes the director modifies the original plan, to make things easier for everybody. If, in a complex situation, the director does not do this, it is usually the sound supervisor who has the trickiest problems to solve – most television directors put picture first and sound second. Sound supervisors who make light work of complex problems are therefore valued members of any production team – but those who accept unsatisfactory compromises and then explain after the programme that poor sound was an inevitable result will not be thanked.

At the planning meeting the sound supervisor starts to build – or re-establish – a working relationship with several other members of the team that will carry on throughout the term of the production.

The sound supervisor and the television director

The sound supervisor meets various types of director (and producer). Those who outline their needs concisely and precisely and then stick to it are easy to work with and are not likely to waste others' efforts or demand extra staff and resources that in the event are not needed. Another director may have few of these obvious virtues, but still be worthy of all the help the sound supervisor can give him. This is the type who likes to maintain as much flexibility as possible up to a late stage in preparation, so can then take full advantage of elements of the production that turn out better than could reasonably have been expected, and cut losses on ideas that do not shape up. In these circumstances the sound supervisor may have to plan by intelligent anticipation, expecting some waste – but with this sort of director the means can be justified by the results.

Inexperienced directors also need a lot of help from the sound supervisor, but should be persuaded toward simpler solutions so that things are less likely to go wrong. Then, when they do, the problem is relatively easy for everyone (including the director) to cope with.

For certain types of production (in particular, plays) the sound supervisor sees the script before the planning meeting, and will already have questions for the director – who will probably be ready with answers without even being asked. The director may start with an outline of the programme content and its intended treatment, and will show a plan (perhaps even a scale model) of the studio layout. The sound supervisor should pay particular attention to descriptions of the visual style: a 'strong' style with fast cutting, mixtures of close and wide shots, unusual angles and shifting focus could require different sound coverage from a simple and more logical progression.

At this stage the director may welcome advice on sound coverage. This may involve pre-recordings or extra expense for equipment or staff; and the director should have the opportunity to decide whether the effect being sought is worth this.

Before the planning meeting, the studio sound supervisor may also have been consulted on links between location material and the studio. Mismatches of background sound or voice quality can sometimes be avoided by the appropriate choice of microphone type or technique. For example, a close balance might be possible on location, but should be avoided because it gives a bad sound cut with a related studio shot – or specially recorded sound effects might be requested.

A running order for rehearsal and recording is also agreed at the planning meeting: here the director needs to hear whether (for example) a recording break will save equipment and staff and therefore perhaps money. But to bring a television studio to a halt for 10 minutes for the lack of one person may be an expensive economy. Note, however, that a sound supervisor's desire to provide perfect results without regard to cost may conflict with the director's responsibility to make the best use of limited resources.

Actors and other performers may have a period of rehearsal before coming into the studio, towards the end of which the sound supervisor should attend and watch. At this stage, scenery is indicated mainly by floor markings, but much of the action will already be set. This includes speed of movement and relative positions, the quality of the voices and direction of projection. It should be clear whether the planned sound coverage is likely to work in practice and also how it may be affected by the evolving plans for lighting and camera movement. Any substantial consequent changes in sound coverage should be discussed with the director and specialist departments, and entered in the studio plan. It should also be possible to visualize how equipment such as property telephones or acoustic foldback will be used in the studio, and get a feel for the timing of speech and action relative to music and effects. Sometimes recorded sound is brought to the rehearsal and tried out with the performers.

Rehearsal-quality sound equipment used outside the studio, for example to work up sequences mimed to sound replay, may be collected from the sound department and operated by an assistant floor manager. But if a sound assistant rehearses its use and timing with the performers, this may save studio time later.

Sound and the television designer

The television designer produces the scale plan of the proposed studio layout, to which symbols for booms, slung and stand microphones and loudspeakers will be added by the director after their positions (boom tracking lines, etc.) have been discussed with the sound supervisor.

Rostra, ceilings, doors, archways and alcoves are all accurately marked on the plan, as are features such as tree-trunks or columns

Studio plan
Corner of set in a television studio, including cameras (1 and 2), boom (A), loudspeaker (LS), monitor (MON) and scenery. Here, camera 1 is a small motorized tracking camera (other cranes require much more space). Camera 2 is a pedestal camera: the circle represents the working area it requires. The boom is capable of further extension (to reach beyond the back wall of this set) if required. Details of set: B, Single-clad flattage. C, Double-clad flattage. D, Door. E, Ceiling over (broken lines). F, Rostrum (height will be marked). G, Columns. H, Steps. J, Arch. K, Drapes on gallows arm. L, Swinger (flat which can be pivoted in and out of position). M, Slung microphone. N, Bracket. P, Lighting barrel over (these will be shown all over the plan).

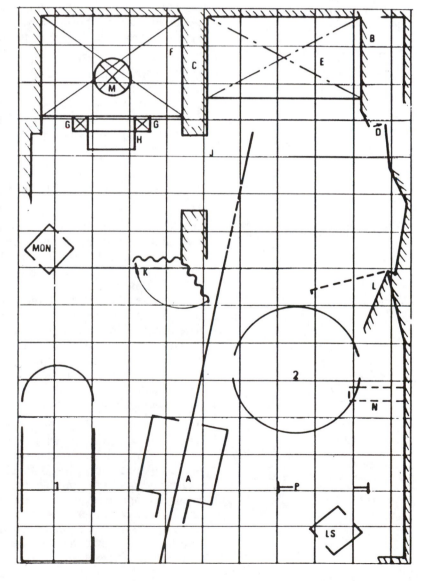

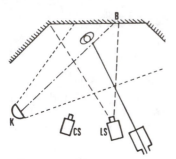

Lighting cameras, and microphone: problems
The actor is too close to the backing, and the microphone, in an attempt to get good sound, has followed. There is a shadow visible at B in the long shot.

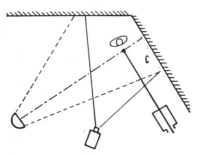

Boom shadow
With a boom close to a long wall there is a danger of a shadow along it (C) on wide shots. If the boom is very close, even soft 'fill' lighting will cast a shadow.

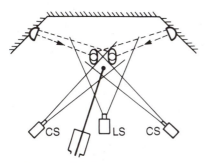

Rear key lighting
Here, two people facing each other are both satisfactorily lit, and there is no danger of microphone shadows as these are thrown to the front of the set.

that may obstruct a boom arm, or areas of floor covered with materials that make it impossible for sound equipment to follow. Stairs and upper working levels are also shown. At this stage the plan is not final: a ceiling may be redesigned using an acoustically transparent material so that a microphone can 'see' down from above; a tree may be chopped off at a calculated height so that a boom can swing over it; or a potted plant may be strategically repositioned so that it can bear improbable fruit in the shape of a concealed microphone.

Certain types of set reflect and therefore reinforce sound: this is particularly true of flattage built of wood. Such flats may help musicians or other performers, but can cause trouble if they are arranged in a curve that will focus sound (whether this be close dialogue or distant noise). Scenic projection equipment may be noisy and require acoustic treatment; this requires cooperation between sound and design. Fortunately, projection has now largely been replaced by electronic colour-separation (chromakey) techniques.

Floor and other surfaces used in television are often painted or made up to look other than they really are. Consequently, footsteps may be audible when they should not be heard at all, or for which the natural sound is inappropriate to picture. Here again the designer can help sound by the suitable choice and treatment of surfaces. This, together with practical wind, rain and fire effects has been described in Chapter 15.

Planning microphone coverage

In planning television sound coverage the first question is almost invariably: can it be covered by a boom? Static or 'personal' microphones may be used. For many types of television programme, microphones may appear in vision. At the planning stage the sound supervisor must decide how many booms are required and how they must be moved about the studio. At this stage, too, it is necessary to anticipate minor changes in the script that could have a big effect on coverage. For example, it may seem reasonable to plan for speeches by two performers who are separated from each other to be covered by a single boom because the script indicates there is time for the boom to be swung. But suppose that at a late stage it becomes apparent that, dramatically, the pause is wrong and must be cut? Plainly this kind of problem should be anticipated.

A boom may require no operator (if it is simply to be placed in relation to a performer who does not move), one operator (for normal operation), or two (if it must be tracked at the same time). The sound supervisor must ensure that an operator can always follow the action, and is also responsible for the operator's safety: as the boom swings the operator must not run out of platform to stand on; with certain types of boom this can happen very easily. Positions and tracking lines must be agreed with cameras, and the angle of the boom with lighting, so that shadows are not thrown on to performers or visible parts of the set.

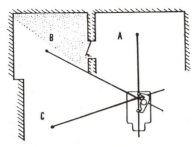

Boom operator in difficulties
If the operator of a boom positioned for use in area A is asked to swing to area B, he is blind to all action which is obscured by the wall, and so cannot judge the angle and distance of his microphone. If the action continues to area C without movement of the dolly, the operator is now supported only by his boom and a toe-hold on the platform. Alternative provision must be made for coverage of B and C before the position for A is decided; this requires planning, as camera positions and lighting are also involved.

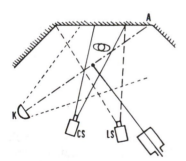

Lighting cameras and microphone: the ideal situation
The close-shot camera has near-frontal key lighting (from K). The microphone shadow is thrown on to the backing at point A where it appears in neither picture. Few cases are, however, as simple as this. Cameras: CS takes close shot; LS takes long shot.

Booms and lighting

To see how shadows may be caused it is necessary to understand a little about television and film lighting.

Performers are generally lit by two frontal lights. One, the *key*, is the main light: it is the strongest, and casts sharp-edged shadows. The second frontal light is the *fill*: ideally this should be from a source that has a large area (e.g. an array of relatively weak lamps) so that it does not cast a hard-edged shadow. In addition there is *rim* lighting from the rear to outline the figure and separate it from the scenery behind (which will be lit separately).

For portrait close-ups the camera and the key light are not widely separated in their angle to the subject (although they must be far enough apart to give satisfactory modelling to a face). If the boom were also to come in from a similar angle, it might cast a shadow on the subject's face. So the first rule is that *the boom should come in from the side opposite to the key light* (perhaps at 90–120° to it).

Provided that the same camera is not used for wide-angle shots, there should be no danger of the shadow being seen against the backing. This leads to the second rule: *long shots should if possible be taken from an angle well away from the key light.* Then, if the boom is coming in from roughly the same angle as the long-shot camera the shadow is thrown to the side of the shot and not into it.

In television, lighting often has to serve several purposes in order that a scene can be shot from a variety of angles. For example, it is often arranged for a light that is used as a key for one person to become the rim for another person facing. But as scenes become more complex in their camera coverage, what may be relatively easy to arrange in shot-by-shot filming may become tricky in television, and compromises must be made. A third rule (which generally makes things easier) is that *plenty of separation should be allowed between the main action and the backing*, on to which a boom shadow could easily fall. This is, in any case, desirable for good back lighting.

As a corollary to this, note that the boom arm should not be taken close to a wall or any other vertical feature of the set that will appear in vision. If the arm is close enough, even the soft fill lighting begins to cause shadows. Slung microphones may also cause shadows, especially if they hang close to a wall.

Sometimes there is no convenient alternative to having a microphone shadow, perhaps from the fill light, in vision: it may be the only way to get close enough for good sound. If so, arrange for the shadow to fall on a point of fussy detail rather than an open, plain surface, and keep it still.

In the planning stage, if the action seems to be getting too close to the backing, the lighting and sound supervisors are likely to join together in an appeal for more room to work. The director may then change the action or the timing of the words spoken; move action props away from the wall; or arrange with the designer for the working area to be physically enlarged.

Preparations for rehearsal

During the first run-through of a complex programme the sound supervisor ('studio manager' in BBC radio) glances ahead down each page of script, quickly estimating what will be needed, and to ensure that the right microphone channels are open and get the feel of the overall and relative levels and such things as fades and mixes. At this first attempt the results are rarely exactly right, but each operator quickly notes down anything out of the ordinary in some sort of personal shorthand: it might be the level of a recorded effect, perhaps, or the setting for the first words that are to be heard in a fade in. In radio, extra words may have to be written for a fade, or an adjustment made to an actor's distance from the microphone: such questions are referred to the producer. In television, the sound department is more inclined to try to solve problems with the minimum of reference to the director. Often the difficulties are more severe; generally revolving around the question of how to get adequate sound coverage without the microphone getting in the picture or throwing shadows.

Nothing should ever be too precisely set prior to the recording or transmission: in a play, for example, actors' emphases and levels may be substantially different on the 'take' as they let loose their final reserves for a performance.

Difficult sequences or unusual techniques require pre-rehearsal and perhaps pre-recording. Awkward material should be broken up into tractable segments.

Line-up

In any studio session time must be allocated for *line-up*: this is the check that for a standard signal at the start of the chain all meters read the same, showing that nothing is being lost or added on the way. When extra equipment is introduced into the system, this too must be lined up. What this means in practice is that when a recording session or transmission follows a period of rehearsal, time must be scheduled for further line-up tests before the 'take' or live show. Television line-up (involving similar procedures for the video signal) lasts longer, so there will be adequate time for sound line-up within that. (However, in the more complex television desks provision can also be made for line-up tone to be sent to line without disturbing normal sound rehearsal circuits.) Where a recording machine is part of the studio equipment, or permanently linked to it, the line-up procedure may be abbreviated.

The procedure adopted for a BBC sound radio recording illustrates the form that such routines may take. At the end of the rehearsal, and a few minutes before the scheduled recording time, the studio manager calls the recording room (if this is separate) on the *control line* (an ordinary telephone circuit). Details of the recording are checked, and the equipment is lined up, using zero-level 1 kHz tone which is sent from the studio. This reads '4' on a mono or 'M' signal PPM. It reads 3 dB below this on the stereo A and B channel meters.

Then *level* is given: someone in the studio is asked to read a few lines. This gives any distant recording engineer an additional check on quality, the last that will be heard directly from the studio before switching to that of the recording.

After a final check that everybody is ready, the studio manager announces over the studio talkback (which also goes to the recording room), 'We'll be going ahead in ten seconds from ... now!' then switches on a red light indicating to people outside the studio that a recording or transmission is in progress, and changes the state of the desk, so that talkback and monitoring circuits are re-routed; then fades up the studio, flicks a green cue-light – and the recording is under way.

In television line-up, the video and sound lines are tested (again the BBC uses zero level tone for sound line-up). At the start of a continuous ('as-live') recording a video clock giving programme details is shown for the 30 seconds or so prior to the start of the required material. (This is used for transmission line-up, and may be added or updated at any editing stage.) On an open microphone the floor manager calls '30 seconds ... 20 seconds ... ten–nine–eight–seven–six–five–four–three ...' and then both sound and picture are faded out for the last three seconds. As an additional sound check, 10 seconds of tone (lined up to zero level at the studio control desk) is sent to line between the counts of 20 and 10, and this, too, is recorded on the tape as a final reference level.

At the end comes the reckoning: the duration is checked, and possible retakes discussed. If these are necessary further cues are given and they are recorded at the end of the main programme tape, to be edited later. Apart from documentation, the recording session is now complete.

A live radio or television transmission differs from a recording mainly in that just before the studio goes on the air the end of the preceding transmission is monitored. The studio manager (in television, sound supervisor) hears the cue and once the transmission is under way switches back to studio output, and since at every stage between studio and transmitter everyone is listening to the programme as it leaves their own point in the chain, the source of any fault can be rapidly located. The studio is switched back to network output just before the end of the transmission.

Programme timing

By *timing* we mean two things: the tyranny of the stop-watch over broadcast programmes, and the artistic relationship in time of the various elements of a programme. In broadcasting, particularly where commercials have to be fitted into the schedules with split-second precision, and also where there is networking, each second must be calculated. An American director once described himself as 'a stop-watch with an ulcer'.

Today the stop-watch is probably no longer in his hands, but for some purposes is built into the operation of an automatic control unit or possibly a computer. For the purely mechanical business of getting programmes, short items, music, commercials, time checks and station identification announcements on the air some stations are now almost entirely automated, leaving the staff formerly occupied in this work free for more useful work, such as (according to the manufacturers of one broadcast automation system) selling more air time to advertisers, or giving news stories on the spot coverage. Or perhaps that director still has both his ulcer *and* his stop-watch – for the segments from which programmes are built still have to be created, and often to a strict overall duration: automation can – so far – only build transmission schedules from material which already exists.

This pre-occupation with the value of time may also remind us to tighten up our programmes and make sure that every part counts. Although 'time' and 'timing' are two separate things – and are often enough at odds with each other – they are closely interconnected. Differences between rehearsal and transmission timings may explain a fault of pace. The relative lengths in time of different scenes or sections of a programme may throw light on faults in overall shape. So we see that the obvious and overwhelming reason for timing programmes is not the only one.

When timing a radio item, the usual practice is to set the stop-watch going *on* the first word and not before it (not, that is, when the cue is given); and the watch is stopped after the last word is complete. In this way, if various timings are added together in sequence, the pause between successive items is counted only once. Television timings normally begin with the first usable picture.

There are various ways of marking the rehearsal and recording timings on a script. It can be done at minute or half-minute intervals, at the bottoms of pages, paragraph by paragraph, or by scenes or parts of scenes. Timings can also be made on a running order, timed cue-sheet or recording form (along with rough editing notes, if necessary). Marking times on a script each minute makes calculation easier; but marking the time at natural junctions in the programme probably has more actual value.

If a programme is being timed in rehearsal or recording and an error occurs, so that it is necessary to break off, first stop the watch and mark in the script the point at which this was done, even if this is after the mistake. Then when the same point is reached again, the watch can be restarted. Even with a substantial overlap, the true duration is probably close to that shown on the stop-watch.

Where a programme consists of separate elements – as in a magazine programme – two (or more) sets of timings may be required, including one for the whole, and one for individual items. In television this is done for recorded inserts, and the producer's assistant keeps everybody informed over talkback how many more minutes there are (useful if moves or microphone changes have to be made in the studio) and counts down the last ten seconds.

Another way in which timings may be used in radio is to give a speaker his or her last minute of air time. A stop-watch is started by someone in the studio and placed in front of the speaker, who knows, as it ticks up to one minute, the time to stop.

Cueing methods

For cueing purposes, there are two principal systems of communication between director and performer. These are the cue-light system (used by BBC radio and adopted by many other radio organizations which are modelled on the BBC and also in film dubbing theatres), and the hand-signal system (favoured in American radio and universally in television).

BBC radio studios are equipped with green cue lights. These are operated by switches on the control desk. A separate cue light is provided for each studio area (or table) and is placed so that the artist does not have to turn away from the microphone to see it. One green flick means 'go ahead'. It is held on long enough for the least observant speaker to see it, but not so long as to hold up the person who thought you said 'when it goes off'. This convention is standard for all types of programme except one, the fast-moving news and actuality programme. For this the narrator may be given a steady green 'coming up' cue 10 or 20 seconds (as agreed) before the actual word cue is expected. The narrator goes ahead when the cue light is switched off.

Some hand signals are also used: the equivalent hand signal for 'go ahead' is to raise the arm for 'get ready' and then to drop the arm, pointing at the performer.

For radio talks, cue lights are also used to help regulate timing and pace. A series of quick flicks means 'hurry up', and a long steady means 'slow down'. A mnemonic for this is 'quicker-flicker, slow-glow'. The equivalent hand cues are: hand and forefinger rotating in a small circle, fairly rapidly – 'speed up'; hand, palm down, gently patting the air – 'slow down'. These are both used extensively even when cue lights are available. In television the director gives the cue via the production talkback circuit to the floor manager, who signals it to the performer.

In discussion programmes the chairman may wish to use these signals and others for precise information about timing, e.g. two steadies at two minutes before the end, one at one minute to go, and flicks at 30 seconds. To give the signals by hand the director must stand where clearly visible, and then indicate 'minutes to go' by holding up the appropriate number of fingers; and 'half' by crossing the forefinger of one hand with the forefinger of the other.

Other hand signals sometimes used are: pulling the hands apart repeatedly – 'stretch'. But have care, as it might also be taken for 'move back from the microphone'; as moving the hands together may be used to indicate 'get closer' – either to the microphone or to another speaker. Waving the arms across each other, palms down in a horizontal plane, can be used for 'stop'; and the side of the hand drawn across the throat predictably means 'cut'.

Crowd noise in a television play may be waved out; but in radio a series of quick flicks may be used to indicate 'the studio is faded out – everybody stop' and can be followed quite soon by the single flick for the start of the next scene.

There is one special case in radio: when the speaker is blind. The director, or an assistant, must stand behind the speaker's chair and give the various cues by laying a hand on the blind person's shoulder or arm. Complex cues can be given in this way if arranged beforehand.

Cueing applause

The management of audience applause is important to many radio and television programmes.

In most ordinary circumstances an audience will wait politely until the end of a sentence and may then respond appropriately. Through uncertainty as to whether they should interrupt the proceedings or not, the applause may be delayed a second or two and apparently, or actually, be half-hearted. They may even completely fail to applaud (particularly if the speaker is inexperienced in this branch of stagecraft). In any of these cases the effect is a lack of warmth and a failure to generate the feeling of an event with unity of location and purpose; as a result, there is no one with whom the viewers or listeners can identify to feel vicariously that they, too, are taking part.

At the other extreme, the studio applause may be excessive to the extent of breaking the flow of a programme. If applause is injected one sentence before the best point, then that point is not applauded at all (and might just as well be dropped), or it is given reduced applause (and appears as an anticlimax) or it receives a totally excessive reaction that simply holds up the programme and alienates the home audience. Excessive applause may be inappropriate in its programme context, or simply unsatisfactory for sound – for example, young people stamping their feet on the wooden or metal rostra that support the seating in many television studios.

In the UK, the overt management of applause is viewed with caution. It is, however, necessary where the programme becomes clumsy without it; it is particularly necessary in the case of small audiences of, say 100 or less, which may lack enough of the people who help to start a crowd reaction; and it may be useful where 'atmosphere' demands that applause continues unnaturally long, such as at the close of a programme, behind closing music and titles or announcements, until sound (and where appropriate, vision) finally fades out. The management of applause is right and proper where it leads the local audience to react in places where reaction would in any case be natural, where such applause helps to interpret the material to which it is a reaction, and where it helps by pacing a programme and allowing the home audience time to absorb and assess what has gone before.

The behaviour of the audience may be partisan, either in the sense of being composed of admirers of a particular performer, of a participating team, or of a political party: in any of these cases it is reasonable to show the outward trappings of an existing genuine relationship. In all of these cases applause must be handled with responsibility towards both truth and intelligence.

Appropriate signals (in television from a floor or stage manager) might be as follows:

For 'start applauding', the stage manager stands in clear view of the audience, raises hands high, and starts clapping (but not too close to a live microphone). A folded script in the hand may help to draw attention.

For 'continue applauding', the same action may continue, or be replaced by waving the script in a small circle, still well above the head (this adds variety to the stimulus and allows the other hand to be used for cueing). Both hands high, palms inwards, will also keep it going. But turning the palms outwards will bring the applause down, and waving the hands downwards at the same time, or if necessary more violently across the body, will bring it to a stop.

Other signs may be used: the only requirement is that the audience understands and follows. Avoid the ludicrous 'applause' signs that are sometimes seen in old films. It is better to lead an audience than to direct it. For natural, warm applause it is often best if there is a slight overlap between speech, or music, and applause. This implies careful timing and clear understanding by the sound supervisor about whether the last words must be intelligible or whether it is permissible for them to be partly or completely drowned. It must be clear whether the first words of a new cue starting over applause require moderate or high intelligibility (ideally, for 'warmth', it should be the former, with redundant information in the words spoken until the applause has almost completely died away).

The 'warm-up' is an essential part of any programme with an audience. Often the sound department play records of a suitable style until the warm-up proper, which consists of speeches giving information about the programme and the audience's important role in it, and, for a comedy show, 'warm-up' jokes.

Documentation

After a recording is complete various details must be written up for later reference. Some of these are technical responsibilities and some production. Typically, they may include:

- Names and addresses of owner and recordist, programme title, any reference or costing number it has been given, and the date and place of recording, etc.
- Technical notes. Besides details of equipment and tape (including reel or cassette number where there is more than one, e.g. 'reel 2 of 5'), the report should indicate the format and tracks used and list the material shot; also details of any technical imperfection, fault in balance, etc.
- Brief details of contents with durations, e.g.:
  (a) names and addresses, etc., of interviewees
  (b) music information: composer, writer, performer, copyright owner, record numbers, etc.
  (c) identification of material that is not original.
- Timed script or cue sheet, with editing notes; and details of the current state of the editing process.
- Details required for programme costing.
- Any extra information related to the programme, e.g. source of material used, contacts, etc.

In addition to this, summaries of the programme details may be written on the cassette or, for quarter-inch tape, on the leader of the tape itself and on the spine of the box in which it is kept. For sound radio programmes the BBC uses printed leader tape: 'British Broadcasting Corporation' followed by spaces for the title of the programme, its reference number, reel number, tape speed, duration and date of recording. On the spine of the box the reference and reel numbers, title and speed are repeated. If a tape is finally wiped and reclaimed the leader is removed, ready for re-use. Leader tapes are not used on film sound tapes recorded in the field; instead, it is more useful for the boxes to have the film slate numbers, etc., listed for quick reference.

Radio and television stations usually log their output as it is broadcast: in the USA this is a requirement of the FCC. Such a log can be used – among other things – to check the total duration of commercials per hour, or the proportion of sustaining (i.e. unsponsored, public service) material in a schedule. Since the effort involved benefits no one at the station directly, it is not surprising that even its storage space is grudged. However, stations using broadcast automation systems may use a very low-speed recorder for the automatic logging of programme details to comply with FCC regulations: the tapes take up relatively little space. In computer-controlled versions the log is printed out on demand.

Chapter 20

Communication in sound

Just how important is technique? Let us seek a perspective.

The mechanics of programme work must be seen in their true relationship to the full process of creation, and they should be given neither too much deference nor too little. Obviously I think technique *is* important, otherwise this book would not have been written. But technique is a means to an end, not an end in itself: and its major object must be to concentrate the audience's attention upon the content, and to do so in a particular way, with particular emphases. In such a supporting role technique is indispensable.

When combined with picture, in television or film, sound should always support, occasionally counterpoint, and rarely dominate; its most effective role is to extend, strengthen and more closely define the information that comes first through vision.

It is relatively easy for the skilled professional to make technique serve the programme; and work is very rarely, if ever, done merely to display the operational ingenuity of those involved. But for the newcomer much of his work must necessarily also serve as exercises in handling sound. To novices, technique is a vital need; perhaps even their most vital need – because they may have no lack of ideas, but little or no knowledge of the means of expressing them. They must learn the techniques, not merely of construction, but also self-criticism and correction. Once the techniques are mastered, they must learn to apply them – to make them serve the needs of each successive production. The medium itself is *not* the message.

What makes a good production? The answer is of importance not only to the writer or the person who is directing a programme but to everyone in the studio. It concerns them all in two ways: first, so that they can see their own contribution in perspective, and to know in what ways this must be subordinated to the whole; and secondly, as the programme's first audience.

Over the years, people in daily contact with a medium become sensitive to virtues or defects that even the director may be unsure of; and this front-line audience can be of immense value in cross-checking the director's calculation or intuition.

The things that matter most in any programme are the ideas within it (its content) and the way they are presented to the audience (its form). Some subjects are better for sound than others, because they make contact more easily through the listener's imagination. Others are more suitable for television or film, demanding the explicitness or possibilities for visual imagery that these media give.

Many subjects can be treated adequately in any medium – adequately but differently, because the medium dictates certain possibilities and impossibilities; or it presents alternatives, where one path presents only difficulties, while another luxuriates in opportunities for attractive or artistic results. To this extent at least the medium does condition and reflect itself upon the message.

But so far as technique is concerned there is one major condition which, in conventional (if not in modern experimental) terms, must be fulfilled: whatever the choice of subject, it must be presented in such a way as to create interest and maintain it throughout the production.

Sound and the imagination

The greatest advantage of the pure sound medium lies in its direct appeal to the imagination. A child once said that he liked sound better than television 'because the scenery is better'. Of course the scenery is better: it is built in the mind of the listener. It is magic scenery, which can at times seem so solid that you feel you could bang your head against it; but in a moment it may dissolve into an insubstantial atmospheric abstraction. It is scenery within which not only people can stand but also ghosts. Ideas, emotions, impressions can be conjured up as easily as the camera can show reality.

Well, perhaps not quite so easily, because sound does require a greater contribution from the listener. To compel this involvement – and to deserve it – the material coming from the loudspeaker must maintain a high level of interest throughout. If it drops for one moment the imagination is turned off like a light, and all real communication is lost. The magic scenery is built and painted by the artists and craftsmen in the studio; but it is illuminated by the imagination of the listener.

The ways in which the imagination can be stimulated are many. To some extent they depend on form, which depends in turn on techniques. Apart from this, what is done, rather than the way it is done, is a matter for individual judgement or intuition.

In our analysis of the qualities of any programme we should perhaps take 'interest' as our starting point, because the one factor that all successful productions have in common is the ability to engage and retain the listener's interest. For a start, it is clear that interest depends on previous experience. What we know, we like; what we do not know leaves us relatively indifferent: this is true even of most people who regard themselves as being 'interested in anything'. For very few people indeed are interested in subjects that cannot be explained in terms of things that are already understood.

The reason is simple enough: any reference to something within the listener's experience can call up an image in his mind. When we listen to somebody talking we soon forget such superficial qualities as accent and voice timbre (unless we are forcibly reminded of them at any point); we do not listen closely to the actual words, but go straight to the meaning. The words we forget almost the moment they are spoken; the meaning we may retain. This meaning, together with its associations in the listener's mind, forms an image that may be concrete or abstract, or a mixture of the two. And if it is possible to present a subject in terms of a whole series of such images a considerable amount of interest may be aroused.

The audience

When a radio or television producer starts work he must bear in mind not only what he wants to say but also who he wants to say it to. This is not merely a matter of the listener's intelligence, but also of what sort of background he has; what sort of things he is likely to have learned at school and in later life; what sort of places he knows and people he has met; what sort of emotions he is likely to have experienced, and so on. It is also related most powerfully to the things he would *like* to do, consciously or not – ranging from taking up a life of meditation and prayer to winning a thousand pounds for saying that the Eiffel Tower is in Paris.

In practice many programmes develop a distinctive personality according to the characteristics of the audiences at which they are aimed, and may be further modified in style according to the time of day at which they are transmitted. But, all in all, the dominant factor in programme appeal is the previous background and knowledge of the listener; and failure to take proper account of this is one of the most serious mistakes any producer can make – if his aim is communication. I am not really concerned here with works of art where the artist is *not* interested in communication. I am certainly not decrying them. But if they are works of importance to anybody but the artist himself, teachers and critics busy themselves with explaining and relating until in the end communication between artist and public is established, just as though it had been intended.

In an investigation into the effect of BBC educational broadcasts, intelligibility was the main issue, but a few sidelights were thrown on 'interestingness'. It was noted that while too many main 'teaching points' (say six or more in a 15-minute programme) were definitely bad for the programme, the presence of what sometimes seemed to be an unduly large number of subsidiary points did not seem to result in the confusion that might have been expected; and also whereas long, complex sentences with difficult vocabulary (for this particular audience) and lots of prepositions (used as an index of complexity of construction) did appear to have a slightly adverse effect on intelligibility, the presence of a large number of adverbs and adjectives did not.

It seems reasonable to conclude that an apparent excess of facts, figures and descriptive terms is no disadvantage, because although they may not be assimilated themselves, they help to maintain interest during the necessary gaps between the teaching points. Apart from any other function they may serve, these are the things that provide 'colour'; they are part of that vital series of images that the listener's imagination needs. And remember here that what is true for educational programmes is also true in the broader sense for a much wider range. On the larger canvas we can take 'adverbs and adjectives' to include any descriptive, or even simply amusing, piece of illustration in sound or picture.

In this survey, concrete subjects were preferred to abstract ones; and, when ideas were being explained, concrete examples – references to people, incidents and actions – were preferred to high-flown 'clever' metaphors. This, too, must be a more general requirement. For apart from the genuine absent-minded professors of this world – the people for whom the world of the mind is more real than the world of reality – there can, I think, be few people for whom abstractions (except those dealing with the emotions) are more telling than are concrete references. Indeed, one danger was noted: that well-developed illustrations can sometimes be so effective as to attract attention away from the main point.

If there's no picture – where is everybody?

Some problems in communication apply to one medium more than another. Before we turn to the more general case, let us consider one or two that apply particularly to the pure sound medium.

We have noted that an advantage of sound lies in its direct appeal to the imagination; and that the really successful programme has one quality above all else – the ability to conjure up an unbroken succession of powerful images in the mind of the listener. These are not just imagined visual images, there to replace the concrete visual evidence that the eye is denied: they are a great deal more. Nevertheless, we must now recognize that some of the main limitations and disadvantages of sound do stem from the confusion caused by the lack of exact pictorial detail.

In our ordinary life we take this detail for granted. When we take part in a conversation with half a dozen other people we are never in any doubt as to who is talking at any time; our eyesight combines with our faculty for judging direction aurally, and no thought is necessary. Indeed, we would have no difficulty in coping with almost any number of people; the ears localize and the eyes fix on the speaker immediately.

Without vision, the number of voices that we can conveniently sort out is reduced very considerably. In the case of stereophonic sound, half a dozen people, spread out and not moving around too much, should be fairly easy to place; but with more people (or with more movement) the picture becomes confusing. When this happens, the imagination fails: the voices no longer appear as identities. Instead of a mental picture of a group of individuals, we merely register the sound as a quantity of depersonalized speech coming from various points in space.

In monophonic sound the picture is even more restricted: we have room for only three or four central participants in any scene at any one time, and even then the voices and characters must be very clearly differentiated. Further voices may come and go, provided that it does not matter if they do so in a fairly disembodied manner. The butler who comes in to say 'My Lord, the carriage awaits' does not count as an extra character to confuse the ear; he is hardly more than a sound effect. But the basic group must be small.

It must also be remembered that the moment a character falls silent, the audience 'loses' him. Leave him out of a conversation for half a dozen or more lines and he becomes invisible, so that when he speaks again his sudden re-entry can come as a shock; he seems to leap in from nowhere.

From this we already have one considerable restriction on subject and treatment. So beware: any feature that involves many-voiced scenes is in danger of sounding vague. From a discussion programme involving half a dozen speakers the listener will probably identify only about two or three clearly. The extreme case is an open debate: this will be merely a succession of voices, though it will help if each speaker in turn is named. For clarity this must generally be done even when a voice has been heard before.

Often, of course, it does not matter to the listener that voices are not identified individually; but the producer must know what is intended – and also what has been achieved. Otherwise the audience may be living in a place of shadows when the artist thinks he is communicating a world of imagined substance.

The next difficulty with sound is that of establishing location. This must be fixed as early as possible, and then reaffirmed by acoustics, effects and references in the text – anything to prevent the scene slipping back into a vague grey limbo and becoming once again no more than a set of voices from a box.

On the other hand, it is often an advantage that the listener is not unnecessarily burdened by pictorial detail. During a television talk the eye is captured by the picture but may wander within it. It may shift to the clock on the shelf behind the speaker; it may be distracted by the shape of a lampshade, or the pattern on wallpaper. Without all that – given sound alone – the speaker becomes part of the atmosphere of the listener's own living-room.

These weaknesses and strengths must be seen before the treatment of a subject is chosen; it is important to allow for the frequent reiteration of locale in the scene of a play, or the formal disregard of surroundings in a straight talk.

Plainly, it is impossible to establish one picture, and one only, in the mind's eye of every single listener. So it is wise to avoid subjects that require precise pictorial detail, or treatments where the impact of a thing that is seen has to be explained in dialogue – in fact, any stories that depend heavily on visual information. Let us take as an example of this, the first meeting of boy and girl in a crowded room: eyes meet, fall, and meet again . . . it needs a Shakespeare to express verbally the magic of such a moment; and even in *Romeo and Juliet* the visual element adds immeasurably to the impact. Suppress it, and the shape of the whole scene is altered.

This example also illustrates another point: that direct emotional effects are best expressed visually. The close-up of the human face, and most particularly the eyes, can touch the audience directly. In terms of pure sound, using only the voice, parallels are not easy to find.

The emotional link between screen and viewer can be very strong. Vision is the dominant faculty, and it must be either fully engaged or completely suppressed if the audience is to identify itself with the performer. With film or television the link is simple, direct and potentially compulsive. In sound only, the link has to be established indirectly but no less firmly through the imagination. The results may be quite as effective, but they are a great deal more difficult to achieve.

The time factor

As a medium, sound has much in common with print. Both the printed and spoken word can be presented as straight argument (or narrative) or as dialogue. There are comparable forms of illustration: the sound effect and the graphic display. And further, because of the restriction to a single channel of communication, neither medium can engulf the senses without first making a strong appeal to the imagination. The analogy is so good that it is worth considering where the differences lie: the more so as there is inevitably a tendency to assume that what can be done in print can also be done in sound.

The great difference lies in the dimension of time. The printed page leaves the time element in the hands of the reader. He can read at any speed he likes. He can read a novel in a few hours, or take weeks over a textbook. He can read one paragraph quickly and the next slowly.

He can stop and go back when the meaning of something is not clear at first reading. In the process of reading, the eye can afford to run on ahead in order to see how a sentence is shaped, and so take in its meaning as a whole.

For the radio (or television) audience the time relationship between successive elements of story is fixed. A recording may be stopped and reviewed, but this disrupts the flow more than for print. Once a listener has decided *when* he is going to play a record he has little control over any other aspect of time: for technical reasons, the interrelation of speed and pitch, he may not play it faster or slower than was originally intended.

To avoid ambiguity and aid understanding (if that is the intention), ideas must be presented in clear logical sequence. It is not only the order of the arguments that must be clear; the style in which they are presented must itself be lucid and lively. If possible, that is, if the subject matter does not militate against it, and sometimes even if it does, speech should be grammatical, as well as vivid in expression. Realistic dialogue, with its broken sentences, repetitions, fractured logic and other inconsistencies, is often regarded as essential. But if it is used carelessly it can also confuse the listener; and when this happens he will hear not meanings but only strings of words, and particularly any imperfections in the way they are spoken. The most frequent complaint about this sort of dialogue is 'inaudibility' – even when the sound would be adequate in other circumstances. This is not to say that it should never be done, only that it should be done only for good reason; and the degree of communication calculated.

At the other extreme, in the 'radio talk' by a single speaker with a prepared script, good delivery is still important. For some listeners, a bad speaker who is obviously reading, and doing so from material written in a flat literary style that makes no concession to colloquial speech, will seriously diminish enjoyment. The listener may well ask, 'Why bother to present this in sound at all, when it is so obviously designed for print?' But this may be a minority view, for the quality of delivery has very little effect on intelligibility, and does not worry most listeners any more than would any other relatively unimportant fault of technique. For example, many of the film-stars of the thirties now seem curiously stilted when we see revivals of their films; but this never prevented huge audiences from adoring them. Only if a programme is thrown completely out of balance, as, for example, by speech that is too fast, is faulty or 'unrealistic' delivery found upsetting.

It is reasonable to demand that the basic message of a piece should be understood in a single hearing, for both sound and television are basically one-shot media. By all means include deeper layers of meaning for the benefit of the people you hope will be shouting for an encore – or for whom your main points may be old hat. But a good basic assumption is that no one will ever hear your creation twice.

If you find yourself telling a baffled audience, 'You'll probably see what I'm getting at if we hear it through again', then you may take it that you are almost certainly on the wrong lines. Only if you recognize this, is it at all likely that anyone will *want* to hear you twice. Of course, there are many important exceptions; and it is always tempting to hope that one's own work is such an exception. This is almost invariably a major error: so aim to make a major impact at the first hearing.

Understanding and remembering

Control over the rate at which ideas are presented is completely in the hands of the programme maker, so he must be particularly careful not to use this power in such a way as to impair intelligibility. There are two ways in which he may be tempted to disregard this. One is the quest for 'pace' – but as we shall see later, pace is not achieved by over-condensing vital information.

The second factor that may influence a producer is more subtle: it is his wish to present a subject in a tidy, well-shaped manner – in itself an unexceptionable aim. But, being literate both by nature and education, he all too often falls into the trap of assuming that what looks well shaped in the form of a script will become so in its finished form, once such minor details as logical presentation and colloquial speech values have been attended to.

Unfortunately, this may not be the case. For it also depends on the rate at which new ideas are being presented. To take an extreme case: on the printed page, a concise, elegant argument can be presented neatly, tautly, as finely shaped as a mathematical theorem – and the eye will dwell on it until it is understood. In a broadcast, such a construction is out of the question.

In order to ensure at least a moderate degree of intelligibility in this constantly progressing medium, the argument must sometimes be slowed down to what in literary terms might seem like an impossibly slack pace. The good magazine programme, play or documentary may appear in script form to have a much looser construction than a well-written article or short story: there may perhaps be an apparent excess of illustration and padding. This is necessary because it seems that the brain will accept and retain material that is new to it only if that material is presented at no more than a certain speed. Even if a listener is intellectually equipped to follow a concise, complex line of argument he needs time – more time than it takes to present the bare bones of the argument – if it is to sink in. He needs time to turn it over in his mind.

There is a clear analogy with the teaching process. But learning – in the sense of the *permanent* retention of fact and argument – may not be the immediate result. *Repetition* plays an important part in learning; if something that is clearly understood in the first place is repeated after a few days, and yet again after a few weeks, there is a reasonable chance of it being retained. Nevertheless, radio or television may help in the learning process, for although very few individual items will be given a second hearing by the audience, their content may be related to what they have read somewhere or heard in a previous programme, or know already. In such a case the content of the programme stands a much better chance of being understood and making some impact.

Clearly, learning is very much a matter of chance, or perseverance on the part of the audience. But the immediate effectiveness of any particular item is, or should be, more predictable: the programme-maker must calculate each effect, allowing sufficient time for the audience to take each point as it is made. Certainly, this will affect the amount that remains in the mind after a programme is over, or the following day – but that is only half the story. If there is a logical argument or a narrative line, a single failure may render the whole programme incomprehensible, or considerably change its meaning. In the action of a play, for example, an understanding of points made early on is often essential to being able to follow developments at a later stage. Fail to establish a single important plot point and the play fails.

This is a matter of vital importance to a broadcasting organization such as the BBC, and it is perhaps not surprising that a number of experiments have been carried out to determine the intelligibility of programmes and the rate at which fact and arugment can be absorbed.

One test was made with news bulletins to determine how many items and how much detail could be recalled by an attentive listener immediately after the end of the news. It turned out that only a small proportion was retained. Anyone with a tape recorder can easily check this. Record ten minutes of news and then, before playing it back, try to write down all the items of news in as much detail as possible – the exact words are not important; the gist will do. Playback will confirm that by far the greater part was missed.

Another way of trying this experiment is in a group, as a sort of party game. One person writes down the subject matter and timings of the items; the others just listen, then try to reconstruct it by discussion and without help from the running order. It may be instructive to record the discussion as well as the original news; and then play them both back together. Such experiments are called aural assimilation tests. Marks may be awarded for major and minor points that are recalled, and are subtracted for errors. Many tests of this sort have been carried out, and the marks have always been low. The ear is not efficient at taking in ideas.

The intelligibility of educational and current affairs broadcasts was also studied. The aim of a particular series of five-minute single subject broadcasts was that they should be understood without undue difficulty by the greater part of the mass audience. At the time of the survey, getting on for a third of Britain's adult radio audience listened to it regularly. It was respected, enjoyed and thought to be interesting and informative. The enquiry was thorough. Each talk was divided up into main and subsidiary points and marks for understanding them were allocated according to their importance. Individual talks scored between 11% and 44%. The average was 28%.

It was quite clear that, simple as these talks already were, they could with benefit be simplified still further. If this could be done without giving an appearance of talking down – always to be avoided – the intelligibility could be increased. Direct faults in presentation were found in some cases: such things as taking too much background knowledge for granted; the use of unfamiliar words, or jargon, or flowery analogies or abstractions; poor logical construction or unnecessarily complex sentences. One of the major conclusions was that the presentation of the more important points should be better organized. Major points should be emphasized and repeated in such a way that it is quite clear which, out of all that is said, is the important part to grasp, and a summary at the end would improve things still further.

This particular survey was carried out for a specific programme aimed at a specific audience, but would be valid for almost any type of programme (not just serious ones) and for any audience, provided account is taken of possible differences in background knowledge, literacy and intelligence.

The communication of ideas

Experience, together with the above experiments, seems to suggest a working rule: that *not less than three or four minutes of programme time should be devoted to any point of importance.*

This is a generalization, but if it errs, it is on the side of allowing too little time: these are likely to be *minimum* times for points of average complexity. Of course, for a simple point it may be possible to get away with spending less than three minutes; while a complex point may need much longer to establish fully. In the latter case it may be better to break the argument down into stages.

In a seven-minute item, for example, I would say that it is possible to make two major points. The rest of the seven minutes must be spent in supporting and linking those points. If a further major point is introduced, then it is likely that one of the three will be lost. To take an extreme case, if the seven-minute piece is made to consist of seven important points, then nearly all of them will be lost, and the listener may be left with a frustrating feeling that he has missed something. Alternatively, he may remember just one of the points that happens to appeal to him, plus the thought that this was part of a pattern that included half a dozen other points which on first hearing seemed plausible. In this highly condensed case there is no exact or predictable control over what the listener picks out. It may be something totally unsuitable from the point of view of the programme-makers.

One can point to Sir Winston Churchill as a broadcaster who triumphantly mastered the technique: a study of recordings of his Second World War speeches shows how he took each point in turn and attacked, analysed and presented it in a variety of ways before moving on to the next. Despite a difficult voice and an almost complete refusal to make any concession to radio (oratory is one of the 'don'ts' of the medium) or even replacement by the voice of an actor the results remain masterpieces.

It seems that all broadcasters who want to get something over to their audience must follow the methods of a schoolteacher – though not too overtly, for unlike the teacher, they have no captive audience. A good teacher, at whatever level, aims to lay out his subject in just the way I have described. However, in the classroom the pace of delivery is subject to immediate correction by the pupils. In contrast, when speaking to an invisible audience, there is a feeling of pressure to 'get on with it', because everyone is behaving as though air time is valuable (which it is) and because there is nearly always more to say than time to say it.

This rule about good layout is more clearly valid for programmes that seek to educate and inform; it is also true for most of those whose outward purpose is solely to entertain, and in particular those with a story to tell.

Pace

The term *pace* has already crept into this discussion: let us attempt to define it more precisely.

To maintain interest, a story or an argument must be kept moving at a fairly brisk rate. The secret of pace is *to allot just enough time to each point for it to be adequately understood – and then move on.* Pace is impossible without intelligibility. It is perhaps odd that it should depend in part on *not* going too fast. But there it is: pace depends on precise shaping of the programme in time, and it provides the basic foundation on which the finer details must be built.

When we talk of lack of pace, we are often thinking of another, secondary, defect which frequently afflicts the work of those without access to skilled critical analysis: a superficial slackness that has as its origin the lack of an adequate sense of timing. But it is the other more subtle fault, lack of organization in the underlying shape, that is more likely to mar professionally produced shows that are slick enough to all outward appearance.

To sum up: in the construction of almost any successful programme we must have a series of basic units each occupying a few minutes. Each of these should contain one major point and sufficient supporting detail to help establish it as clearly as possible. There may also be material that retains the attention of the listener in between the major points that are to be made, but which may not be of direct importance to them, and which must not actively draw the listener's attention away from them. Quite apart from anything else, this sort of construction helps to provide necessary light and shade within a piece – that is to say, some variation in the pressure of attack. And that again is absolutely essential to a feeling of pace in the final product.

All this affects the choice of form for the presentation of a particular subject. An argument that is by its nature concise may have to be filled out by extra material: illustration perhaps, repetition in a different way, anything. Or an extra point that seems to be creeping in may have to be cut, because it is interesting enough to distract the listener from the main line that is being followed, but cannot be allowed its own full development.

Certain types of construction, whether in radio, television or film, simply do not work: they contain too much argument in too little time. The cool, swift logic of a mathematical theorem, or a solid, indigestible, encyclopaedic mass of essential detail are both death to clear, intelligible construction in sound.

Of course, there are exceptions to all such rules. As a television producer, director or writer who enjoys dealing with complex themes, I never construct by rote, but instead work outward from the story, the meaning, the material itself. Often the results may be criticized as overturning some of the tenets of communication, and testing others to the limit, which at least avoids deadening uniformity or predictability. But I believe that the programmes that seem to have succeeded despite breaking all the rules invented are in reality those that stick most closely, if subtly, to the ideal form.

Glossary

As well as short definitions, some indication of the context in which terms may be used is given. Unless otherwise indicated, the term is usually in international use, although local variants may exist. Where usage in Britain (*Br.*) and America (*Am.*) differs markedly, both forms are given. Certain expressions that seem to have originated in BBC usage are so indicated. No attempt is made here or elsewhere in this book to enter into a study of basic electrical theory, or to describe equipment or circuits in technical detail.

'A' signal. In stereo, the signal to be fed to the left-hand loudspeaker of a two-speaker system.

AB. Stereo system in which left and right components are generated and/or processed and transmitted separately, as distinct from MS (middle-and-side).

Absorption coefficient. The fraction of sound absorbed on reflection at any surface. It therefore takes values between 0 and 1, and unless otherwise stated, is for a frequency of 512 Hz at normal incidence. *Soft absorbers* depend for their action on such things as the friction of air particles in the interstices of the material. This friction increases with particle velocity, and the absorption is therefore greatest where particle velocity is greatest, i.e. at a quarter wavelength from the reflecting surface. Soft absorbers (glass wool, fibre board, etc.) in layers a few cm in depth are poor absorbers at low frequencies and are fair to good at middle and high frequencies. If the outer surface is hard (e.g. painted) and unperforated, the high-frequency absorption will deteriorate. *Membrane and vibrating panel absorbers* are, in effect, boxes with panels and cavities that can be tuned to resonate at particular frequencies or broad frequency bands. They readily remove sound energy from the air at their resonant frequency and this is then mopped up within the absorber by various forms of damping. Such absorbers may be efficient at low and middle frequencies. *Helmholtz resonators* respond to narrow bands and if damped internally remove them selectively.

Acetate. (a) Cellulose acetate, a material that has been used as a tape base (i.e. the backing on which the magnetic oxide coating is carried). Tends to break rather than stretch when subjected to excessive stress, and so can be mended easily. (b) Colloquial term for direct-cut lacquer disc.

Acoustic reproduction of effects, etc. (from tape or disc). A feed of pre-recorded sound to a loudspeaker in the studio instead of directly to the mixer.

The sound picked up by the studio microphone depends on the characteristics of the loudspeaker and microphone, and the studio acoustics and balance. It may be noticeably different from the original sound even if a close balance is used (often the loudspeaker is balanced at a separate microphone). Acoustic reproduction techniques may be used either to modify the sound or to provide cues when no other method (headphones, cue lights or hand signals) is possible. In the latter case volume should be kept as low as practicable, otherwise there may be a noticeable change of quality due to acoustic coloration. See **Foldback**.

Acoustics. The study of the behaviour of sound. The acoustics of an enclosed space depend on its size and shape, and the number and position of absorbers. The *apparent acoustics* of a studio depend on the effect of the acoustics on the indirect sound reaching the microphone, and also on the ratio of direct to indirect sound. Thus, apparent acoustics depend on the microphone position, distance and angle from the sound source (i.e. the balance) and also on the level of sound reproduction. Besides this, the actual acoustics of the studio may be modified close to the microphone by the placing of screens, etc.

Actuality. Recording of the actual sound of an event, as distinct from a report, interview or dramatized reconstruction.

ADR. Automatic Dialogue Replacement. A short loop or long digitally recorded sample of the original picture is replayed repeatedly so that new dialogue can be rehearsed in order to attempt to match the original lip movements and duration. When an acceptable version has been recorded, fine adjustments can be made at a later, track-laying stage before the final re-mix.

Ambient noise. In any location, from quiet studio or living-room to busy street, there is a background of sound. In a particular location the ear automatically adjusts to and accepts this ambient noise in all except the loudest (or quietest) case. In a monophonic recording, however, noise at a natural level generally sounds excessive. One purpose of microphone balance is to discriminate against such noise, where necessary. Unexplained noises in particular should be avoided. See also **Atmosphere**.

Amplifier. A device for increasing the strength of a signal by means of a varying control voltage. At each stage the input is used to control the flow of current in a circuit that is carrying much more power. Precautions (feedback) may be necessary to ensure that the degree of amplification does not vary with frequency. In successive stages the magnification of the signal (i.e. ratio of output to input) may be much the same, so that the rise in signal strength is approximately exponential. Relative levels may therefore conveniently be measured in decibels (q.v.).

Amplitude modulation (AM). A method whereby the information in an audio signal is carried on the much higher frequency of a radio wave. The envelope of the amplitude of the radio wave in successive cycles is equivalent to the waveform of the initial sound. Historically, AM was the method used first and such transmissions still crowd the short, medium and long wave bands. This crowding restricts the upper limit of the frequencies that can effectively be broadcast without the transmitter bandwidth overlapping an adjacent channel. The random background of radio noise that a receiver is bound to pick up at all frequencies also appears as minor fluctuations in the amplitude of the carrier wave, and cannot be distinguished electrically from the audio signal. For high-quality transmission, AM has largely given way to FM (frequency modulation) or digitally coded systems such as the British NICAM.

Artificial reverberation ('echo'). Simulation of the natural die-away of sound that occurs in a room or any other enclosed space (e.g. in a cave, down a well, etc.). Such techniques are used when the available studio acoustics are not reverberant enough. Multimicrophone techniques for music need 'echo', as the studios used are normally much less reverberant than those used for natural balance. The technique is used to permit different treatment to be applied to each microphone output.

Atmosphere. The background sound at any location. This can enhance authenticity and give listeners a sense of participation in an event. Even so, it may be advisable to discriminate against background sound in the main recording, and record atmosphere separately for subsequent mixing at an appropriate level (which may not be held constant). When proper monitoring conditions are available at the time of recording (or broadcast), separate microphones may be set up and the mixture again judged by ear. *Studio atmosphere* is the ambient noise and it may at times be obtrusive even in a soundproofed room – particularly if omnidirectional microphones are used for other than close balances. The slightest movement, the flow of air for ventilation or even heavy breathing, may create problems for the balancer. See also **Ambient noise**.

Attenuation. Fixed or variable losses (usually in an electrical signal). As with amplification it is convenient to measure this in decibels (q.v.). See **Fade**.

Axis (of microphone or loudspeaker). Line through centre of diaphragm and at a right angle to it. It is usually an axis of structural symmetry in at least one plane, and thus an axis about which the polar response is symmetrical. Also, it is generally the line of maximum high-frequency response.

Azimuth. The angle that the gap of a recording or reproducing head makes with the line along which the tape moves. For audio tape, this should be exactly 90°. Misalignment may be due to incorrect setting of the head, or to the tape transport system not being parallel to the deck, and may be corrected by adjustment to either. Adjustment of a reproducing head may be made using a standard recording of high-frequency tone (i.e. one recorded with a correctly adjusted head). Maximum output, which may be judged by ear or by meter, occurs when alignment is correct. A separate recording head may then be adjusted by recording tone with it and checking for maximum output from the reproducing head. Azimuth misalignment produces a recording (or reproduction) apparently lacking in top.

'B' signal. In stereo, the signal to be fed to the right-hand loudspeaker of a two-speaker system.

Backing. The base on which the magnetic coating of tape is carried. It gives strength and permits flexibility, and although it has no screening effect its thickness ensures that the physical separation of successive layers of the magnetic coating is sufficient to maintain printing at a low level. Common materials are cellulose acetate, PVC and polyester.

Backing track. Pre-recording of the accompaniment to a singer (etc.) who then listens to a replay on headphones as he contributes his own performance. At this stage the two are mixed to give the final recording. The performer's (and other musicians') time is saved, and adequate separation is achieved without the use of separate studios. *Back tracking* is the technique in which a preliminary recording by the same artist is used as the accompaniment.

Baffle. (a) A rigid screen around a loudspeaker diaphragm, extending the acoustic path from front to rear in order to reduce the flow of air around the edge of the speaker at low frequencies – an effect that causes serious loss of bass. (b) A small acoustic screen that causes a local variation in the acoustic field close to the microphone diaphragm. A baffle of hardboard, card, etc., when clipped to a microphone produces a 6 dB peak at a frequency about 1/1000th of its linear dimension (1/300th in metres). In some microphones, gauze baffles are used as an integral part of the design to improve the response.

Balance. (a) The relative placing of microphones and sound sources in a given acoustic, designed to pick up an adequate signal, discriminate against noise, and provide a satisfactory ratio of direct to indirect sound. This is the responsibility of a *balance engineer, sound supervisor* or (in BBC radio) a *studio manager.* (b) The distribution of a sound between A and B stereo channels.

Balance test. One or several trial balances, judged by direct comparison.

Band. Separately recorded section of a disc, of which there may be several on a side. By extension, the term may also mean an individual section of a tape recording bounded by spacers. Locations may be indicated on a script in shorthand form, e.g. 'S2B3' = side 2, band 3.

Bass. Lower end of the musical scale. In acoustics it is generally taken as the range (below 200 Hz, say) in which difficulties (principally in the reproduction of sound) are caused by the large wavelengths involved. Loudspeakers that depend on the movement of air by cones or plates (e.g. moving-coil or electrostatic types) become inefficient in their *bass response* at wavelengths greater than their dimensions. This is due to too small a piston attempting to drive what is at these frequencies too pliant a mass of air. Certain systems of mounting (or cabinets) extend the bass response by means of a resonance in the octave below the frequency at which cut-off begins, but this results in a sharper cut-off when it does occur (e.g. bass reflex and column resonance types). *Bass lift* may be resorted to, but requires a more powerful amplifier (most power is concentrated in the bass of music, anyway) and a loudspeaker system capable of greater excursions without distortion. An efficient way of producing bass is an acoustic exponential horn with a low rate of flare. This ensures that coupling between the moving elements and the air outside is good, up to wavelengths approaching the diameter of the mouth of the horn.

Bass tip-up (*Br.*). Proximity effect (*Am.*). Increase in bass response of directional microphones when they are situated in a sound field that shows an appreciable loss of intensity in the distance travelled from front to back of microphone. The operation of the microphone depends on the pressure gradient and is exaggerated at low frequencies. At middle frequencies the effect is masked by the phase shift that occurs in this path distance in any case: and at high frequencies the mode of operation is by pressure and no longer by pressure gradient.

Beat. If two tones within about 15 Hz of each other are played together, the combined signal is heard to pulsate, or beat, at the difference frequency. Two tones may be synchronized by adjusting the frequency of one until the beat slows and finally disappears.

Bias. A carrier frequency with which the audio signal is combined in order to reduce distortion during tape recording. Its importance is greatest in recording low frequencies. It is generally of the order of 50–100 kHz and is produced by an oscillator associated with the tape-recorder amplifier. It mostly vanishes from the tape soon after recording, by a process of self-demagnetization. However, a little of the bias signal can still be heard if recorded tape is pulled slowly over the reproducing head. The bias oscillator also supplies the erase head.

Bi-directional microphone. One that is live on the front face and back, but dead at the sides and above and below. The polar response is a figure-of-eight.

Board, clapperboard, slate. In film, a means of marking the start of each shot with scene and take numbers. It is held open for mute scenes (those shot without sound); for sync takes, the scene and take number are identified on sound and then the board is clapped in vision so that sound and picture can subsequently be matched. When large quantities of film are handled in the cutting room, this method of marking is more convenient than electronic techniques.

Boom. A telescopic arm (attached to a floor mounting) from which a microphone is slung.

Boomy. Subjective description of a sound quality with resonances in the low frequencies, or a broad band of bass lift. Expressions with similar shades of meaning are tubby or, simply, bassy.

Bright surface. Strongly reflecting surface, particularly of high frequencies.

Broadcast chain. The sequence: studio–continuity suite–transmitter –receiver, together with any intermediate control or switching points, through which a signal passes.

Butt change-over. A change-over in mid-programme from one tape or disc to another, the second being started as the out cue on the first is about to come up. The cross-over from one reproduction to the other should not be noticeable.

Cancellation. Partial or complete opposition in phase, so that the sum of two signals approaches or reaches zero.

Cans. Headphones.

Capacitance. The ability of an electrical component, or components, to store static charge. Charges present in a conductor attract opposite charges to nearby but not electrically connected conductors; changes of charge induce corresponding changes of charge in nearby conductors. A signal may therefore cross between components where there is no direct path. Capacitors are used in electronics for coupling, smoothing and tuning purposes. In lines, capacitance is a mechanism whereby the signal is gradually lost (capacitance between the wires or to earth). Capacitance varies with distance between plates, or other components: this variation is used in electrostatic microphones and loudspeakers. *Stray capacitances* form the mechanism whereby audio signals or mains hum is transferred unintentionally from one component to another nearby.

Capacitor microphone. See **Electrostatic microphone**.

Capstan (*Am.*: Puck). Drive spindle of tape deck: the tape is driven between the capstan and an idler pulley which presses against it when the machine is switched to record or replay. The diameter of the spindle and the motor speed determine the speed of the tape. If the rotation of the spindle is eccentric there is a rapid fluctuation in tape speed, causing a flutter in the signal.

Capsule. (a) Removable pick-up head. (b) The diaphragm assembly of a condenser microphone capsule.

Cardioid microphone. Microphone with a heart-shaped polar diagram, which may be achieved by adding omnidirectional and figure-of-eight responses together, taking into account the phase reversal at the back of the latter. In electrostatic (condenser) microphones (q.v.) the combination occurs in the principle of construction. See also **Pressure gradient microphone**.

Cassette, cartridge. A fully encapsulated tape system that does not need to be laced up. Cassettes (or cartridges) are also used for endless tape loops, in this case often employing standard-width tape. To avoid confusion it is probably best to use the term cartridge for the endless loop system, and cassette for the encapsulated double-spool system.

CCIR characteristic. In tape recording, the pre-emphasis and subsequent equalization standard used in Britain and Continental Europe (set by the Comité Consultatif International des Radiocommunications). See also **NARTB** and **RIAA**.

Cemented joint (permanent joint). A method of editing tape or film in which the two ends are overlapped and fixed together by an adhesive or bonding agent. As there is a slight step in the tape, this type of joint is more likely to cause a momentary loss of signal on subsequent re-use than is the butt joint, which leaves the surface smooth and substantially continuous.

Ceramic microphone or pick-up. This is similar in principle to a crystal type, but its barium titanate element is less sensitive to temperature and humidity.

Channel. Complete set of (professional) recording equipment. Recording room (or part of recording room that can operate independently to record a programme). The series of controls within an individual signal pathway in a control desk: see **Microphone channel**.

Clean feed (*Am.*: mixed-minus). A cue feed back to a programme source that includes all but the contribution from that source.

Clean sound. Actuality sound of an event, without superimposed commentary.

Clock. See **Videotape clock, Time code**.

Coating (emulsion). The layer of finely divided magnetic material, bonded in plastic and polished to allow smooth flow over the tape heads, that carries the magnetically recorded signal.

Cocktail party effect. The means whereby a listener can concentrate on a single conversion in a crowd. Depends largely on the spatial spread of sound, a quality that is lost in monophonic recording – so that if a cocktail party type of background sound is recorded at a realistic level it can be unpleasantly obtrusive and impede intelligibility of foreground speech.

Co-incident pair (*Am.*: X/Y technique, intensity stereo). In stereo, a pair of directional microphones or a double microphone with two directional elements mounted close to each other (usually one above the other) and at an angle to pick up the A and B signals. This technique avoids the phase interference effects produced by a spaced pair.

Coloration. Distortion of frequency response by resonance peaks. Marked acoustic coloration in a studio may be due to the coincidence of dimensional resonances, to wall-panel resonances, or to frequency-selective excessive absorption of sound. Some mild acoustic coloration may, however, be beneficial. Coloration is often exhibited by microphones and loudspeakers, and in these is generally unwanted.

Comb filter. The effect of adding identical signals with a small time difference, so that there is cancellation, taking the form of 'notches' throughout the audio-frequency range. These are distributed in arithmetic, not logarithmic (i.e. not musical) progression.

Commag. Combined magnetic film sound, i.e. with the sound on a magnetic track running beside the picture, and recorded 28 frames ahead of it. Note: wide-screen combined magnetic prints have their sound recorded after the picture.

Comopt. Combined optical film sound: the sound track is recorded beside the picture, but advanced from it by $19\frac{1}{2}$ or 20 frames (35 mm) or by $25\frac{1}{2}$ or 26 frames (16 mm). See **Optical sound**.

Compact disc. Commercially-available recording which is digitally coded and read by laser. Since the noise level is low, a greater dynamic range is available – which may show up problems at other points in the audio chain.

Compatibility. In stereo, the requirement that a recording or broadcast can also be heard satisfactorily in mono.

Compliance. The ratio of displacement to applied force. It is the inverse of the (mechanical) stiffness of the system. Compliance is the acoustical and mechanical equivalent of capacitance and is a measure of the performance of microphone, loudspeaker, or record-replay components.

Compression. Control of levels in order to ensure that (a) recorded or broadcast signals are suitably placed between the noise and distortion levels of the medium, and (b) the relationship between maximum and minimum volumes is acceptable to the listener who, in his own home, may not wish to hear the greatest volume of sound that he would accept in a concert hall, and who may be listening against background noise. *Manual compression* seeks by various means to maintain as much of the original dynamics as is artistically desirable and possible, and varies in degree with the expected listening conditions. The purpose of *automatic compression*, in short-wave transmissions, is to bring peak volumes consistently to full modulation of the carrier. It is also used in the transmitter of radio microphones to prevent overloading at a point where volume cannot be controlled, and in pop music to limit the dynamic range of the voice or a particular instrument. Above a selected onset volume, say 8 dB below full modulation, compression is introduced in a ratio that may be selected: e.g. 2:1, 3:1, 4:1 or 5:1. *Digital compression* of video or audio signals may be used in off-line editing to speed processing and reduce the cost of hard disc storage. The reversible compression of a digital audio signal increases the capacity of optical discs (CD ROM) or ISDN (Integrated Switched Digital Network) lines: with a ratio of 4:1 the effects should not be perceptible.

Computer mixdown. See **Mix**, also **Digital audio techniques, Pulse code modulation**.

Condenser microphone. See **Electrostatic microphone**.

Cone. A piston of stiff felted paper or possibly of plastics. It should be light and rigid. Paper cones are often corrugated to reduce any tendency to 'break up' radially and produce sub-harmonic oscillations. Elliptical cones give a greater spread of high frequencies out along the minor axis: they should therefore be mounted with the major axis vertical.

Contact microphone. Transducer attached to a solid surface.

Continuity. Linking between programmes, including opening and closing announcements (when these are not provided in the studio), station identification, trailers and other announcements.

Continuity suite. A centre through which programmes are routed or where they are reproduced to build a particular service ready for feeding to a transmitter or to provide source material for other programme services.

Control. The adjustment of programme levels (in the form of an electrical signal) to make them suitable for feeding to recorder or transmitter; where necessary, this includes compression. Mixing involves the separate control of a number of individual sources. At a *control desk* there are faders (individual, group and main control), with their associated sound treatment, feed, cue and monitoring systems, and a variety of communications equipment. A *control line* between one location and another is a telephone circuit on which programme details may be discussed, and is so called to distinguish it from the broad-band (i.e. high-quality) audio 'music line' along which programme is fed. A control line may, of course, be narrow-band, but for outside broadcasts it is safer to use lines of equal quality so that the two are interchangeable.

Control cubicle (*BBC radio*). The soundproof room occupied by production and operational staff and equipped with control desk and monitoring equipment.

Control room. In BBC radio, a switching centre; otherwise, the sound mixing room.

Copyright. The law in relation to the ownershp of creative works, which is initially vested in the author, composer or artist. In Britain this extends to 50 years after the author's death (expiring at the end of the calendar year). For private study, research, criticism or review, 'fair dealing' provisions generally permit extracts without payment. Otherwise payment must always be made, the details being arranged by negotiation with the author or his agent. Sound recordings, films and broadcasts are also protected by copyright and are not subject to the 'fair dealing' exceptions. Thus a gramophone record may have two copyrights, that of the composer and that of the company that made the recording, and the two may extend from different dates. Public performance of music and records, etc., is easy to arrange, and fees are small (usually paid through the Performing Rights Society) but recording or re-recording – except of music recorded specially for the purpose (mood music) – is permitted only with elaborate safeguards, and even so permission may be difficult to obtain. Copying of records and films for private purposes is not permitted. In Britain, the BBC and its contributors' unions do permit the recording of schools and further education programmes by schools and colleges, provided that the recordings are used only for instructional purposes and that they are destroyed at the end of specified periods. There is no copyright in events: e.g. an impromptu speech or interview without a script or casual sound effects (though recordings of them are). Dramatic or musical performers are also protected: their permission must be obtained before a record is made of their performances other than for private or domestic purposes. Copyright laws are easily evaded and owners are justifiably concerned at the losses they undoubtedly sustain. Copyright laws vary considerably in detail from country to country – the above comments relate mainly to British law.

Cottage loaf characteristic. See **Hypercardioid**.

Crab. Move camera or sound boom sideways relative to the performing area.

Crossed microphones. See **Co-incident pair**.

Cross fade. A gradual mix from one sound source or group of sources to another. During this, both faders (or groups) are open at the same time. The rate of fade can be varied manually if desirable for artistic effect.

Crossover. The frequency at which a signal is split in order to feed separate parts of a loudspeaker. *Crossover network:* the filter that accomplishes this.

Crosstalk. In stereo, the breakthrough between channels. The separation (in dB) between wanted and unwanted sound is checked by feeding tone at zero level through one channel and measuring its level in the other. Outputs of $-37\,dB$ at 1 kHz rising to $-30\,dB$ at 50 Hz and 10 kHz are reasonable. Crosstalk also means breakthrough (or 'induction') of signal between any other pair of lines, e.g. on a telephone circuit.

Crystal microphone or gramophone pick-up. This generates a signal by means of a crystal bimorph – two plates cut from different planes of a crystal such as Rochelle salt and held together in the form of a sandwich. Twisting the bimorph produces a voltage (termed a *piezoelectric* voltage) between foil plates on top and bottom surfaces of the sandwich. The principle is well suited to use in a gramophone pick-up, as the frequency response is a fair match to the characteristic needed to reproduce gramophone records. This means that the output level after equalization is higher than it would otherwise be and smaller amplifiers are needed. The whole assembly can be very light in weight and may be cheap to produce. Crystals have been widely used in inexpensive equipment but have given way to electrets.

Cue. Signal to start. This may take the form of a cue light or a hand cue, or may be related to a prearranged point in a script. Headphones or, exceptionally, a loudspeaker may also be used for cueing.

Cue material. Introductory matter supplied with a recording for the scriptwriter to fashion into a cue for announcer or narrator to read.

Cue programme. A feed for cueing purposes.

Cycle. One complete excursion of an air particle when vibrating in a sound, or the corresponding signal in electrical or any other form.

Cycles per second (cps). Now written Hz. See **Frequency, Hertz**.

DAT. Digital audio tape. High-quality rival to compact disc, initially less successful in the domestic market, but more so in broadcasting.

dB. Decibel (q.v.).

Dead acoustic. One in which a substantial loss is introduced at every reflection. For studio work this is the nearest approximation to an outdoor acoustic (in which little or no sound is reflected).

Dead room (anechoic chamber). Has very thick soft absorbers (typically about a metre) and is used for testing the frequency response of microphones and loudspeakers. Such an acoustic is unsuited to use as a studio because of its claustrophobic effect.

Dead side (of microphone). The angles within which the response of a microphone is low compared with the on-axis response. In figure-of-eight microphones this is taken to be about $50°-130°$ from the axis; in cardioids, the entire rear face. In the dead region there may be rapid changes in response with only slight movement of position, and the frequency response is more erratic than on the live side(s). In a studio these effects may be masked by the pick-up of reflected sound on the live sides.

Deadroll (*Am.*). See **Prefade**.

Decay characteristic (of studio). The curve that indicates how the sound intensity falls after a steady note is cut off. It is generally slightly erratic. Plotted in decibels against time, decay should (roughly) follow a single straight line, and not be broken-backed.

Decibel (dB). A measure of relative intensity, power or voltage. Sound intensity is the power flowing through unit area and is calculated relative to a reference level of 2×10^{-5} Pa (Pascals, q.v.). The threshold of hearing at 1000 Hz is at about this level for many young people. However, the aural sensation of loudness is not directly proportional to intensity (I) but to the logarithm of the intensity. It is therefore convenient to measure differences in intensity in units that follow a logarithmic scale, i.e. decibels. Differences of intensity in decibels are calculated as $10 \log_{10} (I_2/I_1)$. Thus, a tenfold increase in intensity is equivalent to a rise of 10 dB (i.e. 1 Bel). Doubling the intensity gives a rise of almost exactly 3 dB. In a circuit the power is proportional to the square of the voltage, so that for a particular impedance a power ratio may be expressed in decibels as $10 \log (V_2^2/V_1^2)$, or $20 \log (V_2/V_1)$. This means that a range of 1:1 000 000 000 000 in intensity is equivalent to 1:1 000 000 in voltage. Note that the voltage scale dB (V) is used for microphones, and that this is not the same as the acoustic intensity scale dB (A). In acoustics the decibel is also convenient in that it can be regarded (very roughly) as the minimum audible difference in sound level. A change in speech or music level of 2 dB or less is not ordinarily noticeable to anyone who is not listening for it. Loudness levels are measured in *phons* (q.v.). For dBA, dBN and PNdB, see **Noise**.

De-gauss or de-flux. Demagnetize (e.g. a tape recording head, or scissors or blades used for cutting tape). Wipe or erase (tape) in bulk.

Delay line. Device that electronically delays a signal – for example by (switched) multiples of 7.5 ms.

Diaphragm. The part of a microphone upon which the pressure of a soundwave acts. In size it should be small enough not to suffer substantial high-frequency phase cancellation for sound striking the diaphragm at an angle, and big enough to present a sufficiently large catchment area to the pressure of the sound wave. In microphones working on certain principles the diaphragm mass may be small. In such designs (e.g. electrostatic microphones) there are minimal inertial effects and therefore a very good transient response.

Difference tone. A perceived frequency 'heard' when two tones are played together, e.g. if 1000 Hz and 1100 Hz are played, 100 Hz may also be perceived.

Diffusion of sound. The degree to which sound waves are broken up by uneven surfaces, and by the placing of absorbers more or less equally in all parts of the studio rather than allowing them to cluster in one particular part. A high degree of diffusion is desirable in a sound studio.

Digital audio techniques. In these, an audio signal which was initially obtained as an electrical *analogue* of the sound is converted into binary form, in which it may be operated upon, fed to random access memory stores, etc. The analogue signal is sampled at a rate which must be well over twice the highest frequency required, e.g. at 32 kbits/s for 15 kHz. Other sampling rates are 48k for DAT, 44.1k for CD, 15k for client presentations of digital video edits, and 38k for applications demanding a signal that is flat and distortion-free to 16 kHz. The signal must be converted back to analogue form before it is broadcast or fed through components designed to operate in that form.

DIN (Deutsche Industrie Norm). German industrial standard, applied to film speeds, tape equalization characteristics and items of equipment such as plugs and sockets.

Directivity pattern. Polar diagram (q.v.).

Dissonance. The sensation produced by two tones in the region of about a semitone or full tone apart. As the tones get closer together they beat together and finally become concordant; as they get further apart the dissonant sensation disappears, and for pure tones does not return with increasing separation. However, dissonances and concords also arise from the presence of harmonics. The sensation of dissonance is sometimes described as unpleasant, but 'astringent' would perhaps be a better antithesis to the sweetness of consonance.

Distortion. Unwanted changes of sound quality, in the frequency response, or by the generation of unwanted products. *Harmonic distortion* is most easily caused by flattening of peaks in an analogue waveform. The human voice and nearly all musical instruments possess a harmonic sound structure that tends to mask the distortion unless it is very severe. 1% harmonic distortion is not usually noticeable. *Intermodulation distortion* is, however, more serious, because it includes sum and product frequencies that are not necessarily harmonically related to those already present.

Drop-out. Loss of signal due to a fault in tape coating.

Dry acoustic. Lacking in reverberation. By way of mixed metaphor, the opposite is 'warm' or 'bright'.

Dubbing. Copying, to arrange for material to be available in convenient form, e.g. transferring music from disc to tape. Also, copying together and combining effects or music with pre-recorded speech. In film, the mixing of all sounds to a single final track (or the several stereo tracks).

Dynamic loudspeaker, microphone or pick-up. See **Moving-coil microphone**.

Dynamic range. The range of volumes in a programme. It may be measured as the range of peak values (i.e. the difference between the highest PPM readings at passages of maximum and minimum volume) or, alternatively, the range of average volume. It may refer either to the range of the original sound, or to what remains of it after compression.

Dynamics. The way in which volume of sound varies internally within a musical work (or in a speech or speech-and-music programme). It may refer to the variation of levels within the work as a whole, or from note to note, or in the envelope of a single note.

Earphone, earpiece. A single headphone (q.v.) or ear-fitting of the hearing-aid type.

Echo. Discrete repetition of a sound, produced by a single reflected sound wave, or a combination of such waves whose return is coincident in time and at least 0.05 second after the original sound. Colloquially, 'echo' is used to mean the sum of such reflections, particularly in artificial reverberation.

Echo chamber. A room for producing or simulating the natural reverberation of an enclosed space. Rarely of sufficient size, and largely superseded by electronic artificial reverberation.

Echo plate. Better known by its proper name, reverberation plate. Like the echo chamber, largely superseded by electronic devices.

Edit decision list (EDL). A list of time codes at which edits are programmed to take place. It may be written down from numbers superimposed on picture or, better, generated from computer memory and stored on floppy disc.

Editing block. A metal plate with a channel the width of magnetic tape running along the centre, with lapped edges to grip the tape, and an angled cutting groove to guide a razor blade across it.

Effects. Simulated incidental sounds occurring (a) in the location portrayed (usually *recorded effects*) or (b) as a result of action (usually *spot* or *foley effects* – those created on the spot). A heightened realism is generally preferred to the random quality of naturally occurring sounds; but effects taken a stage further and formalized into musical patterns are called *radiophonic effects* (*BBC*) or *musique concrète*. *Comedy* or '*cod*' *effects* are those in which some element, a characteristic natural quality of the sound, or a formal element such as its attack or rhythm, is exaggerated for comic effect. In this category are simple *musical effects*, in which a sound is mimicked by a musical instrument.

Eigentone. The fundamental frequency that belongs to any dimensional resonance of a room. Its wavelength is twice that dimension. It can form effectively only between parallel surfaces.

Electret. Electrostatic microphone capsule in which the diaphragm or base-plate has a permanent charge and therefore requires no polarizing voltage.

Electronic music. A work constructed from recorded electronic source materials by arranging them in a formal pattern (which may be beyond the range of conventional instruments or musicians).

Electrostatic loudspeaker. An application of electrostatic principles to the movement of air in bulk. This is done by means of a large charged diaphragm suspended between two perforated plates. As the alternating signal is applied to the outer plates, the diaphragm vibrates; and so long as the excursions are not too large the transfer of power is linear (the use of two plates rather than a single backplate helps considerably in this respect). The problem of handling sufficient power and transferring it to the air at low frequencies is not fully solved in any such speaker of convenient size (though baffles reduce the immediate losses). Electrostatic loudspeakers should not be placed parallel to walls, or standing waves will form behind them. With careful positioning, however, the quality of sound achieved is very clear. As with a number of other transducer principles this can also be applied in reverse, as a microphone.

Electrostatic (condenser or capacitor) microphone. In this, the signal is generated by the variations in capacitance between two charged plates, one of which is rigid and the other flexible, acting as a diaphragm in contact with the air. If the space between the plates is enclosed, the air vibrations can affect one side of the diaphragm only and the microphone is pressure operated (and substantially omnidirectional). If the backplate is perforated the diaphragm is operated by a modification of the pressure gradient principle. It can be arranged that the distance that sound has to travel through the backplate (via an acoustic labyrinth) is the same as the effective path difference from back to front of the plate: in this case the pressure gradient drops substantially to zero for signals approaching from the rear of the microphone, and a good cardioid response is obtained. If two such cardioids are placed back to back, the characteristics depend on the sense of the potential between centre plate and the two diaphragms. The doublet can then become bidirectional, cardioid or omnidirectional by varying the size and sense of the potential on one of the diaphragms. Such microphones have to be fitted with a head amplifier (a single stage is sufficient) in order to convert the signal into a form suitable for transmission by line. A further complication of design of both electrostatic microphone and loudspeaker is that a polarizing voltage has to be available, unless the polarization is permanent, in an electret (q.v.).

End-fire. Orientation of the diaphragm within a long microphone such that the axis of directional pick-up is along the main axis of the casing. An end-fire microphone is therefore 'pointed' toward the sound source, an arrangement easily understood by performers.

Enclosure. A loudspeaker cabinet. Its most important function is to improve bass response. It may do this by acting as a baffle and increasing the distance rear-emitted sound has to travel before it can combine with and cancel or reinforce forward sound (thus lowering the frequency at which cancellation occurs). It may be a ported or bass-reflex cabinet, from which a low-frequency resonance is emitted more or less in phase with the forward output from the cone. In another design the box entirely encloses the rear of the loudspeaker: in this 'infinite baffle' the rear-emitted sound is supposed to be absorbed entirely within the box.

Envelope. The manner in which the intensity of a sound varies with time. Graphical representation of the envelope (or dynamics) of a single note may show separate distinctive features in its attack, internal dynamics and decay. The term may also refer to the envelope of frequency content, or of an imposed frequency characteristic, as in a formant.

Equalization (colloquially, EQ). The use of a filter network to compensate (a) for any distortion of the frequency response introduced by a transducer or other component (e.g. a landline), or (b) for a recording or transmission characteristic employed to ensure an efficient, low-noise use of the medium (as in disc recording or FM transmission).

Erasure. The removal of recorded signals from a tape so that it is ready to re-use. This is done automatically on most machines as the tape passes the erase head, which lies between the feed spool and the recording head. The erase head is similar in structure to the other heads except that the gap is wider. The pure bias signal fed to it magnetizes the tape first in one direction and then the other until it approaches the end of the gap, when the magnitude of the oscillations dies down, and the tape is left demagnetized. With a *bulk eraser*, the tape as a whole is given a similar washing treatment, being physically removed from the region of the alternating current before switching off. In very powerful bulk erasers it is not necessary to remove the tape to complete the wiping action.

Establish. Establishing a sound effect, etc., is allowing it sufficient time (and volume, which may be greater than that subsequently used) for it to register in the listener's mind. Establishing a location may be helped by the above technique, but equally it may be a matter of supplying sufficient 'pointers' in scripted speech.

Exponential. An exponential curve is one that follows the progress of natural unrestrained growth or decay. If, in exponential growth, something has doubled in time t, it doubles again in further time t, is eight times its original size at time $3t$, and so on. Exponential growth inevitably overhauls all other forms of regular growth (unless as happens in nature, some regulating factor intervenes). Exponential decay halves in time t, halves again in further time t, and so on. This must eventually become the slowest regular form of decay. A *logarithmic scale* (e.g. on a graph) is one in which the scale is gradually reduced according to the same principles. The spaces between 1, 2, 4, 16, 32 . . . are all equal. This form of representation reduces an exponential growth to an apparently linear growth. However, the ear judges both changes of volume and changes of pitch by ratio; thus a logarithmic scale is more relevant than any other to perceived sound. The doubling of sound frequency is equivalent to an interval of one octave. A tenfold increase in the intensity of a sound is equivalent to an interval of 1 Bel (10 decibels).

Extinction frequency. Frequency at which there is complete loss of signal when the dimension of a component matches signal wavelength.

Fade. Gradual reduction or increase in the signal. This is accomplished by means of a *fader* (i.e. a potentiometer – or 'pot' for short). A *logarithmic fader* is one in which the ratio of gain or loss is the same for equal movements of the fader throughout its main working range. Such a fader may be marked off in a linear scale of decibels. Faders used in audio work are logarithmic over their main working range.

Feedback. The addition of some fraction of the output to the original; this in turn may contribute to a further addition, and so on. If the feedback is mixed in at a higher level than that of the original signal, a howlround occurs; if at a level that is sufficiently lower, it gradually dies away. See also **Spin**. In amplifiers a signal may be fed back to an earlier point: *negative feedback* (in which the signal fed back reduces output) is used to control distortion introduced by the amplifier.

Field pattern (*Am.*). Polar response. See **Polar characteristic**.

Figure-of-eight. See **Bidirectional microphone**.

Film leader. Section of film used at the head of picture or magnetic track showing a common 'start mark' and footage marks at 16-frame intervals between 12 and 3 after which the leader is black until the first frame of the film. The same intervals (representing feet on 35-mm film) are used for all film footages. Television films with synchronous sound are usually run off 10.

Filter. A network of resistors and condensers (inductances could also be used) that allows some frequencies to pass and attenuates others. The simplest form of filter (one resistor and one condenser) rolls off at 6 dB/octave above or below a certain frequency. There is an elbow in the curve at the turnover point: the nominal cut-off frequency is that at which the loss is 3 dB. For many audio purposes this gentle form of filter is quite as satisfactory as the sharper cut-off that can be obtained with a more complex network. The terms *bass-* and *top-cut filter* and *stop-* and *pass-band filter* are self-explanatory. An *octave* filter is one in which the signal is divided into octaves whose levels may be controlled separately. Again, the simpler circuitry, providing smoother slopes at the boundaries, may be just as satisfactory.

Flanging. Effect produced by playing two recordings almost in sync but with one running slightly slower than the other, so that phase cancellation sweeps through the audio-frequency range. Originally produced by manual drag on the flange of one of the spools, but now usually simulated electronically.

Flat. (a) On a stylus, a surface of wear that appears on the two sides of the tip after some period of use. Seen under the microscope, the surface really is flat. These areas gradually become larger, the angles at their edges sharper, and the tip shape more like that of a chisel moving end-ways on in the groove. (b) On the rubber tyre of an idler wheel, a flat is an indentation that may form if the idler is left parked in contact with drive spindle, etc. It causes a momentary flutter in recording or replay.

Flutter. Rapid fluctuation in pitch having a warble-frequency of, say, 8 Hz or more due to a fault in equipment such as an eccentric drive-spindle.

Foldback. A feed of selected sources to a studio loudspeaker or headphones at suitable levels for the benefit of the performers. It may be fed from outside sources, tape or disc, or from a microphone in a distant part of the studio to overcome problems of audibility or (in music) time lag. See **Acoustic reproduction of effects, Public address**.

Foley artist. US film industry term for specialist in matching footsteps and other sound effects in post production. Tracks are laid to replay of edited picture, and may be further edited, adjusted or treated before the final mix. See **Spot effect**.

Footage. Length of film expressed in feet: either in terms of its own gauge or the equivalent for 35 mm film. One 35-mm foot is 16 frames; one 16-mm foot is 40 frames.

Formant. A characteristic resonance region: a musical instrument may have one or more such regions, fixed by the geometry of the instrument. The human voice has resonance regions associated with the nose, mouth and throat cavities, which are capable of more or less variation in size and shape, permitting the formation of the vowel sounds and voiced consonants. The range of the formant regions is not directly related to the pitch of the sound on which they act. If the fundamental is well below or low in the formant range, the quality of sound produced is rich, as harmonics are clustered close together within the formant; if the fundamental is relatively high, there are less harmonics in the range and the quality is thinner.

Frequency. The number of complete excursions an air particle makes in one second (formerly described as cycles per second, c/s or cps, now as hertz, Hz). The only sound that consists of a single frequency is a pure (i.e. sinusoidal) tone. This corresponds to the motion of an air particle that swings backward and forward in smooth regular cycles (simple harmonic motion). The velocity of sound in air is about 1120 ft/s (340 m/s), depending on temperature. Frequency (f) and wavelength (λ) are related through the velocity of sound (c) by the formula $c = f\lambda$, so any frequency can be represented alternatively, but with a little less precision, as a wavelength. A complex sound can be analysed in terms of its frequency components, and plotting volume against frequency gives the *frequency spectrum*. A sound which is composed of individual frequencies (fundamental and harmonics or partials, or a combination of pure tones) has a *line spectrum*. Bands of noise have a *band spectrum*.

Frequency correction. The change required in the frequency characteristics of a signal to restore it to its original form; but the term is sometimes ambiguously used to indicate the application of desired deliberate distortion of the response (often a peak in the upper middle frequencies). A *linear response* in any process is one in which the frequency distribution is the same at the beginning and end. See also **Equalization**.

Frequency modulation (FM). A method whereby the information in an audio signal is carried on the much higher frequency of a radio wave. The frequency of the audio signal is represented by the rate of change of carrier frequency; audio volume is represented by amplitude of frequency swing. The maximum deviation permitted for FM transmission is set at 75 kHz. Overmodulation does not necessarily cause immediate severe distortion, as with AM. Noise, which appears as fluctuations of carrier amplitude, is discriminated against, though it does produce phase-change effects that cannot be eradicated. Pre-emphasis of top (i.e. prior to transmission), with a corresponding de-emphasis at the receiver, helps to reduce noise further. Unless two carriers on the same wavelength have almost the same strength, the stronger 'captures' the area: there is only a small marginal territory between service areas.

Frequency response. Variation in gain or loss with frequency.

Fundamental. The note associated with the simplest form of vibration of an instrument, usually the first and lowest member of a harmonic series.

Fuzz box. Used in the pick-up and reproduction circuitry of some 'electric' instruments (electric guitar, piano, etc.). This deliberately heavily overloads the system, thereby filling in the texture of the sound with a dense array of distortion products. A buzzing effect is achieved.

FX (indication in script). Effects.

Gain. Amplification: ratio of output voltage to input voltage. It is most conveniently calculated in decibels.

Gap width (*Am.:* Gap-height). The distance between poles of magnetic recording (or reproducing or erasing) head at the point of contact with the tape.

Gate. A switching circuit that passes or cuts off a signal in response to some external control, e.g. to whether or not an applied voltage is above or below a given threshold.

Graphic filter. Filter in which the signal is divided into narrow bands, perhaps of a third of an octave, each controlled by a slide fader. Side by side on the face of the instrument, the fader knobs form a curve approximating to the response of the filter.

Grid. A framework below the roof of a theatre or television studio from which lighting and microphone leads may be suspended. A television studio normally has some of its microphone points in the grid.

Groove. Track on record which carries the mono audio signal in the form of a lateral displacement and stereo in a combination of two 45° displacements. *Coarse groove.* The groove that was used for 78 rpm recordings. The pitch was approximately 100–150 grooves per inch (*Am.:* lines per inch). *Fine groove, microgroove, minigroove:* the groove normally used for $33\frac{1}{3}$ and 45 rpm recordings. The pitch, about 250–330 grooves per inch, varies with modulation.

Group fader (*Am.:* Sub-master fader). One to which the output of several individual faders is fed.

Guide track. A second track on a twin-track recorder which is replayed to artists to assist synchronization, but is not itself included in the finished performance.

Gun microphone. Moving-coil or electrostatic microphone fitted with an interference tube leading out along the axis. Sound enters the tube through a series of ports. That approaching off-axis reaches the diaphragm by a range of paths of different lengths, and mostly cancels out. The microphone is thus highly directional except for sound of wavelengths substantially greater than the length of the gun.

Haas effect. When a sound is heard from two loudspeakers in different directions and there is a time-delay on one path all the sound seems to come from the other loudspeaker unless the delayed sound is increased in volume to compensate. The effect grows rapidly to a maximum as the delay increases to 4 or 5 milliseconds (equivalent to a difference in path-length of about 5 ft or 1.5 m). At the maximum, an increase in volume of 10 dB in the delayed sound re-centres the image, and a greater volume swings it to the louder source.

Half-track recording. A recording occupying the upper 40% of magnetic audio tape (the tape moving left to right across the heads).

Harmonics. A series of frequencies that are all multiples of a particular fundamental frequency. They are produced by the resonances of air in a tube (e.g. woodwind, brass, organ) or of a vibrating string, etc.

Head. Transducer that converts electrical energy into magnetic or mechanical energy or vice versa. Thus we have tape recording and reproducing heads and disc cutter and pick-up heads. The electromagnet used for erasing tape is also called a head.

Headphones. A pair of electro-acoustic transducers (e.g. moving coil) held to the ears by a headband. Alternatively, devices similar to hearing aids may be used in one ear (e.g. for listening to talk-back instructions during a programme). Earphones are efficient and produce the best quality when the channel from transducer to ear is completely closed (as with moulded plastic ear-fittings). But when headphones are used mainly for communication purposes, high quality is of little importance. Some degree of control over volume can be achieved with telephone receiver type phones by moving them a little off the ear, but with the ear-fitting, it is desirable to have a volume control in circuit.

Hearing. Essentially, the ear acts like an instrument for measuring frequencies and volumes of sound logarithmically (so that they appear linear in octave scales and decibels). The subjective aspect of frequency is pitch: judgement of pitch is not entirely independent of volume. The subjective aspect of sound intensity is loudness. Below about 1000 Hz the ear is progressively less sensitive to low volumes of sound: above 3000 Hz the hearing may also become rather less sensitive, but not necessarily following a smooth curve. The upper limit of hearing for young ears may be 16 kHz or higher, but with increasing age this is gradually reduced. No audio system needs to exceed the range 20–16 000 Hz (the signal may be cut off at this upper frequency). The acoustic part of the ear (the outer and middle ear) is relatively simple: essentially, it consists of a group of bones the purpose of which is to provide efficient coupling, i.e. the matching of mechanical impedances, between a diaphragm in contact with the air and a second diaphragm which transmits sound to the liquid of the inner ear. The physiological construction of the inner ear is known, but the mechanism of aural perception is complex.

Hertz (Hz). The measure of frequency (q.v.) formerly expressed as cycles per second. It is more convenient for mathematical analysis to consider frequency as an entity rather than as a derived function.

Hiss. High-frequency noise.

Howlround or **Howlback.** Closed circuit (wholly electrical, or partly acoustic) in which the amplification exceeds the losses in the circuit. In a typical case the loudspeaker is turned up high in a monitoring cubicle and the microphone in the studio faded up high for very quiet speech, and then if the acoustic treatment is inefficient at any frequency (or if both connecting doors are opened) a howl may build up at the frequency for which the gain over the whole circuit is greatest.

Hum. Low-frequency noise at the mains frequency and its harmonics.

Hybrid transformer. A transformer with two secondary windings, so arranged that cross-talk directly between them is as small as possible.

Hypercardioid. Cottage-loaf-shaped polar response of microphone, intermediate between figure-of-eight and true cardioid.

Idler. On a tape deck, the idler presses the tape against the capstan when the drive is switched on (but does not itself transmit rotation).

Impedance. A combination of d.c. resistance with inductance and capacitance, which act as resistances in alternating circuits. An inductive impedance increases with frequency: a capacitative impedance decreases with frequency. Either type introduces change of phase. See **Matching**.

Indirect sound (in microphone balance). Sound that is reflected one or more times before reaching the microphone.

Inductance. The resistance of (in particular) a coil of wire to rapidly fluctuating alternating current. The field built up by the current resists any change in the rate of flow of the current. And this resistance increases with frequency.

Intensity of sound. The sound energy crossing a square metre. Relative sound intensities, energies, or pressures, may all conveniently be measured in decibels. See **Decibel, Wave**.

Intermodulation distortion. See **Distortion**.

Joint. Point on a tape at which two physically separate butt ends have been joined together.

Jointing tape. A specially prepared, non-oozy adhesive tape, slightly narrower than magnetic tape, used to back two pieces of tape that have been butted together.

Jump cut. Originally a cut in replaying a disc, made by lifting the stylus and replacing it in a later groove (at which the cue has been checked in advance). There is a momentary loss of atmosphere while the replay is faded out. In film, a cut between scenes or two separate parts of the same scene in such a way that continuity of action is deliberately lost.

kc/s. Kilocycles per second. Now written kHz. See also **Frequency, Hertz**.

Lavalier. Personal microphone originally worn suspended around the neck, like pendant jewellery; now a microphone in any similar position.

Lazy arm. Simple form of boom consisting of an upright and a balanced cross-member from which a microphone may be slung.

Leader. White uncoated tape that may be cut on to the start of a spool of recorded audio tape, and on which may be written brief details of the contents of the sound recording. See also **Film leader**.

Level. Volume of electrical signal as picked up by the microphone(s) and passed through pre-amplifiers and mixer faders. At the BBC this volume is calculated in decibels relative to a *reference level* of 1 milliwatt in 600 ohms (also called *zero level*). Zero level corresponds to 40% modulation at the transmitter: 100% modulation is 8 dB above this. To *take level* is to make a test for suitable gain settings, in order to control the signal to a suitable volume for feeding to transmitter or recorder. This test may sometimes be combined with the balance test. *Voice level* is the acoustic volume produced by a voice in the studio.

Limiter. An automatic control to reduce volume when over-modulation occurs, e.g. to prevent dangerous heavy peaks of power reaching a transmitter that might be damaged by it. One form of operation uses feedback: the signal is monitored by the limiter and any wave that exceeds a certain volume causes a corresponding increase in feedback, which reduces the signal. After a while, the operation of a recovery device allows the gain to return to normal.

Line. A send and return path for an electric signal. In its simplest form a line consists of a pair of wires. A *landline* is a line (with equalizing amplifiers at regular intervals) for carrying a programme across country (c.f. **Radio link, Broadcast**).

Line microphone. Gun microphone (q.v.)

Line up. Arrange that the programme signal passes through all components at the most suitable level. In a broadcast chain this is normally at or about zero level (see under **Level**).

Line-up tone. Pure tone, usually at 1000 Hz, fed through all stages of a chain. It starts at zero level, and should read the same on a meter at any stage. A drop or jump in the level of the tone between successive points in the chain may be due to a fault in the line or in other equipment.

Lip-ribbon microphone. A noise-cancelling microphone placed close to the mouth, the exact distance being determined by a guard resting against the upper lip. The microphone's directional properties discriminate in favour of the voice; close working also discriminates against unwanted sound, and low frequency noise is reduced still further by the compensation necessary when working so close to a directional microphone.

Live angle. Angle within which reasonable sensitivity is obtained. See **Cardioid microphone, Bidirectional microphone**.

Live side (of microphone). The face that must be presented to a sound source for greatest sensitivity.

Logarithmic scale. See **Exponential**.

Loop. Continuous band of tape made by joining the ends of a length of tape together. A tape loop may be used (a) to provide a repeated sound structure or rhythm (in radiophonics), (b) for an 'atmosphere' track where this is regular in quality, (c) in tape delay techniques. Magnetic film loops have been used for continuous 'atmosphere' in film dubbing, but have been superseded by electronic sampling devices.

Loudness. Subjective aspect of sound intensity. See **Meter, Decibel**.

M signal. The combined A + B stereo signal. Corresponds to the signal from a single 'main' microphone. See **S signal, AB**.

M and E. A pre-mix, nominally of music and effects. In practice, in documentary it is a mix of all but commentary.

Main gain control. (*Am.*: Grand Master or Overall Master Control). Final fader to which the combined outputs of all group (*Am.*: sub-master) or individual faders are fed.

Matching. Arranging that the impedance presented by a load is equal to the internal impedance of the generator. Unless this is done there will be a loss of power, and the greater the mismatch the greater the loss. Often, when there is adequate gain in hand, some degree of mismatch is not critical, but the input impedance should be lower than that of the following circuit: 'low into high will go'. However, in the case of microphones and similar generators, the signal is low to start with and any immediate loss will result in a poorer signal to noise ratio. Matching is done by means of a small transformer. Where a microphone impedance is strongly capacitive (e.g. in an electrostatic microphone) its output voltage controls current in an external circuit.

Meter. Device for measuring voltage, current, etc. In audio, several types of meter are used for measuring programme volume. The *VU (volume unit) meter* is used on much American equipment. Over its main working range it is linear in 'percentage modulation'. It is not, therefore, linear in decibels, and this means that for all but a very narrow range of adjustments of level the needle is either showing small deflections, or flickering over the entire scale. The VU meter is not very satisfactory for high-quality work where a comparison or check on levels (which are not always close to the nominal 100% modulation) is required. In Britain the *peak programme meter* (PPM) is used. It is linear in decibels over the main working range.

Micron (μ). A thousandth of a millimetre.

Microphone. Electro-acoustic transducer. A microphone converts the power in a sound wave into electrical energy, responding to changes in either the air pressure or the pressure gradient (q.v.). Principal response patterns are omnidirectional, figure-of-eight, cardioid; principal types are moving-coil (dynamic), electrostatic (condensers) and ribbon (q.v.). *Microphone sensitivity* is measured in dB relative to 1 volt Pa. See **Pascal**.

Microphone channel. The pre-amplifier, equalization circuit, fader, etc. in the mixer which are available for each microphone. The channel may also include facilities for 'pre-hear', and feeds for echo, foldback and public address.

Midlift. Deliberate introduction of a peak in the frequency response in the upper-middle frequency range (somewhere between 1–8 kHz). A group of midlift controls may be calibrated according to the frequency of the peak and degree of lift.

Mil. A thousandth of an inch. About 25 microns.

Mix. Combine the signals from microphones, tape and gramophone reproducers, and other sources. To *mix down* or *reduce* is to combine separately recorded tracks in a multimicrophone balance. *Computer mixdown* is an aid to mixing multitrack recordings. A trial mix is fed to the memory and then successive changes are made to improve it. This ensures that each individual stage is operationally simple: complexity does not grow, and so does not become a limiting factor.

Mixed-minus (*Am.*). Clean feed (q.v.).

Modulation. Superimposition (of sound wave) on a carrier, which may be a high-frequency signal (e.g. by amplitude modulation or frequency modulation, q.v.) or (on a record) a smooth spiral groove which when it carries no recorded signal, is described as *unmodulated*. 100% *modulation* is the maximum permissible amplitude for any recording or transmission system.

Monaurai sound. Alternative term for monophonic sound suggested by the analogy that listening with one microphone is like listening with one ear – which is not strictly accurate. See **Mono**.

Monitoring. (a) Checking sound quality, operational techniques, programme content, etc., by listening to the programme as it leaves the studio (or at subsequent points in the chain, or by a separate feed, or by checking from a radio receiver). A *monitoring loudspeaker* in the studio is usually as good as the best that listeners might be using (but it must be remembered that adverse listening conditions might radically alter the listeners' appreciation of sound quality – also they may be listening at a level lower than that at which the monitoring loudspeaker is set). (b) Listening to other broadcasts for information.

Mono (monophonic sound). Sound heard from a single channel. This is defined by the form of the recording or transmission, and not by the mumber of loudspeakers. A number of microphones may be used and their outputs mixed; several loudspeakers may be used, and their frequency content varied, but this is still mono unless there is more than one channel of transmission. Multiplex radio transmissions and the single groove of a stereo disc each contain more than one channel. In mono the only spatial movement that can be simulated is forward and backward. There is no sideways spread; in particular there is no spatial spread of reverberation, or of ambient noise (see **Cocktail party effect**). The balance of sound in mono is therefore not natural but worked out in terms of a special convention. NB. In television, 'mono' is also used to mean monochromatic, i.e. black and white pictures.

Mouse. Microphone laid flat on floor (e.g. of stage) inside neutrally coloured foamed plastic shield that is flat on the underside. As the capsule is very close to a hard reflecting surface, interference effects are limited to very high frequencies. See **Pressure zone microphone**.

Moving-coil microphone, loudspeaker or pick-up. (*Am.*: Dynamic). These all use a small coil moving in the field of a permanent magnet. In the microphone or pick-up the movement generates a current in the coil; in the loudspeaker the current causes movement that is transmitted to a cone (q.v.) which drives the air.

MS. Middle-and-side (*Am.*: mid-side): a stereo system in which the components are generated and/or processed and transmitted as 'M' and 'S' signals.

Multiplex stereo. Radio transmission carrying A − B stereo information on a subcarrier above the A + B signal, or other combined signals of this type.

Music Line. Full audio-frequency cicrcuit for carrying programme (including speech) as distinct from a telephone line which may occupy only a narrow band.

Musique concrète (*French*). A work in musical form constructed from natural sounds that are recorded and then treated in various ways.

Mute film (*Br.*). Scenes shot without sound. (*Am.*: MOS).

NARTB characteristic. In tape recording, the pre-emphasis and subsequent equalization standards used in America and Japan. (Standard set by the American National Association of Radio and Television Broadcasters.) See also **CCIR** and **RIAA**.

Network operation. A broadcasting system involving many local stations and transmitters that may join together or separate at will.

Newtons per square metre (N/m^2). See **Pascal**.

NICAM. Near-Instantaneous Companded Audio Multiplex. A complex, compressed digital transmission system developed by the BBC. NICAM 728 is a version of this adopted for stereophonic television broadcasting in the UK. With a sampling rate of 32 kbits/s it is limited to a bandwidth of 15 kHz, but is otherwise comparable in quality to CD reproduction. There is no perceptible lag behind picture, but enough to cause severe comb-filter effects if decoded NICAM is mixed with synchronous stereo.

Noise. This is generally defined as unwanted sound. It includes such things as unwanted acoustic background sounds, unwanted electrical hiss (e.g. caused by the random flow of electrons in a transistor, or the random placing and finite size of oxide particles in tape coating), or rumble, hum or unwanted electromagnetic noise (background noise picked up on a radio receiver). In all electronic components and recording or transmission media the signal must compete with some degree of background noise, and it is vital in radio and recording work to preserve an adequate signal-to-noise ratio at every stage. In general, noise consists of all frequencies or a band of frequencies rather than particular frequencies (hum is an exception to this). *White noise* contains all frequencies in equal proportion. *Coloured noise* is a defined frequency-band of noise (the term is used by analogy with coloured light). Acoustic noise levels are often measured in dBA (or SLA). 1 dBA = 40 dB relative to 2×10^{-5} Pa at 1 kHz, and at other frequencies is weighted (electrically) to be of equal loudness. PNdB (perceived noise decibels) use a weighting function that can be used to place aircraft and other noises in order of noisiness. The analysis of sounds in PNdB can be complex; a similar analysis uses an electrical network giving adequate results in units that are designated dBN.

Noise reduction systems. Applied mainly to analogue tape recordings, and essential to multitrack, these add some 10 dB to the signal-to-noise ratio, of which part is used to hold down recording levels that might cause print-through. The signal is encoded by splitting into frequency bands that are separately compressed. Digitally encoded audio signals are intrinsically less susceptible to noise and distortion.

Obstacle effect. Obstacles tend to reflect or absorb only those sounds with a shorter wavelength than their own dimensions; at greater wavelengths the object appears to be transparent to the sound. For example, a screen 27 in (67 cm) wide is substantially transparent to frequencies below 500 Hz, but will absorb or reflect higher frequencies. 15 kHz is equivalent to a wavelength of about 0.9 in (just over 2 cm): a microphone containing parts of about this size is subject to effects that begin to operate when this frequency is approached.

Off-microphone. On the dead (i.e. insensitive) side of the microphone, or at much more than the normal working distance on the live side.

Omnidirectional microphone. One that is equally sensitive in all directions. In practice, there is a tendency for this quality to degenerate at high frequencies, with the top response progressively reduced for sounds further away from the front axis. Microphones with small capsules are less affected.

Optical sound. A film recording system that is replayed by scanning a track of variable width (or, sometimes, density) by means of a lamp, slit and photocell. See **COMOPT**.

Oscillator. A device that produces an alternating signal, usually of a particular frequency (or harmonic series). An audio oscillator produces a pure sine tone at any frequency in the audio range. *Square wave* and *sawtooth generators* produce waveforms of roughly those shapes (these correspond to harmonic series). Other oscillators (e.g. tape bias and r.f. oscillators) provide signals at higher frequencies.

Outside source (*Br.*). Remote (*Am.*). A source of programme material originating outside the studio to which it is fed, and appearing on an individual fader on the mixer panel just as any local source does (except that no pre-amplifier is needed as it is fed in at about zero level).

Overdub. Added recording (usually on a separate track) made while listening to replay of tracks already recorded.

Overlap changeover. A type of changeover from one recorder to the next in sequence which may be used when the two have half a minute or so of material in common. The overlap is used to adjust synchronization before crossfading from one to the other: before changing over, the first recording is heard on the loudspeaker, and the second prefaded on headphones, or on a loudspeaker of different quality.

Overmodulation. Exceeding the maximum permissible amplitude for recording or transmission. Distortion may be expected, and in certain cases damage to equipment or to the recording itself (limiters are used where this could occur).

Overtone. A partial in a complex tone, so called because such tones are normally higher than the fundamental.

Pad. Attenuator of fixed loss.

Panning or **Steering.** Splitting the output from a monophonic microphone between stereo A and B channels. *Panpot:* potentiometer (fader) to do this.

Parabolic reflector. A light rigid structure (of aluminium or other material) that reflects sound to a focus at which a microphone is placed. The assembly is very strongly directional at frequencies for which the corresponding wavelengths are less than the aperture of the reflector. It differs from gun microphone systems in that it is not degraded by interference from reflecting surfaces.

Parity. Whether the number of 1s in a binary number is odd or even. This is designated by a 1 or 0 which is attached to the end of the number as an error-check. If the parity of each row and column of a matrix of digital numbers is recorded, any single bit (a 0 or 1) that is missing from a row or column can be detected and restored.

Partial. One of a group of frequencies, not necessarily harmonically related to the fundamental, appearing in a complex tone. (Bells, xylophone blocks and many other percussion instruments produce partials that are not harmonically related.)

Pascal. International unit of sound pressure, equal to 1 newton per square metre or (in older units) 10 dynes/cm^2. A common reference level

2×10^{-5} Pa approximates to the threshold of hearing at 1 kHz. Microphone sensitivity is calculated in dB relative to 1 V/Pa: this gives values 20 dB higher than those using the older reference level, 1 V/dyne/cm^2. Characteristically, microphones have a sensitivity of the order of -50 dB relative to 1 V/Pa or -70 dB relative to 1 V/dyne/cm^2.

Patch (*Am.*). Cross-plug.

PCM. Pulse code modulation (q.v.).

Peak. A period of high volume.

Peak Programme Meter (PPM). A device for measuring the peak values of programme volume. In Britain it is the main programme control aid. See **Meter**.

Perspective. Distance effects, which may be simulated by varying the levels and the proportions of direct and indirect sound. In a studio, this can be done by an actor moving on and off microphone. In dead acoustics the ratio of direct to indirect sound cannot be varied, as indirect sound must be kept to a minimum. In this case it may help if an actor 'throws' his voice, simulating raising it to talk or shout from a distance. In television, sound and picture perspectives are often matched by the adjustment of boom microphone distance, or by placing a fixed microphone sufficiently close to the camera that as a performer moves toward the camera the sound also appears closer.

PFL. Prefade-listen. See **Pre-hear**.

Phase. The stage that a particle in vibration has reached in its cycle. Particles are *in phase* when they are at the same stage in the cycle at the same time.

Phase-shift. The displacement of a waveform in time. If a pure sine tone waveform is displaced by one complete wavelength this is described as a phase-shift of 360°. If it is displaced by half a wavelength (i.e. through 180°) it has peaks where there were troughs and vice versa. If two equal signals 180° out of phase are added together, they cancel completely; if they are at any other angle (except when in phase) partial cancellation occurs. Some electronic components introduce phase-shift into a signal, and the shift is of the same angle for all frequencies. This means that the displacement of individual component frequencies in the wave is different (depending on their wavelength) and distortion of the waveform results. This will not normally be of importance, because the ear cannot detect changes in phase relationship between components of a steady note. Interference between two complex signals, similar in content but different in phase, results in the loss of all frequencies for which the two signals are 180° out of phase. Electrical cancellation may occur when two recordings are played slightly out of synchronization (this is used creatively in popular music), or when a signal has to be sent by two different paths (this can make the transmission of stereo signals by landline difficult, when lines with broad transmission characteristics are not available). It occurs in radio reception when a sky wave interferes with a direct signal, etc. In a studio, cancellation may occur when one of a pair of bidirectional microphones has its back to a sound source. Care must always be taken when there is any possibility of pick-up on two such microphones at the same time. A check can be made by asking someone to speak at an equal distance from both, fading both up to equal volume, and mixing. If the sound appears thin and spiky, or direct sound is completely lost, one of the microphones should be reversed. (Studio reverberation and random sound generally are not affected by cancellation.) Pairs of loudspeakers may also be out of phase. If a monophonic signal is fed to an in-phase pair the sound appears to come from behind them: if they are out of phase the sound appears to be thrown forward to a point between speakers and listener. Correct phasing of speakers is, of course, vital to true stereo reproduction.

Phon. A unit of loudness. Phons are the same as decibels at 1000 Hz, and at other frequencies are related to this scale by contours of equal loudness.

Pick-up. The electromechanical transducer of a gramophone. The movement of a stylus in the record groove gives rise to an electrical signal.

Pilot tone. See **Sync pulse**.

Pilot tone system. Stereo broadcasting system in which the A − B signal is transmitted on a subcarrier, and a pilot tone is also transmitted to control the phasing of the two signals.

Pitch. The subjective aspect of frequency (in combination with intensity) that determines its position in the musical scale. The pitch of a group of harmonics is judged as being that of the fundamental (even if this is not present in the series). Dependence of pitch on intensity is greatest at low frequencies. Increase in loudness may depress the pitch of a pure tone 10% or more at very low frequencies, although where the tone is a member of a harmonic series the change is less apparent. Pitch does not depend on intensity at frequencies between 1–5 kHz, and at higher frequencies increase in loudness produces a slight increase in pitch.

Polar characteristic, polar diagram. (*Am.*: field pattern). The response of a microphone, loudspeaker, etc., showing sensitivity (or volume of sound) in relation to direction. Separate curves are shown for different frequencies, and separate diagrams are necessary for different planes. Such diagrams can also be used to indicate the qualities of reflecting screens or the radiation pattern of transmitter aerials, etc.

Popping. Breakup of the signal from a microphone caused by blowing the diaphragm far beyond its normal working range. This may occur when explosive consonants such as 'p' and 'b' are directed straight at the diaphragm at close range.

Portamento. Used in musical performance, a slide in frequency to reach a note, or from one note to another. Also provided in some synthesizers. Measured in octaves per second.

Pot (Potmeter, Potentiometer). Fader (q.v.).

Pot-cut. Dropping a short segment of unwanted material out of a programme without stopping the replay, by quickly fading out and fading in again. Results in a momentary loss of atmosphere. Editing is preferable if time is available.

Power (of sound source). The total energy given out by a source (as distinct from intensity, which is energy crossing unit area). Power is the rate of doing work, and in an electrical circuit equals voltage times current.

PPM. Peak Programme Meter. See **Meter**.

Practical. A prop (property) in film or television that works partly or completely in its normal way, e.g. a telephone that rings, or an office intercom that is used normally.

Pre-amplifier. Amplifier in circuit between a source and the source fader.

Pre- and post-echo. An 'echo' of a particular programme signal which may appear before or after the signal, On tape it occurs one turn of tape before or after the normal signal, and is caused by printing (see **Print-through**); on discs it may appear one groove before and after as molecular tensions in the groove walls relax.

Pre-emphasis. Increasing the relative volume of part of the frequency response (usually h.f.) in order to make the best use of some segment of a recording or transmission system (e.g. record, VHF radio). This is compensated by matched de-emphasis after the link or component in question.

Pre-fade, deadroll. Playing closing music from a predetermined time in order to fit exactly the remaining programme time, and fading up at an appropriate point during the closing words or action.

Pre-hear. A facility for listening to a source either on headphones or on a loudspeaker of a quality characteristically different from the main monitoring loudspeaker. The source can thereby be checked before fading it up and mixing it in to the programme.

Premix. A mix made of several but not all components of a final mix, e.g. a music, effects and dialogue mix for a documentary (the commentary being added subsequently). Sometimes used for complicated effects sequences or dialogue with a lot of cuts and quality matching.

Preparation, 'prep'. In film and video sound editing, the stage in which tracks are converted to the format required for sound mixing. In hard disc editing, this becomes an integral (and major) stage of the sound editing.

Pre-recording. Recording made prior to the main recording or transmission, and replayed into it.

Presence. A quality described as the bringing forward of a voice or instrument (or the entire composite sound) in such a way as to give the impression that it is actually in the room with the listener. This is achieved by boosting part of the 1–8 kHz frequency range. In fact, emphasis of a single component (or several, at different frequencies) in this band give greater clarity and separation, although at the expense of roundness of tone. Presence applied to the entire sound (e.g. by a distorted loudspeaker response) achieves stridency but little else.

Pressing. Gramophone records have been made by stamping a plastic material such as polyvinyl chloride (PVC) in association with other components. PVC-based pressings have low surface noise (provided they have not been played with a heavy pick-up) and are not breakable, but are easily damaged by scratching or heat. Formerly, shellac was used for pressing records and, having much greater elasticity, was suitable for record materials when only very heavy (low-compliance) pick-up heads were available. The shellac was combined with a large proportion of inexpensive filler (e.g. slate dust) which made them hard, served to grind the stylus to the shape of the groove, and also contributed the characteristic surface noise of 78s.

Pressure gradient microphone. One with a diaphragm open to the air on both front and back, and which therefore responds to the difference in pressure at successive points on the sound wave (separated by the path difference from front to back). The microphone is 'dead' to sound approaching from directions such that the wavefront reaches front and back of the diaphragm at the same time. (See **Figure-of-eight** and **Cardioid microphone**.) A pressure gradient microphone measures the rate of change of sound pressure, i.e. air particle velocity. This is not quite the same thing as measuring intensity (as the ear does) but differs from it essentially only in phase, and as the ear is not sensitive to differences in phase, this does not matter. Pressure gradient operation begins to degenerate to pressure operation for wavelengths approaching the dimensions of the microphone. This may help to maintain the response of the microphone at frequencies such that cancellation would occur due to the wavelength being equal to the effective path difference from front to back of the diaphragm.

Pressure microphone. One with a diaphragm open to the pressure of the sound-wave on one side and enclosed on the other. If the diaphragm and casing is sufficiently small, the microphone is omnidirectional (q.v.).

Pressure Zone Microphone (PZM). Proprietary name for a range of boundary microphones which are attached or very close to a hard reflecting surface, so that interference effects are virtually eliminated.

Print-through. The re-recording of a signal from one layer to another in a spool of tape. The recorded signal produces a field that magnetizes tape separated by only the thickness of the backing. (Other factors affecting printing are temperature, time and physical shock.)

Programme. (a) Self-contained and complete item. (b) The electrical signal corresponding to programme material, as it is fed through electronic equipment.

Proximity effect. See **Bass tip-up**.

Public address (PA). A loudspeaker system installed for the benefit of an audience. The output of selected microphones is fed at suitable levels to directional loudspeakers.

Pulse Code Modulation (PCM). An efficient means of using available bandwidth. The signal is sampled at frequent intervals and encoded in digital form. The coded signal is less susceptible to noise and distortion. On a multitrack recording it can be used for noise reduction (q.v.). A coded signal can travel considerable distances, and television sound may be interleaved with the picture signal: the signal is reconstructed at its destination and broadcast in the normal way.

PVC. Polyvinyl chloride. Plastic commonly used in disc pressings and some tape backings and electrical insulators.

Quad, quadraphony. A system providing four-channel sound surrounding the listener. Two speakers are nominally to the front and two to the rear. The latter may be used to enhance the apparent acoustics of the listening room, or all may carry direct sound as well as the acoustic enhancement.

Quarter-track recording. One that occupies about 15% of the full width of the tape. The tape runs from left to right and the top track is recorded first, and then the third. To record the other two tracks the tape is turned upside down and once again fed left to right. For stereo, the first and third tracks are recorded at the same time, using a stacked head (one with both gaps in line). Quarter-track recording has a signal-to-noise ratio poorer than that of half or full track recording in proportion to the relative widths.

Radio link. A radio transmission focused into a narrow beam and directed toward a directional receiver placed in line of sight. Used in place of a landline.

Radio microphone. Microphone and small transmitter sending a signal that can be picked up at a distance of up to perhaps several hundred metres (provided that the two are not screened from each other).

Radio transmission. A system for distributing audio information by modulating or encoding it on to a carrier at a particular frequency that is then amplified to a high power and broadcast toward the receivers by means of an aerial. Transmitter and receiver aerials act in a similar way to the two windings of a transformer; they are coupled together by the field between them. (The field at the receiver is, of course, very tiny: it diminishes not only as the square of the distance, but also by the action of other 'receivers' along the path. In passing over granite, for example, a great deal of a medium-wave signal is mopped up. For VHF, which does not penetrate so deep into the ground, the nature of the terrain does not matter – but high ground casts a 'shadow'.) This highly inefficient transformer action is improved if the receiving aerial is so constructed as to resonate at the desired frequencies of reception, and it will discriminate against other signals if it is made directional. It is important that the aerials should be parallel (both vertical or both horizontal). Systems of carrier modulation include amplitude modulation and frequency modulation (q.v.).

Radiophonics (*BBC*). Electronic sounds or effects that are formally arranged, as in *musique concrète* or electronic music, but which are not usually intended to be heard independently of other radio or television material.

RAM. Random access memory: an electronic component storing digital information that can be retrieved at any time and in any order.

Ramp. Fade, in programmed (hard disc) editing.

Recording. An inscription of an audio signal in permanent form, usually on magnetic tape or on disc. The term *record* implies commercial gramophone record.

Reduce. See **Mix**.

Reinforcement (in sound balance). The strengthening of direct sound reaching a microphone by the addition of indirect sound. This adds body without adding appreciably to volume. See **Resonance, Reverberation**.

Remote (*Am.*). Outside broadcast or outside source (q.v.).

Resistance. The ratio of e.m.f. (electromotive force) to current produced in a circuit; the ratio of voltage drop to current flowing in a circuit element.

Resonance. A natural periodicity or the reinforcement associated with it. The frequencies (including harmonics) produced in musical instruments (e.g. vibrating strings or columns of air) are determined by resonance.

Response. Sensitivity of microphone, etc. See **Frequency response, Polar characteristic**.

Reverberation. The sum of many reflections of sound in an enclosed space. This modifies the quality of a sound and gives it an apparent prolongation after the source stops radiating. *Reverberation time* is the time taken for sound to die away to a millionth of its original intensity (i.e. through 60 dB).

Reverberation plate. A metal plate, held under stress and fitted with transducers, one to introduce sound vibrations, and the others to detect them at other points on the plate. By this means the decay of sound in an enclosed space is simulated. This technique has been superseded by digital reverberation.

RIAA characteristic. Recording characteristic used on gramophone records (standard set by the Record Industry Association of America). See also **CCIR** and **NARTB**.

Ribbon microphone (or **loudspeaker**). The ribbon is a narrow strip of corrugated aluminium alloy foil suspended in a strong magnetic field provided by a powerful horseshoe magnet with polepieces extending along the length of the ribbon. It is made to vibrate by the difference in pressure between front and back of the ribbon. If both sides are open to the sound wave, the resulting motion is in phase with the velocity, and not with the amplitude of the sound wave. (See **Figure-of-eight, Pressure gradient microphone**.) In the loudspeaker the same principle is used in reverse, but is not suitable for handling considerable power, and so has been used only in a tweeter. A flared (exponential) horn is provided to improve coupling.

Ring. Undamped resonance.

Rock and roll. A facility for discontinuous film sound re-recording (dubbing) in which the recording is stopped after an error is made, the picture and tracks run back in synchronization and a new recording is begun from an earlier point. An essential characteristic is that the join must not be perceptible.

Rumble. Low-frequency mechanical vibration picked up by an audio system.

S signal. The stereo difference signal A − B. The S does not stand for stereo (of which it is only part) but for 'side' response, such as may be obtained by a side-fire bidirectional microphone − used sometimes in combination with a 'main' microphone to produce stereo.

Satellite. In this context, a stage in the transmission of audio or television signals from one part of the globe to another by two line-of-sight paths. Using a synchronous satellite (i.e. one for which the orbital speed is the same as the Earth's rotational speed, so that the satellite remains stationary with respect to a point on the surface of the Earth) there will be a delay of nearly a quarter of a second.

Scale. Division of the audio-frequency spectrum by musical intervals (i.e. frequency ratios). An octave has the ratio 1:2, a fifth 2:3, a fourth 3:4, and so on: common musical intervals are derived from series of ratios of small whole numbers. The *chromatic* or *twelve-tone scale* is a division of the octave into 12 equal intervals (semitones in the equal tempered scale). On this scale most of the small-whole-number intervals correspond (though not exactly) to an integral number of semitones, which may, of course, be measured from or to any point. Certain scales omit five of the 12 notes of the octave, leaving seven (plus the octave) into which most of the small-whole-number ratios still fit − but it is no longer possible to start arbitrarily from any point. Instead, the interval must always be measured from a *key* note, or a note simply related to it. Music that for the most part observes the more restricted scales is called *tonal*, while that which ranges freely over the chromatic scale is called *atonal*. In *twelve-tone music* an attempt is made to give each of the 12 notes equal prominence.

Screen (*Am.*: flat, gobo). A free-standing sound-absorbent or reflecting panel that may be used to vary the acoustics locally, or to cut off some of the direct sound travelling from one point to another in the studio. It may be moved about the studio at will. An object (such as a script) is said to be *screening the microphone* if it lies in the path of sound coming directly from a source.

Screening. A protection from stray electrical fields. It may take the form of an earthed mesh wire surrounding a conductor carrying a low-level signal. Valuable tapes may be screened in a metal box when sent by air.

Script rack. An angled rack on which a script may be placed in order to encourage a speaker to raise his head. It should be transparent to sound.

Segue (pronounced 'segway'). Musical term meaning follow on.

Self-drive studio. One that is operated by the user without continuous technical assistance.

Sensitivity. The ratio of response to stimulus, usually measured in decibels relative to some given reference level. See **Pascal**.

Sepmag. Separate magnetic film sound, i.e. on a separate spool, but recorded with frame-for-frame synchronization. See **Commag, Comopt**.

Separation. Degree to which each of several microphones discriminate in favour of the sources or groups of sources associated with it, and against unwanted sound or that to be picked up by other microphones. The purpose of separation is to allow individual control and treatment of the sources.

Shelf. A dip in frequency response; in equalization, the exact opposite of a peak.

SI. International system of units, adopted also by the British Standards Institution. See, e.g. **Pascal**.

Sibilance. The production of strongly emphasized 's' and 'ch' sounds in speech. These may in turn be accentuated by microphones having peaks in their top response.

Side-fire. Orientation of diaphragm within a microphone such that the axis of directional pick-up is at a right-angle to its length. The front axis is usually marked in some way.

Signal. The fluctuating current that carries audio information (programme).

Signal-to-noise ratio. The difference in decibels between signal and noise levels. A standard reference level is 1 Pa (94 dB at 1 kHz).

Sine tone (pure tone). A sound (or electrical signal) containing one frequency, and one alone. This corresponds to an air particle executing simple harmonic motion, and its graphical representation (waveform) is that of s.h.m. (i.e. a sine wave). In principle, all sounds can be analysed into component sine tones. In practice, this is only possible for steady sounds, or for short segments of irregular sounds: noise contains an unlimited number of such tones. Sine tones are useful for studying frequency response and for lining up equipment.

Slew. Frequency shift, portamento (q.v.).

Slung microphone. One that hangs by suspending wires or by its own cable from a ceiling or grid fitting (or from a boom or lazy arm).

Solid-state device. A circuit element such as a transistor (q.v.) or a complete microcircuit combining a number of circuit elements and their connections. It is characterized by small size, low power consumption and high reliability.

Sound. A series of compressions and rarefactions travelling through air to another medium, caused by some body or bodies (*sound sources*) in vibration. At any place it is completely defined by the movement of a single particle of air (or other material). The movement may be very complex, but there are no physically separate components due to different sound sources. Such contributions may be isolated by mathematical analysis, and the brain, too, acts as a mathematical computer which perceives the different components as separate entities. See **Hearing, Decibel, Level, Frequency, Wave, Pitch, Phase, Formant, Sine tone, Velocity of sound.**

Sound effects. See **Effects**.

Sound subject (in *musique concrète*) A segment of treated sound material, to be assembled with others into the finished work.

Soundfield microphone. Proprietary name for microphone with a tetrahedral array of diaphragms, the output of which can be combined to create any desired (primary) polar response in three dimensions.

Spaced pair. Two separated microphones used to pick up stereo. Phase distortion effects occur but many balancers regard these as tolerable, or even as adding to the richness of the stereo sound.

Spacer. Uncoated tape, which may be yellow or other colours (other than white or red which are used mainly for leaders and trailers), cut into a spool of tape to indicate the end of one segment and the start of another.

Spill (*Am.*: leakage). Sound reaching a microphone other than that intended, thereby reducing separation.

Spin. The combination of original and delayed signals with multiple repetitions. Decay (or continuation of the effect) is controlled by a fader in the feedback circuit. Eventually the quality changes to express the frequency characteristics of the feedback circuit itself.

Spin-start. Quick-start technique for disc, requiring no special equipment, in which the record is set up with the stylus in the groove and the motor switched off. On cue, the motor is switched on and the turntable boosted to speed by hand. This boost is not usually necessary for turntables with rim drive.

Spool. Reel for carrying tape. *Cine spools* for tape are similar in design to 8 mm film spools. The most common types are made of clear plastic in a variety of sizes. Metal spools are available in larger sizes, and are more often used on professional decks. Some also have larger hub size than the cine spool. Transcription tape decks are normally equipped to play either type. (A large hub is used in order that the tension does not vary so much as it would with a large diameter cine spool. Some recorders can be switched to provide reduced tension when cine spools are used.) Another design consists of a hub and a one-side-only backing plate, but this gives the operator less security against spillage than the double-sided spool.

Spot effect. (*Am.:* foley) A sound effect created in the studio. It may be taken on a separate microphone or on the same microphone as the main action.

Spotting. Reinforcement of a particular element in a stereo balance using a monophonic microphone (*Am.:* an accent microphone). The balance is usually very close, to avoid 'tunnel' reverberation effects and other problems.

Standing waves. See **Wave**.

Steering. See **Panning**.

Stereo (stereophonic sound). A form of reproduction in which the apparent sources of sound are spread out. The word 'stereo' implies 'solid', but in fact the normal range of sound in two-channel stereo is along a line joining the loudspeakers – to which an apparent second dimension is added by perspective effects. This may be used to simulate the spread of direct sound of an orchestra or a theatre stage very well. Whereas each individual source may be fairly well defined in position, the reverberation associated with it is spread over the whole range permitted by the position of the loudspeakers. But even this range is small in comparison with the situation in a concert hall, in which reverberation reaches the listener from all around. *Stereophonic sound in the cinema* uses a multitrack, multi-loudspeaker system, partly to ensure that members of the audience who are not ideally seated for two-channel stereo still hear sound that is roughly coincident with the image, but also (in more sophisticated systems) from the sides of the auditorium.

Sting. Musical punctuation pointing dramatic or comic mood.

Stops. A term referring back to the stud positions on a stud fader, and the markings associated with them, and so, by extension, to the corresponding arbitrary divisions of a continuous fader. Thus, to lift programme level 'a stop' is to increase it by moving the fader (potentiometer) about $1\frac{1}{2}$–2 dB.

Stray field. Unwanted a.c. field that may generate a signal in some part of the equipment where it should not be. The use of balanced wiring discriminates against this, and so also, where necessary, does screening (in which an earthed conductor surrounds parts that might be affected).

Stroboscope. Since a.c. electric lighting pulsates in intensity (at twice the mains frequency) such light can be used to illuminate a disc with bars (or spots) on it. When the speed is at an appropriate setting these move a distance equal to their spacing in the time from one pulse of light to the next. The number of bars for $33\frac{1}{3}$, 45 and approximately 78 rpm can easily be calculated – the European mains frequency being 50 Hz and the American, 60Hz. Stroboscopic indicators can also be devised for checking tape speed. Stroboscopes work best with fluorescent tubes, and not at all in sunlight. When the dots move onward the speed is too fast; when they move backward it is too slow. Stroboscopes are very sensitive to the slightest deviation of speed from the normal.

Studio (sound studio). Any room or hall that is primarily used for microphone work. Its most important properties lie in its size and its acoustics – the way in which sound is diffused and absorbed, and the reverberation time (q.v.). A *studio manager* in BBC radio is the balancer in operational charge of the broadcast or recording. (At the BBC the studio manager is not an engineer, which is a separate function.)

Stylus. The needle of a pick-up. Materials most used are diamond and sapphire: diamond lasts about 20 times as long. Wear depends on playing weight (and record material) and is most rapid in the earlier hours of playing.

Subharmonic. A partial of frequency $\frac{1}{2}f$, $\frac{1}{3}f$, $\frac{1}{4}f$, etc., lying below the fundamental f. The subharmonic $\frac{1}{2}f$ can be generated in the cone of a moving coil loudspeaker.

Sync pulse (pilot tone). A signal related to camera speed that is fed to the tape on which the associated sound is recorded. When the sound is subsequently transferred to magnetic film, the pulse controls the speed of re-recording.

Sync take. Film shot with sound, using a synchronous recording system.

Talkback. Communication circuit from control to studio, used mainly for the direction of performers. *Reverse talkback* is a secondary circuit for communication from studio to control. Both forms of talkback are independent of live programme sound circuits, although in unfavourable circumstances breakthrough (or cross-talk) can occur.

Tape. Magnetic recording medium consisting of a magnetic coating on a plastic backing (see **Coating** and **Backing**). *Long play tape* gives 50% more recording time on a spool than *standard play*, and *double play*, of course, gives double. Thinner tape is more susceptible to printing, and its use does not save money (for comparable quality); but it does save storage space, and permits fewer reel changes for a given spool size. *Videotape* is a similar medium. Metric equivalents: the standard width, $\frac{1}{4}$ in = 6.25 mm, $\frac{1}{2}$ in = 12.5 mm approx. and so on. Audio cassettes use a narrower tape, 3.8 mm wide.

Tape deck. The mechanical part (plus heads) of a tape recorder or reproducer.

Tape joiner. Film joiner using a guillotine and polyester tape. As the film is not lapped, no frames are lost. This type has replaced the cement joiner.

Tape speeds. These are all based on the early standard of 30 in/s (inches per second) for coated tape. Successive improvements in tape, heads and other equipment have permitted successive reductions in speed to 15 and $7\frac{1}{2}$ in/s (used professionally for music and speech) and $3\frac{3}{4}$ and less (used domestically). Metric equivalents: 15 in/s = 28 cm/s. $7\frac{1}{2}$ in/s = 19 cm/s, $3\frac{3}{4}$ in/s = 9.5 cm/s and so on.

Telephone adaptor. A pick-up coil that may be placed in the field of the line transformer of a telephone or in some similar position where a signal may be generated.

Telephone quality. A frequency band between 300 and 3000 Hz. Simulated telephone quality should be judged subjectively according to programme needs.

Tent. A group of screens, arranged to trap (and usually to absorb) sound in the region of the microphone.

Timbre. Tone quality. The distribution of frequencies and intensities in a sound.

Time code. A recorded time signal that can be electronically read and displayed. It may be the real time at which the recording was made or, alternatively, tape time (duration) from an arbitrary starting point. It is used for automatic control and synchronization, and also for timing duration, automatic logging, etc.

Time constant. For a capacitor of C farads, charging or discharging through a resistance of R ohms (as in a PPM, limiter, etc.), the time constant, $t = CR$ sec. (Similarly, for an inductance of H henries and resistance, R ohms $t = L/R$ sec.).

Tone. Imprecise term for sound considered in terms of pitch (of frequency content). A pure tone is sound of a particular frequency. See **Sine tone**.

Tone control. Pre-amplifier control for adjusting frequency content of sound (usually bass or treble).

Top. High frequencies in the audio range, particularly in the range 8–16 kHz.

Top response. Ability of a component to handle frequencies at the higher end of the audio range.

Track. (a) To move a camera or boom toward or away from the acting area. (b) An individual recording among several on a record. (c) One of several (typically, up to 8, 16 or 24) recordings made side-by-side on multitrack tape.

Trailer. (a) An item in the continuity (link) between programmes that advertises future presentations. (b) Uncoated tape, usually red, that is cut into a tape to indicate the end of wanted recorded material.

Transducer. A device for converting a signal from one form to another. The system in which the power is generated or transmitted may be acoustic, electrical, mechanical (disc), magnetic (tape), etc. Thus, microphones, loudspeaker, pick-ups, tape heads, etc., are all transducers.

Transient. The initial part of any sound, before any regular waveform is established. The transient is an important part of any sound and in musical instruments helps establish an identifiable character. *Transient response:* Ability of a component to handle and faithfully reproduce sudden, irregular waveforms. See **Diaphragm**.

Transistor. A semiconductor device that performs most of the functions of a valve (vacuum tube), but differs from it principally in that (a) no heater is required, so that the transistor is always ready for immediate use and no significant unproductive power is consumed, (b) it is much smaller in size, (c) input and output circuitry are not so isolated as in a valve, and (d) there is normally no phase reversal in a transistor, as the control voltage is used to promote flow of current, and not to reduce it. A range of designs and functions of transistors (as with valves) is available. The circuit associated with a transistor is somewhat different from that for a valve. Transistors are more sensitive to changes of temperature than valves. Power transistors, which generate heat, need to be well ventilated. See **Tube**.

Tremolo. A regular variation in the amplitude of a sound, e.g. of electric guitar, generally at a frequency between 3 and 30 Hz. Sometimes confused with *vibrato* (q.v.).

Tube (*Am.*) Vacuum tube, or valve (q.v.) When used as a component of a head amplifier of archaic design in an early electrostatic microphone or replica, the term has been applied to the microphone itself.

Turntable. The rotating plate of a record player. Besides supporting and gripping the record, it acts as a flywheel, checking any tendency to wow and flutter. It should therefore be well balanced, with a high moment of inertia (this is highest for a given weight if most of the mass is concentrated at the outer edge).

Tunnel effect. Monophonic reverberation associated with an individual source in stereo.

Tweeter. High-frequency loudspeaker used in combination with a low-frequency unit or 'woofer' and possibly also a mid-range unit (sometimes called a 'squawker'). The problems of loudspeaker design are different at the two ends of the audio spectrum and in some are easier to solve if handled separately. A single unit becomes progressively more directional at higher frequencies: a small separate unit for h.f. can maintain a broad radiating pattern.

Undermodulation. Allowing a recorded or broadcast signal to take too low a volume, so that it has to compete to an unnecessary extent with the noise of the medium (and has to undergo greater amplification on reproduction, risking greater noise at this stage also).

Undirectional microphone. This may refer either to a cardioid (q.v.) or near-cardioid type of response (live on one face and substantially dead on the other), or a microphone with a more strongly directional forward lobe. See also **Parabolic reflector, Gun microphone**.

User bits. Parts of a digital code which are reserved for allocation by the user. In time code they can be used to indicate a reel number.

Valve (*Am.*: vacuum tube). Used in electronic equipment before the advent of the transistor, and still found in one archaic design of microphone, this is an almost completely evacuated glass envelope within which electrons released by a heated electrode (the cathode) are collected by a second positively charged electrode (the anode). In this, its simple form, the valve is a *diode* and it conducts electricity whenever there is a flow of electrons into the cathode (on every other half-cycle of an alternating signal). In the *triode* a grid is placed in the path of the electrons and variation of voltage on this controls corresponding fluctuations in the flow of electrons, producing an amplified version of the signal on the control grid. Additional grids may be added (*tetrode, pentode*, etc.) and these further modify the characteristics of the electron flow.

Velocity microphone (*Am.*). Ribbon microphone (q.v.).

Velocity of sound. In air at room temperature this is approximately 1120 ft/s (or 340 m/s). It can be calculated roughly as $1087 + 2T$ ft/s where T is the temperature in degrees Celsius. Humidity also makes a slight difference: in fully saturated damp air it is 3 ft/s faster than in dry air. In liquids and solids it is much faster than in air.

VHF (Very High Frequency). See **Radio transmission**.

Vibrato. Rapid cyclic variation in pitch at a rate of about 5–8 Hz, used by musicians to enrich the quality of sustained notes.

Videotape clock. Electronic clock on a video screen showing programme details, in vision before the start of a television recording between about minus 30 seconds and minus 3 seconds, after which the picture cuts to black. Reference tone is recorded between minus 20 and minus 10 seconds, with tone pips on each second from nine to three.

Vocal stop. A short break in vocalized sound that precedes certain consonants. These help in the exact location of editing points on tape, and sometimes provide a useful place to cut. It is sometimes possible to lose the sound after a stop completely, e.g. when cutting 'bu' (for 'but') on to the beginning of a sentence.

Volume control. See **Control. Compression, Fader**.

Volume meter, VU Meter. See **Meter**.

Volume of sound. See **Decibel, Level**.

Wave (sound wave). A succession of compressions and rarefactions transmitted through a medium at a constant velocity. See **Velocity of sound**. In representing this graphically, displacement is plotted against time, and the resulting display has the appearance of a transverse wave (like the ripples on a pool). This diagram shows the *waveform*. The distance between corresponding points on successive cycles (or ripples) is the *wavelength* (λ) and this is related to the frequency (f) (the rate at which ripples pass a particular point) and sound velocity (c) through the relation $f\lambda = c$. Wavelength can thus be roughly calculated as 1120 ft/s or 340 m/s. As sound radiates out from a source it forms a *spherical wave* in which intensity is inversely proportional to the square of the distance. When the distance from the source is very large compared with the physical dimensions of any objects encountered this effect becomes less important: the loss in intensity is small and the wave is said to act as a *plane wave*. See **Bass tip-up** for illustration of this. In a *standing wave* the nodes occupy fixed positions in space. In a *progressive wave* the whole waveform moves onward at a steady rate.

Weighted and **unweighted**. Different ways of indicating levels of noise or hum relative to signal, with particular reference to their bass content. Programme meters do not normally give any frequency information, and low frequencies do not sound as loud as a VU meter or PPM indicates, as the ear discriminates against bass. For this reason, quoted noise levels are sometimes 'weighted' against bass according to standard loudness contours. Weighted and unweighted measurements may differ by 20 dB or more at low frequencies.

White noise. See **Noise**.

Wiggle or **Flicker**. The momentary displacement of a stereo image that occurs when audio tape is cut at an angle or when one of the faders on the individual sound channels is not making perfect contact. In a mixer with studded faders it occurs when the sliders do not make simultaneous contact with successive studs.

Wildtrack. Film sound recorded without picture.

Windshield (*Am.*: windscreen). Shield that fits over the microphone and protects the diaphragm from 'rattling' by wind, and also contours the microphone for smoother airflow round it.

Wipe (tape). Erase.

Woofer. Low-frequency unit in loudspeaker. See **Tweeter**.

Woolly. Sound that lacks clarity at high frequencies and tends to be relatively boomy at low.

Wow. Cyclic fluctuation in pitch due to mechanical faults in recording or reproducing equipment (or physical fault in a disc). The frequency of the variation is below, say, 5 Hz. See **Flutter, Vibrato, Tremolo**.

'X' and **'Y'** signals (*Am.*: 'X/Y technique' means coincident microphone pair.). Sometimes used instead of A and B. See also **'A' signal, 'B' signal**.

Zero level. See **Level**.

Bibliography

Alkin, Glyn *Sound Recording and Reproduction* (Second Edition), Focal Press, Oxford (1991). A comprehensive guide to all aspects of studio sound recording, reproduction systems and their use.

Alkin, Glyn *Sound Techniques for Video and TV* (Second Edition), Focal Press, Oxford (1989). A clear account based on BBC practice.

Bartlett, Brian *Stereo Microphone Techniques*, Focal Press, Oxford (1991). Guide to the theories behind the use of free-field, boundary layer and binaural stereo pairs.

Borwick, John *Microphones, Technology and Technique* (Second Edition), Focal Press, Oxford (1990). A good all-round account, if slightly lacking in detail on microphone performance.

Gayford, Michael (Ed.) *Microphone Engineering Handbook*, Focal Press, Oxford (1994). Offers detailed technical and, where appropriate, mathematical descriptions of design and resulting performance characteristics for a wide range of types – even including a section on the promise of optical microphones. Many chapters are contributed by the specialist engineers of particular market leaders, which inevitably somewhat limits the selection of examples.

Huber, David Miles *Microphones Manual, Design and Application*, Focal Press, Oxford (1988). Most directly useful when describing the characteristics of individual instruments and how to exploit them in 'microphone placement'; less so when discussing the complexities of microphone design. American terminology.

Millerson, Gerald *The Technique of Television Production* (Twelfth Edition), Focal Press, Oxford (1990). A highly analytical study of the medium, including the contribution of sound, videotape, electronic effects and on-location news collecting.

Mott, Robert L. *Sound Effects, Radio, TV and Film*, Focal Press, Oxford (1990). Anecdotal (and in this field none the worse for that), but limited to a history of the American experience.

Nisbett, Alec *The Use of Microphones* (Fourth Edition), Focal Press, Oxford (1993). A concise handbook on microphones, balance and control.

Oringel, Robert *Audio Control Handbook* (Sixth Edition), Focal Press, Oxford (1988). A simple and useful introduction to the subject with a strong emphasis on American equipment.

Potter, Kopp and Green-Kopp *Visible Speech,* Dover Publications, New York, and Constable, London (1966). Useful when considering how to record, treat and edit speech and song.

Robertson, A. E. *Microphones* (Second Edition), Iliffe Books, London (1963). Describes the engineering principles behind many types of microphone.

Rumsey, Francis *Digital Audio Operations,* Focal Press, Oxford (1990). A comprehensive handbook, but more on the bewildering variety of equipment and standards than on their operation.

Rumsey, Francis *MIDI System and Control,* Focal Press, Oxford (1990). A sharply-focused account of a specialized digital field.

Rumsey, Francis *Stereo Sound for Television,* Focal Press, Oxford (1989). A short conversion course (including digital equipment) for those familiar with the basics.

Stockhausen, Karlheinz *Study II* (Score) Universal Edition. See this and the scores of other works of a similar nature for illustration of the problems (and purpose?) of the notation of electronic music.

Wood, A. B. *The Physics of Music* (Seventh Edition), Methuen (1975). A useful primer.

Index